AA001090

2009 International Symposium on VLSI Technology, Systems, and Applications

(VLSI-TSA)

Hsinchu, Taiwan
27 – 29 April 2009

IEEE Catalog Number: CFP09846-PRT
ISBN: 978-1-4244-2784-0

Copyright © 2009 by the Institute of Electrical and Electronic Engineers, Inc
All Rights Reserved

Copyright and Reprint Permissions: Abstracting is permitted with credit to the source. Libraries are permitted to photocopy beyond the limit of U.S. copyright law for private use of patrons those articles in this volume that carry a code at the bottom of the first page, provided the per-copy fee indicated in the code is paid through Copyright Clearance Center, 222 Rosewood Drive, Danvers, MA 01923.

For other copying, reprint or republication permission, write to IEEE Copyrights Manager, IEEE Service Center, 445 Hoes Lane, Piscataway, NJ 08854. All rights reserved.

***This publication is a representation of what appears in the IEEE Digital Libraries. Some format issues inherent in the e-media version may also appear in this print version.**

IEEE Catalog Number: CFP09846-PRT
ISBN 13:   978-1-4244-2784-0
ISSN: 1930-885X

Additional Copies of This Publication Are Available From:

Curran Associates, Inc
57 Morehouse Lane
Red Hook, NY 12571 USA
Phone: (845) 758-0400
Fax: (845) 758-2633
E-mail: curran@proceedings.com
Web: www.proceedings.com

2009 International Symposium on VLSI Technology, Systems, and Applications (VLSI-TSA)

Hsinchu, Taiwan
27-29 April 2009

IEEE Catalog Number: CFP09846-POD
ISBN: 978-1-42442-784-0

TABLE OF CONTENTS

PLENARY SESSION

Has The Sun Finally Risen on Photovoltaics? 1
  Mark Pinto

The Future of Semiconductor Industry - A Foundry's Perspective..2
  Fang-Churng Tseng

From Living Faster to Living Better 3
  René Penning de Vries

Carrier Transport and Stress Engineering in Advanced Nanoscale MOS Transistors 5
  Ken Uchida

SESSION 1: PROCESS MODULE

The Impact on Device Characteristics with STI Formed by Spin-on Dielectric in High Density NAND Flash Memory 7
  W. Z. Wong, J. J. Fan, J. D. Jiang, C. H. Huang, C. Y. Chen, H. H. Chen, C.C. Hsu, Rex Young, P. Y. Wang, H. Fujita, H. Kobayashi

Properties of Very Thin Adenine Layer with High Inhibition for 32nm Node Cu/Low-K Interconnection..9
  M. Hara, H. Aoki, T. Masuzumi, D. Watanabe, C. Kimura, T. Sugino

Boron Carbon Nitride Film Containing Hydrogen for 22nm Node Low-K Interconnection....................11
  Hidemitsu Aoki, Takuroh Masuzumi, Makoto Hara, Daisuke Watanabe, Chiharu Kimura, Takashi Sugino

Ge Shallow Junction Formation by As implantation and Flash Lamp Annealing......................13
  Kosei Osada, Tetuya Fukunaga, Kentaro Shibahara

Impact of Lithography Variations on Advanced CMOS Devices 15
  J. Lorenz, C. Kampen, A. Burenkov, T. Fühner

SESSION 2: STRAIN TECHNOLOGY

Stress-Enhancement Technique in Narrowing NMOSFETs with Damascene-Gate Process and Tensile Liner..17
  S. Mayuzumi, S. Yamakawa, Y. Tateshita, M. Tsukamoto, H. Wakabayashi, T. Ohno, N. Nagashima

Additive Mobility Enhancement and Off-State Current Reduction in SiGe Channel pMOSFETs with Optimized Si Cap and High-k Metal Gate Stacks 19
  Jungwoo Oh, Prashant Majhi, Raj Jammy, Raymond Joe, Anthony Dip, Takuya Sugawara, Yasushi Akasaka, Takanobu Kaitsuka, Tsunetoshi Arikado, Masayuki Tomoyasu

Dopant and Thermal Interaction on SPE Formed SiC for NMOS Performance Enhancement 21
  P. W. Liu, T. F. Kuo, C. I. Li, Y. R. Wang, R. M. Huang, C. H. Tsai, C. T. Tsai, G. H. Ma

Successful Integration Scheme of Cost Effective Dual Embedded Stressor Featuring Carbon Implant and Solid Phase Epitaxy for High Performance CMOS 23
  M. Nishikawa, K. Okabe, K. Ikeda, N. Tamura, H. Maekawa, M. Umeyama, H. Kurata, M. Kase, K. Hashimoto

An Investigation about the Limitation of Strained-Si Technology 25
M. H. Liao, Lingyen Yeh, J. C. Lu, M. H. Yu, L. T. Wang, J. Wu, P. R. Jeng, T. L. Lee, Simon Jang

SESSION 3: NON-VOLATILE MEMORIES

Characterization of Poly-silicon Emitter BJTs as Access Devices for Phase Change Memory 27
B. Rajendran, M. Breitwisch, R. Cheek, M-H. Lee,Y-H. Shih, H-L. Lung, C. Lam

Low Current and Voltage Resistive Switching Memory Device Using Novel Cu/Ta$_2$O$_5$/W Structure 29
S. Z. Rahaman, S. Maikap, C. H. Lin, T. Y. Wu, Y. S. Chen, P. J. Tzeng, F. Chen, C. S. Lai, M. J. Kao, M. J. Tsai

A Novel Multi - Nitridation ONO Interpoly Dielectric (MN-ONO) for Highly Reliable and High Performance NAND Flash Memory 31
C. H. Liu, Y. M. Lin, Y. Sakamoto, R. J. Yang, D. Y. Yin, P. J. Chiang, H. C. Wei, C. Y. Ho. S. H. Chen, H. P. Hwang, C. H. Hung, S. Pittikoun, S. Aritome

Forming-free HfO$_2$ Bipolar RRAM Device with Improved Endurance and High Speed Operation 33
Yu-Sheng Chen, Tai-Yuan Wu, Pei-Jer Tzeng, Pang-Shiu Chen, Heng-Yuan Lee, Cha-Hsin Lin, Frederick Chen, Ming-Jinn Tsai

Multi-Level Phase Change Memory Using Slow-Quench Operation: GST vs. GSST 35
Der-Sheng Chao, Frederick T. Chen, Yen-Ya Hsu, Wen-Hsing Liu, Chain-Ming Lee, Chih-Wei Chen, Wei-Su Chen, Ming-Jer Kao, Ming-Jinn Tsai

SESSION 4: GREEN DEVICES (ALL INVITED TALKS)

The Standby Power Challenge: Wake-up Receivers to the Rescue 37
Jan Rabaey

Realizing Steep Subthreshold Swing with Impact Ionization Transistors 38
Yee-Chia Yeo

Optimizing Tunnel FET Performance – Impact of Device Structure, Transistor Dimensions and Choice of Material 40
Joachim Knoch

Nanoelectromechanical Systems for Ultra-Low-Power Computing and VLSI 42
Philip Feng

SESSION 5: HIGH-K/METAL GATE

Single-Metal Dual-Dielectric (SMDD) Gate-First CMOS Integration Towards Low VT and High Performance (Invited) 43
L-Å Ragnarsson, T. Schram, E. Röhr, F. Sebaai, P. Kelkar, M. Wada, T. Kauerauf, M. Aoulaiche, M. J. Cho, S. Kubicek, A. Lauwers, T. Y. Hoffmann, P. P. Absil, S. Biesemans

Low Capacitance Approaches for 22nm Generation Cu Interconnect (Invited) 45
T. I. Bao, H. C. Chen, C. J. Lee, H. H. Lu, S. L. Shue, C. H. Yu

A V$_{FB}$ Tunable Single Metal Single Dielectric Approach Using As I/I into TiN/HfO$_2$ for 32nm Node and Beyond 51
J. Petry, G. Boccardi, R. Singanamalla, C. S. Liu, K. Xiong, P. Escanes, J. L. Huguenin, J. Tseng, L. Van Nimwegen, F. Voogt, C.W.T. Bulle-Lieuwma, M. Müller

La-doped metal/High-K nMOSFET for Sub-32nm HP and LSTP Application 53
C. S. Park, J. W. Yang, M. M. Hussain, C. Y. Kang, J. Huang, P. Sivasubramani, C. Park, K. Tateiwa, Y. Harada, J. Barnett, C. Melvin, G. Bersuker, P. D. Kirsch, B. H. Lee, H. H. Tseng, R. Jammy

Extending Spectroscopic Ellipsometry for Identification of Electrically Active Defects in Si/SiO$_2$/High-k/Metal Gate Stacks 55
J. Price, G. Bersuker, P. S. Lysaght, H. H. Tseng

22nm CMOS Approaches by PVD TiN or Ti-Silicide as Metal Ga 57
C. S. Liu, G. Boccardi, H. Y. Wang, C. T. Lin, J. Petry, M. Müller, Z. Li, C. Zhao, C. H. Yu

Reliability Assessment of Low |V$_t$| Metal High-k Gate Stacks for High Performance Applications 59
C. D. Young, G. Bersuker, P. Khanal, C. Y. Kang, J. Huang, C. S. Park, P. Kirsch, H. H. Tseng, R. Jammy

High Performance Metal/Insulator/Metal Capacitors Using HfTiO as Dielectric 61
Hsiao-Hsuan Hsu, Bing-Yue Tsui

High-K/ Metal-Gate Stack Work-function Tuning by Rare-Earth Capping Layers: Interface Dipole or Bulk Charge? 63
H.Y. Yu, S.Z. Chang, M. Aoulaiche, B. Kaczer, P. Absil, C. Adelmann, T. Hoffmann, S. Biesemans, C. Wann, Y. J. Mii

SESSION 6: CMOS

Scaling Challenges of MOSFET for 32nm node and Beyond (Invited) 65
Yasuo Nara

P-FinFETs with Al Segregated NiSi/p$^+$-Si Source/Drain Contact Junction for Series Resistance Reduction 67
Mantavya Sinha, Rinus T. P. Lee, Sivasubramaniam Nandini Devi, Guo-Qiang Lo, Eng Fong Chor, Yee-Chia Yeo

Impacts of NBTI on SRAM Array with Power Gating Structure 69
Hao-I Yang, Ching-Te Chuang, Wei Hwang

CMOS Technology Roadmap Projection Including Parasitic Effects 71
Lan Wei, Frédéric Boeuf, Thomas Skotnicki, H. S. Philip Wong

The Device Architecture Dilemma for CMOS Technologies: Opportunities & Challenges of FinFET over Planar MOSFET 73
B. Parvais, A. Mercha, N. Collaert, R. Rooyackers, I. Ferain, M. Jurczak, V. Subramanian, A. De Keersgieter, T. Chiarella, C. Kerner, L. Witters, S. Biesemans, T. Hoffman

Investigation of Low Frequency Noise in Uniaxial Strained PMOSFETs 75
Jack J. Y. Kuo, William P. N. Chen, Pin Su

FDSOI CMOS with Dual Backgate Control for Performance and Power Modulation 77
Jeng-Bang Yau, Jin Cai, Leathen Shi, Robert. H. Dennard, Arvind Kumar, Katherine L. Saenger, Alexander Reznicek, Paul. M. Solomon, Qiqing Ouyang, Steven Koester, Wilfried E. Haensch

Sub-32nm CMOS Technology Enhancement For Low Power Applications 79
R. M. Huang, P. W. Liu, E. C. Liu, W. T. Chiang, S. H. Tsai, Jonas Tsai, Tzermin Shen, C. H. Tsai, C. T. Tsai, G. H. Ma

Highly Performant FDSOI pMOSFETs with Metallic Source/Drain 81
T. Poiroux, M. Vinet, F. Nemouchi, V. Carron, Y. Morand, B. Previtali, S. Descombes, L. Tosti, O. Cueto, L. Baud, V. Balan, M. Rivoire and S. Deleonibus, O. Faynot

High-k/Metal Gate Low Power Bulk Technology - Performance Evaluation of Standard CMOS Logic Circuits, Microprocessor Critical Path Replicas, and SRAM for 45nm and Beyond 83

D. G. Park, K. Stein, K. Schruefer, Y. Lee, J. P. Han, W. Li, H. Yin, C. Pacha, N. Kim, M. Ostermayr, M. Eller, S. Kim, K. Kim, S. Han, K. von Arnim, N. Moumen, M. Hatzistergos, T. Tang, R. Loesing, X. Chen, D. Jaeger, H. Zhuang, J. Chen, W. Yan, T. Kanarsky, M. Chowdhury, Jens Haetty, D. Schepis, M. Chudzik, V. Y. Theon, S. Samavedam, V. Narayanan, M. Sherony, R. Lindsay, A. Steegen, R. Divakaruni, M. Khare

NGL Overview 86

Burn Lin

Double Patterning Interactions with Wafer Processing, OPC and Physical Design Flows 87

Kevin Lucas

Multiple Electron Beam Maskless Lithography for High-Volume Manufacturing 88

Jack J.H. Chen, S.J. Lin, T.Y. Fang, S.M. Chang, Faruk Krecinic, Burn J. Lin

Status and Challenges of Extreme UV Lithography 90

Kurt Ronse, Eric Hendrickx, Mieke Goethals, Rik Jonckheere, Geert Vandenberghe

Recent Developments in NAND Flash Scaling (Invited) 92

Krishna Parat

Modeling and Scaling Evaluation of Junction-Free Charge-Trapping NAND Flash Devices 94

Yi-Hsuan Hsiao, Hang-Ting Lue, Kuang-Yeu Hsieh, Rich Liu, Chih-Yuan Lu

Gate Last MOSFET with Air Spacer and Self-Aligned Contacts for Dense Memories 96

Jemin Park, Chenming Hu

A Physics-Based Compact Model of Quantum-Mechanical Effects for Thin Cylindrical Si-Nanowire MOSFETs 98

Bastien Cousin, Olivier Rozeau, Marie-Anne Jaud, Jalal Jomaah

A New Technique to Extract the Gate Bias Dependent S/D Series Resistance of Sub-100nm MOSFETs 100

Dominique Fleury, Antoine Cros, Grégory Bidal, Hugues Brut, Emmanuel Josse, Gérard Ghibaudo

Analytical Modeling of Accumulation-Mode Suspended-Gate MOSFET and Process Challenges for Very Low Operating Power Devices 102

M. Collonge, M. Vinet, M. Ribeiro, J.-M. Pedini, B. Previtali, T. Ernst, S. Bécu, G. Ghibaudo

A New Robust Non-local Algorithm for Band-to-band Tunneling Simulation and its Application to Tunnel-FET 104

C. Shen, L. T. Yang, E. H. Toh, C. H. Heng, G. S. Samudra, Y. C. Yeo

Investigation of Static Noise Margin of Ultra-Thin-Body SOI SRAM Cells in Subthreshold Region using Analytical Solution of Poisson's Equation 106

Vita Pi-Ho Hu, Yu-Sheng Wu, Ming-Long Fan, Pin Su, Ching-Te Chuang

The Promise and Implementation of Three Dimensional Integration (Invited) 108

Subramanian Iyer

Optimization of the Channel Lateral Strain Profile for Improved Performance of Multi-Gate MOSFETs 109

L. De Michielis, K. E. Moselund, D. Bouvet, P. Dobrosz, S. Olsen, A. O'Neill, L. Lattanzio, M. Najmzadeh, L. Selmi, A. M. Ionescu

A Novel Double-Gated Nanowire TFT and Investigation of its Size Dependency 111

Wei-Chen Chen, Chuan-Ding Lin, Horng-Chih Lina, Tiao-Yuan Huang

Fermi Level Depinning for the Design of III-V FET Source/Drain Contacts 113

Jenny Hu, Ximeng Guan, Donghun Choi, James S. Harris, Krishna Saraswat, H. S. Philip Wong

Influence of Gate Misalignment on the Electrical Characteristics of MuGFETs................................115
Chi-Woo Lee, Aryan Afzalian, Ran Yan, Nima Dehdashti, Weize Xiong, Jean-Pierre Colinge

FinFET Resistance Mitigation through Design and Process Optimization................................117
Cindy Wang, Josephine Chang, Chung-Hsun Lin, Arvind Kumar, Andreas Gehring, Jin Cho, Amlan Majumdar, Andreas Bryant, Zhibin Ren, Kevin Chan, Thomas Kanarsky, Xinlin Wang, Omer Dokumaci, Michael Guillorn, Marwan Khater, Qingyun Yang, Xi Li, Munir Naeem, Judson Holt, Yongsik Moon, John King, John Yates, Ying Zhang, Dae-gyu Park, Christine Ouyang, Wilfried Haensch

Sub-100µW Low Power Operation of Vibrating Body FETs................................119
Daniel Grogg, Adrian Mihai Ionescu

Inversion-channel GaN MOSFET Using Atomic-layer-deposited Al_2O_3 as Gate Dielectric................................121
Y. C. Chang, W. H. Chang, H. C. Chiu, Y. H. Chang, L.T. Tung, C. H. Lee, M. Hong, J. Kwo, J. M. Hong, C. C. Tsai

A Comparison Study of Silicon Nanowire Transistor with Schottky-Barrier Source/Drain and Doped Source/Drain................................123
Zhaoyi Kang, Liangliang Zhang, Runsheng Wang, Ru Huang

High Mobility SiGe Shell-Si Core Omega Gate pFETS................................125
Hemant Adhikari, Harlan R. Harris, Casey E. Smith , Ji-Woon Yang, Brian Coss, Srivatsan Parthasarathy, Bich-Yen Nguyen, Paul Patruno, Tejas Krishnamohan, Ian Cayrefourcq, Prashant Majhi, Raj Jammy

Metal-Oxide-Semiconductor Devices with $UHV-Ga_2O_3(Gd_2O_3)$ on Ge(100)................................128
L. K. Chu, T. D. Lin, C. H. Lee, L. T. Tung, W. C. Lee, R. L. Chu, C. C. Chang, M. Hong, J. Kwo

Self-aligned Inversion Channel $In_{0.53}Ga_{0.47}As$ N-MOSFETs with $ALD-Al_2O_3$ and MBE $Al_2O_3/Ga_2O_3(Gd_2O_3)$ as Gate Dielectrics................................130
H. C. Chiu, T. D. Lin, P. Chang, W. C. Lee, C. H. Chiang, J. Kwo, Y. S. Lin, Shawn S. H. Hsu, W. Tsai, M. Hong

Inversion-Type Surface Channel $In_{0.53}Ga_{0.47}As$ Metal-Oxide-Semiconductor Field-Effect Transistors with Metal-Gate/High-k Dielectric Stack and CMOS-Compatible PdGe Contacts................................132
Hock-Chun Chin, Xinke Liu, Leng-Seow Tan, Yee-Chia Yeo

Sub-100nm High-K Metal Gate GeOI pMOSFETs Performance: Impact of the Ge Channel Orientation and of the Source Injection Velocity................................134
C. Le Royer, A. Pouydebasque, K. Romanjek, V. Barral, M. Vinet, J.-M. Hartmann, E. Augendre, H. Grampeix, L. Lachal, C. Tabone, B. Previtali, R. Truche, F. Allain

Tri-gated Poly-Si Nanowire SONOS Devices................................136
Hsing-Hui Hsu, Ta-Wei Liu, Chuan-Ding Lin, Chiu Kuo-Jung, Tiao-Yuan Huang, Horng-Chih Lin

Overall Operation Considerations for a SONOS-based Memory................................138
C. H. Lee, W. H. Tu, L. H. Chong, S. H. Gu, K.F. Chen, Y. J. Chen, J. Y. Hsieh, I. J. Huang, N. K. Zous, T. T. Han, M. S. Chen, W. P. Lu, K. C. Chen, Tahui Wang, C. Y. Lu

Reliability Study of MANOS with and without a SiO_2 Buffer Layer and BE-MANOS Charge-Trapping NAND Flash Devices................................140
Chien-Wei Liao, Sheng-Chih Lai, Hang-Ting Lue, Ming-Jui Yang, Chin-Yen Shen, Yi-Hsien Lue, Yu-Fong Huang, Jung-Yu Hsieh, Szu-Yu, Wang, Guang-Li Luo, Chao-Hsin Chien, Kuang-Yeu Hsieh, Rich Liu, Chih-Yuan Lu

Reliability of Planar and FinFET SONOS Devices for NAND Flash Applications – Field Enhancement vs. Barrier Engineering................................142
Tzu-Hsuan Hsu, Hang-Ting Lue, Sheng-Chih Lai, Ya-Chin King, Kuang-Yeu Hsieh, Rich Liu, Chih-Yuan Lu

Band Engineered Tunnel Oxides for Improved TANOS-type Flash Program/Erase with Good Retention and 100K Cycle Endurance 144
David C. Gilmer, Niti Goel, Sarves Verma, Hokyung Park, Chanro Park, Gennadi Bersuker, Paul D. Kirsch, Krishna C. Saraswat, Raj Jammy

Author Index

FOREWORD

This year, with the aim of stimulating the interactions between the technology and design communities, the executive committees of 2009 VLSI-TSA symposium and 2009 VLSI-DAT symposium have decided to have both symposia overlap for two days. A joint plenary session with two keynote speeches has been organized accordingly for topics of common interest to the VLSI-TSA and VLSI-DAT attendees. The joint plenary session features the following keynote speeches: "Efficient Analog Signal processing in nm CMOS Technologies" by Dr. Fang-Churng Tseng from TSMC, Taiwan and "From Living Faster to Living Better" by Dr. Rene Penning de Vries from NXP, The Netherlands.

Aside from the joint plenary session, the 2009 VLSI-TSA symposium program includes two additional plenary talks, two special sessions (one on Green Devices and the other one on Next Generation Lithography), nine technical sessions, and two short courses. The Technical Program Committee has again made an effort to put together a program with invited and contributed papers of high quality. These papers will be presented by industrial and academic experts and students from more than 10 countries.

The two VLSI-TSA plenary talks feature the following keynote speeches: "Has The Sun Finally Risen on Photovoltaics?" by Dr. Mark Pinto from Applied Materials, USA and "Carrier Transport and Stress Engineering in Advanced Nanoscale MOS Transistors" by Dr. Ken Uchida from Tokyo Institute of Technology, Japan.

Two special emerging technology sessions feature the progress of Green Devices and Next Generation Lithography. Eight experts are invited to present their cutting-edge results. In the regular sessions of the symposium, invited and contributed papers will be presented in the following nine sessions: Process Module; Strain Technology; Non-Volatile Memories; High-k/Metal Gate; CMOS; Scaling Trends and Related Modeling; Non-Traditional CMOS; High Mobility Channel, Memories and SONOS.

Two short courses will cover "SIP & 3D IC" and "Non-Volatile Memory". We are happy that four distinguished experts have accepted to be an instructor in these short courses: Dr. Fred Roozeboom of NXP, Dr. Makoto Motoyoshi of ZyCube, Mr. Toshiharu Watanabe of Toshiba, and Dr. William J. Gallagher of IBM.

We would like to thank the keynote speakers, invited speakers and contributing authors for their efforts to contribute to the success of the conference. Appreciation is also extended to all the Technical Program Committee members who solicited, reviewed, and selected the papers. Our special thanks go to the regional Sub-Committee Chairs for their assistance in putting together an excellent technical program. Finally, we are most appreciative of the dedication and tireless efforts of the Local Organizing Committee members, who not only edited this Technical Digest but also took great care in preparing all logistic aspects of the 2009 VLSI-TSA symposium.

Technical Program Committee Chair

H.-S. Philip Wong
Professor of Electrical Engineering
Stanford University

2009 International Symposium on VLSI Technology, Systems and Applications Organization

Symposium Chair

Roger De Keersmaecker   IMEC, Belgium

Symposium Co-Chair

Lewis Terman   IBM, USA
Yi-Jen Chan   ITRI, Taiwan

Steering Committee

Morris Chang   TSMC, Taiwan
John Y. Chen   Wafertech, USA
Tze-Chiang Chen   IBM, USA
Alice M. Chiang   Teratech, USA
Ben M.Y. Hsiao   IBM Emeritus, USA
Genda Hu   FocalTech Systems, Taiwan
G. P. Li   UC Irvine, USA
T. P. Ma   Yale Univ., USA
Tak H. Ning   IBM, USA
Jyuo-Min Shyu   NTHU, Taiwan
Hsing-Huang Tseng   SEMATECH, U.S.A
Paul P. Wang   Pacific Venture Partners, Taiwan
Clement H. Wann   TSMC, Taiwan
Ran H. Yan   RealTek, USA
Ping Yang   TSMC, Taiwan
Hwa Nien Yu   IBM Emeritus, USA

Technical Program Committee

Chair

Philip H.S. Wong   Stanford University, USA

Co-Chair

Sheng-Fu Horng   ITRI, Taiwan

USA Subcommittee

Stefan K Lai	Being AMC, USA **(Chair)**
Ko-Min Chang	Freescale, USA
Kin P. Cheung	National Institute of Standards & Technology, USA
Nathan Cheung	University of California at Berkeley, USA
Raj Jammy	SEMATECH, USA
Derchang Kau	Intel, USA
Chung Lam	IBM, USA
Chun-Yung Sung	IBM, USA
Hong Tang	Yale University, USA
Chin-Yu Tsai	Texas Instruments Inc., USA
Been-Jon K Woo	Numonyx, USA

Asia Pacific Subcommittee

Ming-Jinn Tsai	EOL/ITRI, Taiwan **(Chair)**
Mansun Chan	Hong Kong University of Science and Technology
Steve Chung	National Chiao Tung University, Taiwan
Sanford Chu	Chartered Semiconductor, Singapore
Bau Tong Dai	National Nano Device Laboratory, Taiwan
Daewon Ha	Samsung Electronics, Korea
C. S. Hwang	Seoul National University
Meikei Ieong	TSMC, Taiwan
Wen-Yueh Jang	Winbond, Taiwan
Jin-Feng Kang	Peking University, China
Byoung-Hun Lee	Gwangju Institute of Science and Technology, Korea
Ming-Fu Li	Fudan University Zhang Jiang Branch, China
Jengping Lin	Nanya Technology, Taiwan
Rich Liu	Macronix International, Taiwan
Chee-Wee Liu	National Taiwan University, Taiwan
Oliver Lue	Macronix International, Taiwan
Mike Ma	UMC, Taiwan
Hiroshi Matsuo	Powerchip Semiconductor Corp., Taiwan
Jae Chul Om	Hynix, Korea
Elgin Quek	Chartered Semiconductor, Singapore
John Sudijono	Chartered Semiconductor, Singapore
Luan Tran	TSMC, Taiwan
Cheng-Tzung Tsai	UMC, Taiwan
Bing-Yue Tsui	National Chiao Tung University, Taiwan

Yee-Chia Yeo	National University of Singapore, Singapore
Makoto Yoshida	Samsung Electronics, Korea
C. H. Yu	TSMC, Taiwan
Zhiping Yu	Tsinghua University, China
Peter Zhao	ProMOS Technology, Taiwan
Wei Zhang	Fudan University Zhang Jiang Branch

Japan Subcommittee

Seiichiro Kawamura	JST, Japan **(Chair)**
Hidemitsu Aoki	Osaka University, Japan
Atsushi Hori	Matsushita Electric Industrial Company
Hajime Kurata	Fujitsu, Japan
Eiji Morifuji	Toshiba, Japan
Kenji Noda	NSCore, Japan
Hiroyuki Ota	AIST, Japan
Kunihiro Sakamoto	AIST, Japan
Nobuyuki Sano	University of Tsukuba, Japan
Kentaro Shibahara	Hiroshima University, Japan
Kazuo Terada	Hiroshima City University, Japan
Hitoshi Wakabayashi	Sony, Japan
Heiji Watanabe	Osaka University

European Subcommittee

Michel Brillouët	CEA-LETI, France **(Chair)**
Livio Baldi	ST Microelectronics, Italy
Casper Juffermans	NXP Semiconductors, Netherlands
Malgorzata Jurczak	IMEC, Belgium
Stephane Mofray	ST Microelectronics, Italy
Wolfgang Müller	Qimonda, Germany
Paolo Pavan	Universita di Modena e Reggio Emilia, Italy
Herbert Reichl	IZM, Germany
Heiner Ryssel	Fraunhofer Institute IIS-B, Germany
Klaus Schruefer	Infineon, Germany

Local Organizing Committee

Fang-I Chen **(Chair)**
Alen Hsieh **(Co-Chair)**
Fang-Ru Hsu **(Co-Chair)**
Yao-Chou Lu **(Co-Chair)**
Yu-Chuan Wei **(Co-Chair)**

Secretary	Clara Wu
	Elodie Ho
Art Work	Yuansu Hsu
Conference Assistance	Emily HM Cheng
IT Support	Zeng-Chin Chang
	Tzyy-Kae Chang
	Kun-Shu Gorg
	Tzong-Ying Jan
	Ching-Hsiao Yu
Public Relations	Su-Jen Hsiao
Registration	Yvonne Chen
Treasurer	Yu-Jen Wu

Has The Sun Finally Risen on Photovoltaics? (Keynote)

Mark Pinto

Applied Materials, USA

The idea of solar generated electricity dates to discovery of the photovoltaic (PV) effect in 1839 through to the first silicon solar cell in 1954. But even with concerns about oil and the environment, PV currently generates less than 0.1% of the world's electricity. We present here the case that PV is on the verge of becoming a major source of electrical power through a principle similar to that which underlies VLSI – the reduction of unit cost through nanomanufacturing.

The Future of Semiconductor Industry - A Foundry's Perspective

Fang-Churng Tseng

Vice chairman of Taiwan Semiconductor Manufacturing Co., Ltd and Chairman of Global Unichip Corp., Taiwan

The semiconductor industry has been in a path of steady growth tolerating waves of economical ups and downturns. We are currently facing a severe one that very few of us have encountered in our lifetime. Nevertheless, it is not a question whether the growth curve will resume. It is always a matter of when it does.

The adoption of communication, computer and consumer products is spreading across the world in both advanced and developing economies. Opportunities are global and the need for technologies are defined and shaped by different levels of economy but never stopped by the variances of economy.

Meantime, the semiconductor industry has been facing design challenges. Design technology struggles to keep up the pace and to consume all the new possibilities opened up by the new advanced process technologies. It is not a matter to complete the product design, it is equally important to create economical and competitive design. Designers have never cease the dream of creating "just right" designs "just in time".

The knowledge to create product and the resource to complete the design is a necessary condition. While the knowledge and the resource to create optimized design will become the major differentiation. Design and process co-optimization reduces design waste and increase design productivity. TSMC is committed with Open Innovation Platform (OIP) to deliver this in addition to our long-term investment in leading edge technology development and GigaFab capacity expansion needed to enable industry's total innovation.

From Living Faster to Living Better

René Penning de Vries
SVP & CTO, NXP Semiconductors
High Tech Campus 60
5656 AG Eindhoven, The Netherlands

ABSTRACT

For decades semiconductor developments have been driven by Moore's Law productivity gains. This led to extremely fast digital processors, increase in bandwidth and huge memories that boost productivity of PCs, mobile phones, and other applications demanding heavy data traffic and storage. The economics of the IC industry as well as developments in society, however, will cause the trend to turn, leading to a paradigm shift in the semiconductors world.

Our industry is confronted with sky-rocketing cost levels for System-On-Chip development in advanced CMOS technology. Secondly, ongoing shrinking the physical dimensions will eventually cause Moore's Law to come to an end.

At the same time, consumer demand shifts from ever more productive consumer electronics to a variety of new smart products adding value to daily life. A globally graying population demands for innovations in medical devices and food safety. Raising environmental awareness demands for smart "green" solutions. In transportation IC innovations help to bring safety, to overcome traffic congestion, and to deliver all thinkable information to consumers on the move.

These and many more applications are typically realized by integration of existing CMOS technologies with "More-than-Moore" technologies, like Analog/Mixed-Signal, High-Voltage, and MEMS.

In his keynote presentation we will elaborate on the paradigm shift in the semiconductor industry, and present a number of "More-than-Moore" applications and underlying enabling technologies.

PC POWER SUPPLY EFFICIENCY

With vast and growing numbers of PCs and domestic appliances spending much time in standby mode, the global drain on energy has shifted attention to achieving high efficiency right across the load range. IEA estimates of standby power imply this is 5 – 10% of total electricity use in most homes and an unknown amount in industry.

Legislation is now in force or imminent in many territories (e.g., CA, USA [1]) to require AC mains power blocks to operate at high efficiency at minimum load, effectively banning classical transformer supplies. To achieve these savings at an economic cost, special technologies must be deployed where there is intelligence in the operating mode as well as intrinsically high efficiency in customized power devices.

The NXP "GreenchipPC" family [2] is a good example, meeting the Energy Star "80plus" target [3], and reducing losses in a PC power supply by more than a factor of two for all modes from full power to standby by a combination of optimized architecture and design together with special semiconductor technology (Figure 1). Integration of features, as typically done in Moore's law applications, is not applicable here; three different IC technologies are employed to achieve optimized overall performance, with integrated HV devices on the primary controller and power factor correction, a general purpose 12V technology for the main controller and a 5V device to control the synchronous rectifiers in the high current secondaries.

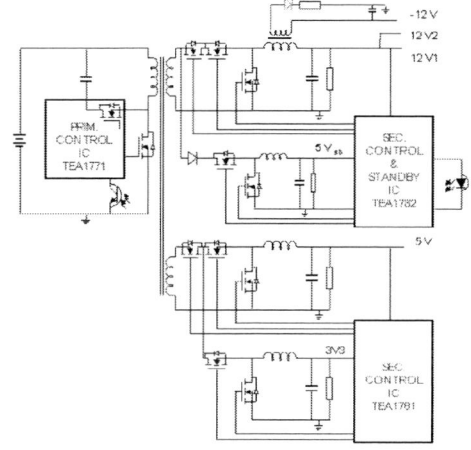

FIGURE 1. GREENCHIP PC POWER SUPPLY

LED FLASH DRIVER

LEDs promise lighting solutions that are small, adaptable, energy-saving, controllable in both color and intensity, and with a long life-time. The widespread adoption of LED lighting is determined by the quality of low-cost, highly integrated driver electronics.

LED lamps have similar efficiency gains as incandescent lamps, but operate at relatively low voltages, placing new demands on efficient AC/DC conversion, and include features such as output and colour balance control.

Today, dimming of LEDs through standard TRIAC dimmers is already available. Further developments in LED drivers and controllers are aimed at ongoing energy consumption reduction, reduced heat generation allowing smaller form factors, and power factor correction (PFC) for harmonic distortion regulation at powers > 25W.

978-1-4244-2784-0/09 $25.00 © 2009 IEEE

Further transition from conventional incandescent, Halogen, and CFL lamps to LED lighting is expected, but is hampered by cost levels rather than technology maturity.

Intelligent Car Keys

Today's car keys allow for immobilization and for remote access through one-way communication. The latest developments have turned car keys into systems combining many functions and features. Equipped with LEDs or a display and two-way communication functionality, within the operating range the car owner can remotely check the status of the vehicle, e.g. locked or unlocked.

A standardized interface in the key compliant to Near-field communication technology will enable further functionality. For example, the car location and also the fuel level or the tire pressure can be displayed by NFC mobiles. In case the mobile contains GPS, it can even lead the user back to the car. Based on the same standardized interface, adding a secure element (like NXPs SmartMX) allows for payments via the car key.

The options for further extension of features in the car key are numerous, and can be realized by smart integration of existing technologies in mature CMOS, high-voltage, analog, RF, MEMS and advanced packaging.

Silicon Based TV Tuners

Until recently TV tuners were constructed with hundreds of discrete components and were therefore bulky, needed many manual alignments and were manufactured for a single or just a few standards. Classical TV tuner solutions depended heavily on discrete RF and Intermediate Frequency (IF) filtering.

The NXP silicon tuner was the first TV tuner with all selectivity integrated. This was possible thanks to the novel low-IF direct-conversion architecture. This architecture incorporates a full complex mixer in combination with both RF and IF poly-phase filtering for high image rejection and prevention of unwanted down-conversion with Local Oscillator (LO) harmonics.

The following low-IF demodulation of analogue modulated TV signals has been implemented completely in the digital domain with CMOS technology. This has enabled the integration of analogue TV demodulation in digital baseband processing ICs and requires no external components for receiving all standards.

MEMS

Being around for more than 50 years already, micro-electromechanical systems (MEMS) are far from new. But their widespread application in consumer and mobile electronics is really taking off lately. MEMS devices are usually combined with components to form complete systems in one advanced package, making them ultimate examples of "More-than-Moore" technology.

MEMS oscillators form an attractive alternative to conventional quartz timers in ICs. Apart from being cheap and small, they are accurate and designs are scalable. Moreover, MEMS oscillators can easily be integrated with an IC into a System-In-Package (SIP). Because of their CMOS compliant process technology they can also be fully integrated into the system-on-chip (SoC) design. However, this will not in all cases be the most cost effective solution.

Ultra Low Power Applications

Medical electronics is one of the fastest growing areas of semiconductor development. NXP's ultra low power solutions, based on magnetic induction radio technology and CoolFlux DSP, form the basis of the latest hearing aid products. The NXP chip supports a data rate of up to 298 kbps and bi-directional communication, enabling novel applications such as stereo audio streaming and binaural processing. Use of a hub allows easy and wireless connectivity of an MP3 player or mobile phone to the hearing aid. Also, on short range (<50 cm) the magnetic induction radio signal is much more energy efficient than an RF signal. It travels through the body tissue with low degradation of the signal strength. Magnetic induction radio signals also interact far less with human tissue than conventional RF signals, and hence cause less risk of tissue damage.

Other ultra-low power medical applications are implanted sensors monitoring vital body functions, devices for applications in neurostimulation and pain management, and cardiac rhythm management devices. These are all good examples of More-than-Moore technologies enabling living better, rather than living faster.

Conclusion

Above examples show that many innovations in semiconductors today are not driven by Moore's Law, but yet enable breakthroughs in applications and products in many areas. Although Moore's Law might continue to drive further scaling in advanced CMOS for some time for specific applications, we have come to the stage where application of state-of-the-art CMOS processes are no longer the default technological solution.

Moore's Law was primarily driven by the need to gain productivity: speed and capacity. Today, to meet the needs of a changing society semiconductors have to be smart, customized, cost effective, and easy to integrate in systems. A variety of "More-Than-Moore" technologies are available and under development, to provide the optimal semiconductor solution.

References

[1] "2007 appliance efficiency regulations", California Energy Commission, Dec 2007

[2] www.NXP.com/greenchip

[3] http://www.80plus.org/

Acknowledgements

Jan van Sinderen, NXP Research

Dirk Wenzel, NXP Automotive

Edwin Kluter, NXP Multi Market Semiconductors

Joost van Beek, NXP Research

Bill Redman-White, NXP Automotive

Antoine Delaruelle, NXP Multi Market Semiconductors

Philippe Maugars, NXP Analog/Mixed-Signal

978-1-4244-2784-0/09 $25.00 © 2009 IEEE

Carrier Transport and Stress Engineering in Advanced Nanoscale MOS Transistors

Ken Uchida[1] and Masumi Saitoh[2]

[1]Department of Physical Electronics, Graduate School of Engineering, Tokyo Institute of Technology
2-12-1 Ookayama, Meguro-ku, Tokyo 152-8552, Japan, Phone/Fax: +81-3-5734-3591, E-mail: uchidak@neo.pe.titech.ac.jp
[2]Advanced LSI Technology Laboratory, Research and Development Center, Toshiba Corporation

Abstract

This paper reviews the carrier transport mechanisms and stress engineering in advanced nanoscale MOSFETs. First, carrier transport in bulk (100) and (110) MOSFETs is reviewed. Subband structure engineering to enhance mobility as well as ballistic current is also examined.

Introduction

Mobility booster technologies, such as stress [1-3] and crystal-orientation engineering [4,5] (the optimization of surface orientations and channel directions), have attracted growing interests, because the scaling of device sizes is not enough to meet performance requirements against present and future FETs. In order to fully utilize the booster technologies and to design high-performance nanoscale MOSFETs, the precise understanding of carrier transport mechanisms is indispensable.

This paper reviews the carrier transport mechanisms and stress engineering in advanced nanoscale MOSFETs. First, carrier transport in bulk (100) and (110) MOSFETs is reviewed. We then discuss mobility in three-dimensional MOSFETs. Subband structure engineering to enhance mobility as well as ballistic current is also examined. Furthermore, future possible directions for new channel materials are addressed.

(100) and (110) MOSFETs on Bulk Substrates

A. Mobility in unstrained (110) MOSFETs

First, carrier transport in unstrained (110) MOSFETs is examined [6]. Note that larger energy splits between valleys or bands generally result in better mobility, since scattering events between valleys (bands) are suppressed. **Fig. 1** illustrates the equivalent energy valleys for (100) and (110) Si. As shown in **Fig.2**, electron mobility in (110) nFETs, $\mu_e^{(110)}$, is lower than that in (100) nFETs, $\mu_e^{(100)}$. The lower $\mu_e^{(110)}$ is mainly attributable to smaller energy split between 4-fold valleys, $\Delta_4^{(110)}$, and 2-fold valleys, $\Delta_2^{(110)}$ (**Fig.3**), as well as larger density-of-states (DOS) of the lower-energy valleys. Since both smaller split and larger DOS than those in (100) nFETs result in higher scattering frequency, lower $\mu_e^{(110)}$ than $\mu_e^{(100)}$ is reasonably explained. $\mu_e^{(110)}$ decreases more rapidly as N_s increases. The rapid decrease of $\mu_e^{(110)}$ at higher N_s is attributable to the non-parabolicity in E-k relationship along <110> at the energy minima.

For pFETs, as shown in **Fig. 4**, $\mu_h^{(110)}$ is much better than $\mu_h^{(100)}$. The larger energy split between the heavy-hole (HH) and light-hole (LH) bands in (110) pFETs, $\Delta E_{HH\text{-}LH}^{(110)}$, than that in (100) pFETs, $\Delta E_{HH\text{-}LH}^{(100)}$ contributes to better mobility in (110) pFETs. As shown in **Fig. 5**, $\Delta E_{HH\text{-}LH}^{(110)}$ is larger than optical phonon energy when substrate impurity concentration (N_{sub}) is high. As a result, optical phonon scattering, which is the dominant scattering mechanism in pFETs, can be suppressed. Whereas $\Delta E_{HH\text{-}LH}^{(100)}$ is much smaller than the optical phonon energy, and thus the optical phonon scattering is not suppressed. Consequently, higher $\mu_h^{(110)}$ than $\mu_h^{(100)}$ is reasonably explained. An interesting property in (110) pFETs is that mobility at a fixed N_s increases with an increase in N_{sub}, which is strongly contradict to our experience in (100) FETs (mobility always decreases as N_{sub} increases). However, this can be also explained by the increase in $\Delta E_{HH\text{-}LH}^{(110)}$ (decrease in optical phonon scattering) with an increase in N_{sub}.

B. Stress Engineering in (100) and (110) MOSFETs

Stress engineering can be understood in terms of energy split between valleys (bands) and change in effective mass [7]. For electrons, biaxial stress on (100) surface and uniaxial stress along <100> are understandable in terms of energy split. **Fig. 6** and **Fig. 7** show the energy split of valleys under (100) biaxial and <100> uniaxial stresses, respectively. For <110> uniaxial stress, both energy split (**Fig. 8**) and effective mass change (**Fig. 9**) need to be considered. **Fig. 10** shows electron mobility enhancement as a function of strain for various stress conditions. Because of the effective mass change, <110> uniaxial stress is most effective to enhance mobility in (100) nFETs at strains of greater than 0.7 %. In (110) nFETs, <110> uniaxial stress is most effective for mobility enhancement.

For pFETs, <110> uniaxial compressive stress is most important for mobility enhancement. Under <110> uniaxial stress, effective mass (m^*) change is great [8], while stress-induced $\Delta E_{HH\text{-}LH}$ in bulk is negligible. In (100) pFETs, it is reported that the change of m^* parallel and perpendicular to the stress (transport) direction is responsible for mobility enhancement. On the other hand, in (110) pFETs, m^* normal to the surface in the HH band gets heavier as the compressive stress increases. The heavier normal effective mass in the HH band results in wider $\Delta E_{HH\text{-}LH}$, leading to better hole mobility [9].

Three-dimensional Transistors

In three-dimensional FETs such as FinFETs and nanowires FETs, nanoscale Si channels and multi-gate structures are utilized. For nanoscale Si, **a)** potential fluctuations due to differences in quantum confinements from one part to another [10] and **b)** an increase in the form factors are concerns. The potential fluctuations form scattering centers and severely degrade mobility (**Fig. 11**) [11]; in UTB SOI FETs mobility is proportional to T_{SOI}^6, where T_{SOI} is the thickness of SOI. In addition, the increase in the form factor lowers mobility. Therefore, mobility degradation is observed in thinner-body FETs both for electrons and holes. However, at a condition, where the effect of energy split induced by quantum confinements is larger than that of the form factor, mobility enhancement can be observed (**Fig. 12**). In other words, there is a space to engineer the subband structures by quantum effects.

In the case of multi-gate FETs, the interaction of carriers at each gate and the redistribution of carriers within the channel take place As a result, the form factor is expected to decrease in double-gate FETs (volume inversion), resulting in mobility enhancement in general (**Fig. 13**).

Conclusions

The carrier transport mechanisms and stress engineering in (100) and (110) MOSFETs and 3D FETs are reviewed. The mobility difference between (100) and (110) FETs are attributable mainly to differences in energy split and density-of-state masses. The effect of stress on mobility enhancement is also understandable in terms of energy split and effective mass change. In future three-dimensional FETs such as FinFETs and nanowires FETs, nanoscale Si channels and will be utilized. In such a thin Si channel, Si thickness fluctuations should be avoided. The quantum effects in those thin films provide the chance to enhance mobility.

Acknowledgement: The authors would like to thank N. Fukushima, A. Nishiyama, and S. Oda for continuous supports. They also appreciate Y. Nishi, K. C. Saraswat, T. Krishnamohan for valuable discussions.

References:
[1] S. Takagi et al., IEDM, p57, 2003.
[2] T. Ghani et al., p978, IEDM, 2003.
[3] K. Uchida et al., IEDM, p229, 2004.
[4] H. C.-H. Wnag, IEDM, p67, 2006
[5] B. Yang et al., VLSI Tech., p126, 2007
[6] K. Uchida et al., IEDM, p1019, 2006.
[7] K. Uchida et al., IEDM, p135, 2005.
[8] S. Thompson et al., IEDM, p221, 2004.
[9] M. Saitoh et al., IEDM, p711, 2007.
[10] K. Uchida et al., IEDM, p47, 2002.
[11] K. Uchida et al., IEDM, p633, 2001.

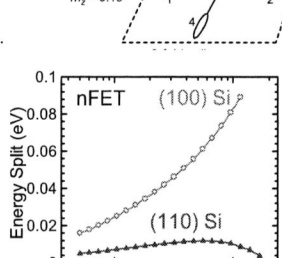

Fig. 1: 6-fold valleys in silicon are projected to (100) and (110) surfaces. In both cases, 6-fold valleys are split into 4-fold and 2-fold valleys. However, whereas 4-fold valleys are energetically lower than 2-fold valleys in (110), 2-fold valleys are lower than 4-fold valleys in Si (100). Between 2-fold and 4-fold valleys, difference in normal mass is larger in (100) Si, resulting in larger energy split in (100) Si. Although 2-fold and 4-fold valleys are isotropic on Si (100), they are anisotropic on Si (110).

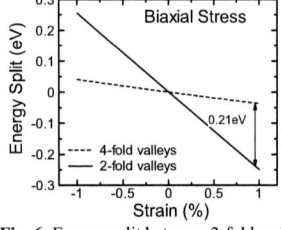

Fig. 2: Electron mobility, μ_e, as a function of surface carrier concentration, N_s, for (100) and (110) nFETs. μ_e of (110) nFETs depends on channel direction. The substrate impurity concentration, N_{sub}, for both substrates is approximately $1 \times 10^{15} \mathrm{cm}^{-3}$.

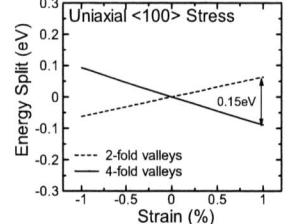

Fig. 3: Energy split between 2-fold and 4-fold valleys in (100) and (110) nFETs. The energy split in (110) nFETs is much smaller than that in (100) nFETs and it decreases at N_s of greater than $5 \times 10^{12} \mathrm{cm}^{-2}$.

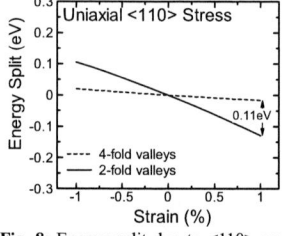

Fig. 4: Hole mobility, μ_e, as a function of Effective field, E_{eff}, for (100) and (110) pFETs. N_{sub}'s are $1.5 \times 10^{15} \mathrm{cm}^{-3}$ and $2.4 \times 10^{15} \mathrm{cm}^{-3}$ for (100) and (110) respectively. To calculate E_{eff}, η of 1/3 is utilized for both (100) and (110). The cannel direction is <110>.

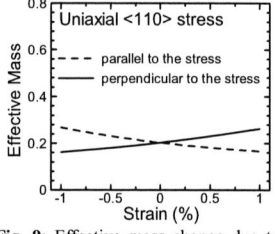

Fig. 5: Energy split, $\Delta E_{HH\text{-}LH}$, between heavy-hole (HH) and light-hole (LH) bands in (100) and (110) pFETs. In (110) pFET, $\Delta E_{HH\text{-}LH}$ is larger than optical phonon energy when N_{sub} is greater than $1 \times 10^{17} \mathrm{cm}^{-3}$.

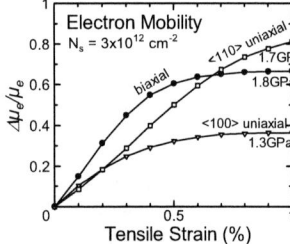

Fig. 6: Energy split between 2-fold and 4-fold valleys as a function of strain induced by biaxial stress on (100) Si. The energy split of 0.21 eV was obtained at the strain of 1 %. The energy split was calculated using the empirical pseudopotential method.

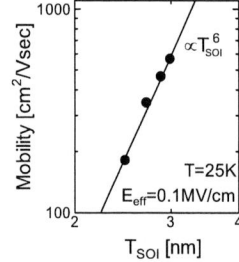

Fig. 7: Energy split due to <100> uniaxial stress as a function of strain induced by the stress. The <100> uniaxial stress also splits 6-fold valleys into 2-fold valleys and 4-fold valleys. The energy split of 0.15 eV was obtained at the strain of 1 %.

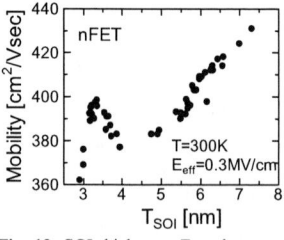

Fig. 8: Energy split due to <110> uniaxial stress as a function of strain induced by the stress. The <110> uniaxial stress also splits 6-fold valleys into 2-fold valleys and 4-fold valleys. The energy split of 0.11 eV was obtained at the strain of 1 %.

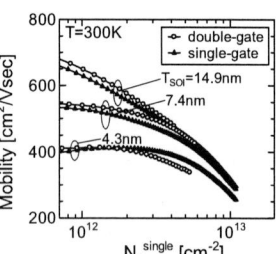

Fig. 9: Effective mass change due to <110> uniaxial stress. The effective mass parallel to the <110> stress is reduced as tensile strain increases, whereas the effective mass perpendicular to the stress is increases as tensile strain increases.

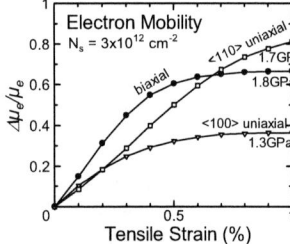

Fig. 10: Electron mobility enhancement ratio as a function of tensile strain for various stress conditions. Because of effective mass change, <110> uniaxial stress shows highest mobility enhancement ratio at tensile strain greater than 0.7%.

Fig. 11: Electron mobility in ultra-thin-body SOI MOSFETs as a function of SOI thickness, T_{SOI}, at low temperature. T_{SOI}^6 dependence indicates that mobility is limited by scattering induced by SOI thickness fluctuations.

Fig. 12: SOI thickness, T_{SOI}, dependence of electron mobility at room temperature. Electron mobility increases as T_{SOI} decreases, when T_{SOI} is in a range from 3.5 nm to 4.5 nm.

Fig. 13: Electron mobility in single-gate and double-gate nFETs as a function of N_s in single-gate FETs, which is the half of N_s in double-gate FETs. Double-gate mobility is generally higher than single-gate mobility.

978-1-4244-2784-0/09 $25.00 © 2009 IEEE

The Impact on Device Characteristics with STI formed by Spin-on Dielectric in High Density NAND Flash Memory

W. Z. Wong, J. J. Fan, J. D. Jiang, C. H. Huang, C. Y. Chen, H. H. Chen, C.C. Hsu, Rex Young, P. Y. Wang, H. Fujita and H. Kobayashi

Technology Development Division I, Powerchip Semiconductor Corp.,

NO. 12, LI-HSIN RD. 1, HSINCHU SCIENCE PARK, HSINCHU, TAIWAN, R.O.C.

E-Mail: weizhe@psc.com.tw

ABSTRACT

The electrical impact from adopting spin-on dielectric (SOD) for shallow trench isolation is demonstrated in this paper. Although perfect STI gap filling and suppressed re-oxidation of tunneling oxide near the active area (AA) edge are achieved through SOD process, some unexpected side effects occur. In peripheral area, severe corner thinning of thick gate oxide and positive fixed charge inside STI are observed, leading to distorted transistor I-V characteristics and deteriorated junction/well isolation capability. They are attributed to the mechanical stress from volume shrinkage when SOD material is transformed into pure silicon dioxide. [1]

INTRODUCTION

The application of high density flash memory, such as memory card and SSD, has been growing up rapidly in recent years. This demand for the flash memory with higher density and lower cost has pushed the memory fabricating technology into the nano-scale field. Different from advanced CMOS technology, more critical layers are required in whole process flow of memory technology. This paper is focused on the STI re-filling process, which is in the very early stage of whole fabricating steps. As shown in Fig. 1(a), obvious void defect is observed in the narrow STI region (memory array matrix) while conventional high density plasma CVD (HDPCVD) oxide is used. To overcome this problem, several methods had been proposed, like HARP [2] and SOD. [3] Figure 1(b) shows the re-filling result by SOD process in the narrower STI region. Compared with HDPCVD process, excellent gap filling and void free STI structure are obtained. However, there are two major concerns. One is the mechanical stress induced from volume shrinkage of SOD material during cure and annealing process. These cure and annealing steps play a very important role to transform SOD material into the pure and dense silicon dioxide. The other is the released solvent and bi-product, such as NH_3, from the above-mentioned process. They may penetrate into the active region and impact the device performance. In this paper, we will explore the electrical impact from adopting SOD process for STI formation in flash memory technology.

EXPERIMENTAL

Conventional NAND strings, along with peripheral devices, were fabricated by 50nm design rule technology. SOD process was adopted for STI formation. The cross-sectional views of memory array matrix along BL- and WL- directions are illustrated in Fig. 2(a), (b) respectively. Both the WL pitch and BL pitch are 100nm. Figure 2(c) also shows the cross-section of the peripheral transistor along its width direction since it is important for the later discussion. Another sample with identical process flow except STI formation by HDPCVD process was prepared for comparison.

RESULT AND DISCUSSION

(1) The impact on memory cell device

As the channel width of memory cell is scaling down, the control on re-oxidation phenomenon of tunneling oxide near the AA edge becomes more and more critical. It was reported that the programming efficiency and operating reliability of the memory cell would be seriously affected by such oxide encroachment. [4] Figure 3 shows the amount of oxide encroachment in the tunneling oxide capacitor with SOD and HDPCVD process respectively. Both less increment in oxide thickness and shorter encroaching distance are found for SOD process. The reason is that apparent oxide thickening near the AA edge already happened during the STI gap re-filling by HDPCVD process. The corresponding endurance and retention results of the memory cell are presented in Fig. 4. As expected, significant improvement in reliability performance for SOD process is achieved.

(2) The impact on peripheral devices

To check the gate oxide condition near the AA edge in the peripheral area, the I-V characteristics of the field-edge-intensive MOS capacitor with thick gate oxide are measured. (Fig. 5) The tunneling current of SOD case is ten times higher than that of HDPCVD, even though their electrical oxide thickness extracted by C-V method is very close. The amount of oxide encroachment in the thick gate oxide capacitor is also checked and the result is shown in Fig. 6. It is amazing that the oxide profile of SOD process already shows corner thinning, possibly resulted from the accompanied mechanical stress. That's why the tunneling current of SOD case is larger. But so far we still can't explain the difference of oxide profile in the tunneling oxide capacitor and the thick gate oxide capacitor.

As for the isolation problem revealed in [3], we also observe that n+-to-n+ and n+-to-Nwell isolations become much weaker for SOD process. To clarify the model, we monitor the I-V characteristic of the n-type field transistor. As seen in Fig. 7, the turn-on characteristic of SOD case occurred ten volts earlier than that of HDPCVD. Different from Choi's model, we suspect that some positive fixed charge may exist in the interface between SOD and linear oxide. Many dangling bounds may be generated at the time when volume shrinkage of SOD material occurred.

Finally, the I-V characteristics of the nMOS transistor with thick gate oxide are presented in Fig. 8. Compared with HDPCVD case, serious kink effect appears in the I-V curve for SOD process. It seems reasonable because the above-mentioned phenomena, oxide corner thinning and positive fixed charge inside STI, will force the corner transistor to turn on earlier, and thus induces obvious kink effect. To make SOD process feasible for flash memory technology, we further optimize the SOD-related process, for example pre-cure condition, and add extra well engineering. The improved results are also shown in Fig. 8. After optimization, the characteristics of the nMOS transistor are comparable to that of HDPCVD case.

CONCLUSIONS

SOD process is adopted for narrow STI formation in advanced NAND Flash technology due to its excellent gap filling capability. The reliability of memory cell is also improved due to less tunneling oxide re-oxidation near the AA edge. However, the accompanied side effects, severe corner thinning of thick gate oxide and positive fixed charge inside STI in peripheral area, will seriously degrade peripheral device characteristics. Therefore, the process optimization and extra well engineering for SOD process are essential to its practicability.

REFERENCES

[1] H. Fujita, et al., "The Method of Gap Filling for Shallow Trench Isolation by Spin-on Dielectric Material for High Density NAND Flash device," IEEE VLSI-TSA, submitted paper

[2] Armin T. Tilke, et al., "Shallow Trench Isolation for the 45-nm CMOS Node and Geometry Dependence of STI Stress on CMOS Device Performance," IEEE Trans. Semiconductor Manufacturing, vol. 20, NO. 2, pp.59-67, May 2007

[3] J. S. Choi, et al., "A Shallow Trench Isolation using Novel Polysilazane-based SOG for Deep-Submicron Technologies and Beyond" IEEE ISSM,2003, pp. 419-422

[4] Jae-Duk Lee, et al., "Degradation of Tunnel Oxide by FN Current and Its Effects on Data Retention Characteristics of 90-nm NAND Flash Memory Cells," in proceeding of the IRPS, 2003, pp.497-501

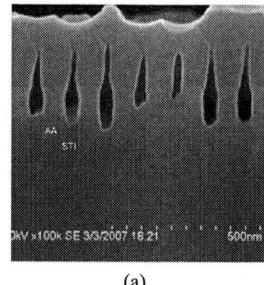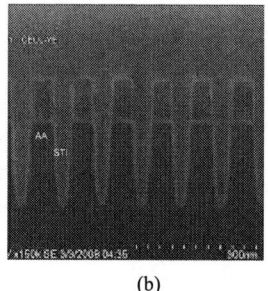

(a) (b)

Fig. 1 SEM photos of the narrow STI patterns filled by HDPCVD oxide (a) and SOD oxide (b).

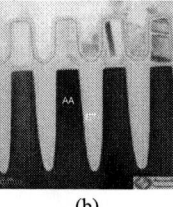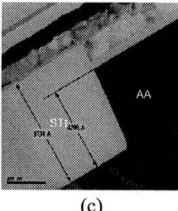

(a) (b) (c)

Fig. 2 TEM photos of memory array matrix along BL-(a), WL-(b) directions, and of peripheral MOSFET along its width direction (c).

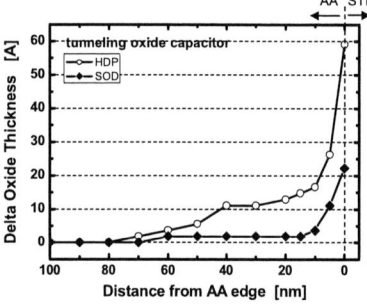

Fig. 3 The amount of oxide encroachment monitored in the MOS capacitor with tunneling oxide.

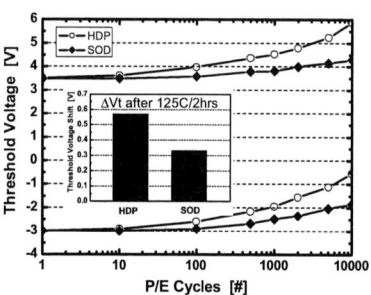

Fig. 4 The P/E cycling characteristics of the memory cell. The inset shows the data retention characteristics.

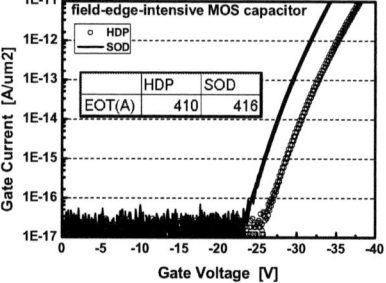

Fig. 5 I-V characteristics of the field-edge-intensive MOS capacitor with thick gate oxide.

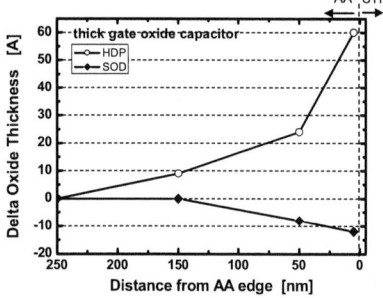

Fig. 6 The amount of oxide encroachment monitored in the MOS capacitor with thick gate oxide.

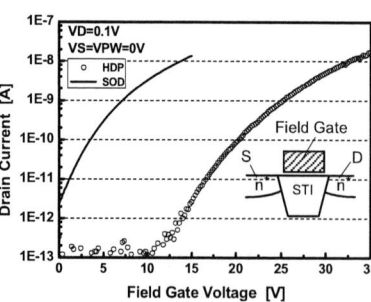

Fig. 7 Id-Vg characteristics of the n-type field transistor in peripheral Pwell.

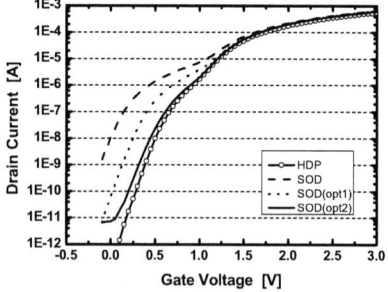

Fig. 8 Id-Vg characteristics of the peripheral nMOS transistor with thick gate oxide.

978-1-4244-2784-0/09 $25.00 © 2009 IEEE

Properties of Very Thin Adenine Layer with High Inhibition for 32nm Node Cu/Low-K Interconnection

M.Hara, H.Aoki, T.Masuzumi, D.Watanabe, C.Kimura and T.Sugino

Department of Electrical, Electronic and Information Engineering, Osaka University,
2-1 Yamadaoka, Suita-Shi, Osaka 565-0871, Japan

ABSTRACT

An effective inhibition with very thin layer is required for Cu/Low-K interconnection of next generation devices. We have achieved an effective suppression of Cu oxidation using adenine as an environmentally friendly material. By using electrochemical measurements, we find that the adenine layer can inhibit Cu oxidation by forming the very thin layer compared with Benzotriazol (BTA) as a conventional Cu inhibitor.

INTRODUCTION

The suppression of Cu oxidation is one of serious issues in the reliability of Cu interconnection for 32nm node device and beyond. BTA has been used as the conventional Cu corrosion inhibitor. However, it is difficult to remove BTA layer on the Cu surface by a light pre-sputtering before barrier layer formation because BTA has a tendency to form thick layer on the Cu surface. (Fig. 1) Especially, as a long time pre-sputtering process to remove the thick inhibitor layer causes damages of Cu interconnection, a thin inhibitor layer is needed for next generation devices. In addition, BTA has a large environmental impact because it creates mutagenicity and biodegrades poorly. Thus, a very thin layer with high inhibition composed of environmentally friendly material is required. As environmentally friendly Cu corrosion inhibitors to replace BTA, there are several typical natural heterocyclic nitrogen compounds such as adenine. It is reported that adenine is an environmentally friendly material from biodegradability examination.[1,2] Especially, adenine is a compound occurring in living organisms. In this study, we have investigated properties of inhibition of adenine and BTA layer and thickness of inhibitor layer estimated by electrochemical measurement.

EXPERIMENTAL

Figure 2 shows the chemical structures of BTA ($C_6H_5N_3$) and adenine ($C_5H_5N_5$). The XPS measurements were carried out to examine the oxidation of the Cu surface after treated with Cu corrosion inhibitor (BTA or adenine). Impedance of inhibitor layer was measured by using an electrochemical system with three electrodes. The impedance measurements were performed in the frequency range from 0.1Hz to 20MHz. The Cu working electrode was treated with the inhibitors (BTA, adenine: 0.1wt%, 1-5min, 25°C) by dipping immediately after treatment with oxalic acid (0.1wt%, 1min, 25°C). Oxalic acid was used to remove the natural oxide layer on the Cu surface.

RESULTS AND DISCUSSION

Figure 3 shows the XPS spectra from the Cu (2p) core level of the Cu surfaces with BTA or adenine treatments after the accelerated exposure test. These samples kept for the ambience of high temperature (100℃) and high humidity (100%) for an hour. The Cu-O peak of BTA at 933.6 eV is clearly observed compared with adenine. It is found that the adenine has superior inhibition than BTA.

Figure 4 shows the XPS spectra of the N (1s) on the Cu surface with adenine treatments before and after first sputtering by Ar gas. This indicates that the adenine layer can be removed by Ar first sputtering. Thus, it is thought that adenine layer on the Cu surface can be easily removed by the light pre-sputtering before barrier layer formation.

Figure 5 shows an equivalent circuit on the Cu surface covered with inhibitor layer during impedance measurements. The circuit constants can be determined from Nyquist plot as shown in Fig.6. The thickness d of the inhibitor layer was estimated by calculating $d =(\varepsilon\varepsilon_0 S)/C_2$, where ε is the dielectric constant of the inhibitor layer, ε_0 is the dielectric constant of free space, S is the exposed area of the working electrode, C_2 is the capacitance of the inhibitor layer. Here, the inhibitor layer thickness was calculated assuming that the dielectric constant of adenine is similar to that of BTA because chemical structure of these inhibitors is similar to that of BTA.

Figure 7 shows the dependence of inhibitor layer thickness on the dipping time in the inhibitor solution. In this measurement, the Cu electrode was treated with the constant concentration of inhibitor solution (0.1wt%). It is found that the thickness of BTA layer increases with increasing the dipping time. On the other hand, the thickness of adenine inhibitor layer does not depend on the dipping time. Moreover, adenine inhibitor layer is much thinner compared with BTA inhibitor layer. It is thought that adenine layer is very thin because adenine molecules may adsorb approximately horizontally on the Cu surface compared with BTA.

Figure 8 shows the stable placement of BTA and adenine molecules on the Cu surface using chemical simulation. Adenine and BTA molecules show different adsorption on the Cu surface. Adenine adsorbs approximately horizontally on the Cu (100) surface. On the other hand, BTA adsorbs with slight inclination to Cu (100) surface. As a result, it is thought that adenine molecule may adsorb more orderly and cover the Cu surface more effectively than BTA. Therefore, the effective coverage of the Cu surface can be achieved by thinner adenine layer than BTA.

CONCLUSION

We have succeeded in an effective inhibition of Cu oxidation with very thin layer using adenine as environmentally friendly material. It is also found that adenine inhibitor forms the very thin layer compared with BTA inhibitor. Adenine is readily applicable to the system LSI devices for 32nm node devices and beyond.

ACKNOWLEDGEMENTS

We are indebted to Mitsubishi Chemical Corporation for supporting the electrochemical measurements and discussion.

REFERENCES

1) T.Koito, K.Hirano, M.Iji, H.Aoki and Y.Kasama, J.Surface Finishing Society of Japan, Vol. 56 №7(2005) 57.

2) M.Hara, D.Watanabe, C.Kimura, H.Aoki and T.Sugino, Extended abstract of 2008 International Conference on Solid State Devices and Materials (SSDM 2008) 68.

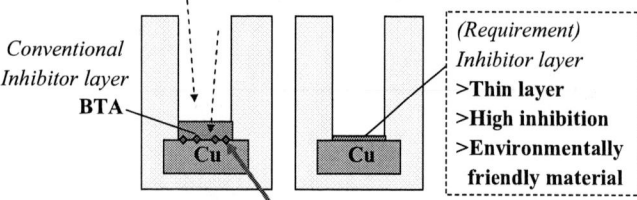

It is difficult to remove the inhibitor by pre-puttering before barrier layer formation.

Conventional Inhibitor layer BTA

(Requirement)
Inhibitor layer
>**Thin layer**
>**High inhibition**
>**Environmentally friendly material**

Cu damage by long pre-sputtering

Fig.1 Requirements to Cu corrosion inhibitor for the Cu/Low-K interconnection.

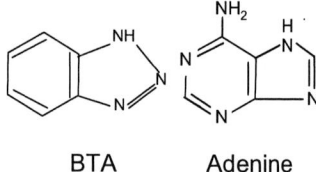

BTA   Adenine

Fig.2 Chemical structure of Cu corrosion inhibitors.

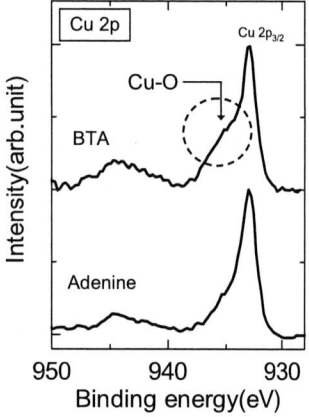

Fig.3 XPS spectra of Cu 2p of the Cu surfaces with BTA or adenine treatments after the accelerated exposure test.

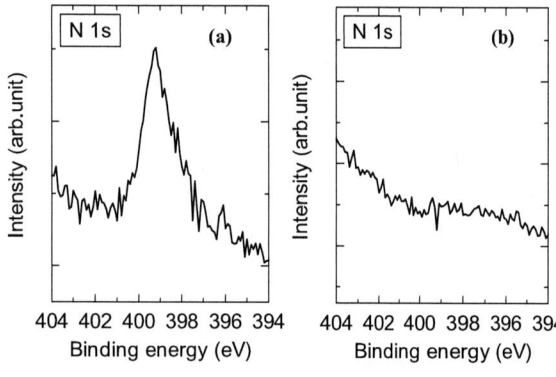

Fig.4 XPS spectra of N (1s) on the Cu surface (a) before and (b) after first sputtering by Ar gas. (N is contained in adenine layer.)

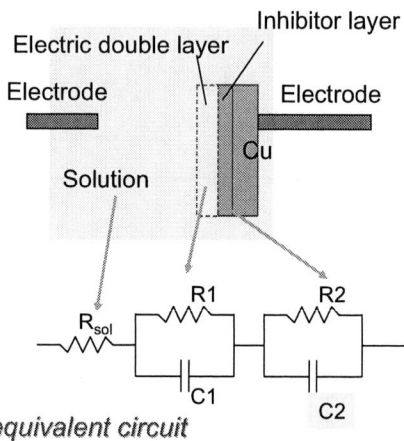

equivalent circuit

Fig.5 An equivalent circuit on the Cu surface covered with inhibitor layer during impedance measurements.

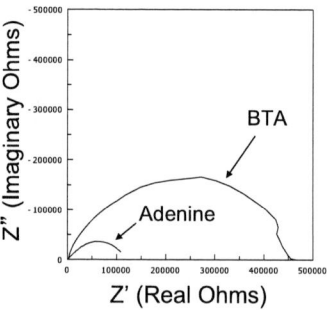

Fig.6 Nyquist plot of the Cu-inhibitor layer by impedance measurements.

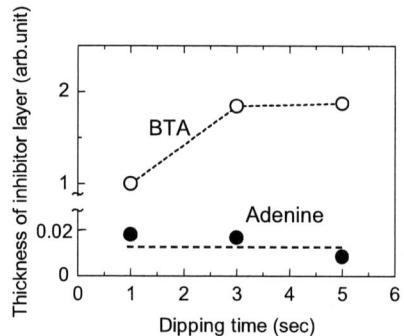

Fig.7 Dipping time dependence of inhibitor layer thickness. (Inhibitor solution: 0.1wt%.)

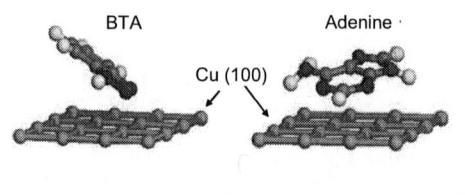

Fig.8 Stable placement of BTA and adenine molecules on the Cu surface using chemical simulation.

Boron Carbon Nitride Film containing Hydrogen for 22nm Node Low-K Interconnection

Hidemitsu Aoki, Takuroh Masuzumi, Makoto Hara, Daisuke Watanabe, Chiharu Kimura and Takashi Sugino
Department of Electrical, Electronic and Information Engineering, Osaka University,
2-1 Yamadaoka, Suita-Shi, Osaka 565-0871, Japan

ABSTRACT

We have investigated the properties of boron carbon nitride containing hydrogen (BCNH) film deposited by using tris (dimethylamino)boron (TMAB) gas. The dielectric constant (k) of the BCNH film was achieved as low as 1.8 by deposition with a low RF power (10W). The film has a sufficient Young's modulus as high as 26 GPa. In addition, k-value of BCNH film is more stable compared with conventional BCN film.

INTRODUCTION

In order to achieve high-performance interconnections with small RC delays requires the integration of a low dielectric constant interlayer (low-k) and Cu interconnection. The use of porous low-k materials is a promising solution for reducing interconnection capacitance. However, most porous low-k materials faced serious issues, such as weak mechanical stress, water incorporation, residual water in the pores and Cu diffusion into the low-k films. Fig.1 shows the relationship between dielectric constant and mechanical strength of various low-k films. Recent effort has been devoted to the development of porous low-k materials with a SiO-based porous structure.

On the other hand, boron nitride (BN) and boron carbon nitride (BCN) are well known as hard materials. The BN and BCN are used in a variety of mechanical applications, such as edged tools. In addition, the BCN film composed of N atoms with high electronegativity has a low dielectric constant without porous structure. We previously reported a minimum dielectric constant value of 1.9 for BCN film.[1] Nonetheless, the k-value of the BCN film deposited by using BCl_3 gas is unstable. Thus, we have reported more stable methyl-BN film with a methyl group and without chlorine.[2] This paper reports the properties of BCN containing hydrogen (BCNH) film deposited by using tris(dimethylamino)boron (TMAB) gas.

EXPERIMENTAL

BCNH films were also deposited by using PACVD with TMAB $(B[N(CH_3)_2]_3)$ gas and N_2 gas mixture. The TMAB chemical structure is shown in Fig. 2. The deposition temperature was 350°C. The RF power was controlled from 10 to 90W and films about 100~200 nm thick were deposited on the Si substrate. On the other hand, conventional BCN films were deposited by using BCl_3, CH_4, H_2 and N_2 mixture gas. We analyzed film bonding from FTIR. XPS measurements were mainly carried out to examine the composition ratio of the constituent atoms of the film. Hydrogen in the film was detected by using thermal desorption spectroscopy (TDS). Current–voltage (I–V) and capacitance–voltage (C–V) characteristics were measured with a metal-insulator-semiconductor (MIS) structure of metal/BCNH/Si at room temperature. The dielectric constant was estimated from the capacitance in the accumulation region (1MHz) of the MIS sample. The strength (Young's modulus) of the films was measured by nano-indentation.

RESULTS AND DISCUSSION

Fig.3 shows the composition of B, N and C atoms plotted as a function of the RF plasma power. The composition ratios of the B, N, and C atoms were estimated to be 35–42%, 33–38%, and 20–31%, respectively. The B and N composition ratios increase with increasing RF plasma power. The composition ratio of C increases with when the power decreases below 50 W. In the case of TMAB and N_2 mixture gas only at 10W of RF power, the BCNH film containing more than 30% carbon can be deposited without CH_4 gas. The BCNH film has good adhesion on the Si substrate and exhibits no physical change after rinsing with deionized water.

TDS measurements examined the hydrogen atoms in the BCN film. Fig.4 depicts the TDS spectra of H_2 (m/z = 2) for the TMAB and BCl_3 samples. A large hydrogen peak is observed for the TMAB sample. However, the conventional BCN sample has barely any TDS spectrum of hydrogen in spite of using H_2 gas. This indicates that the BCN film deposited with TMAB gas contains many hydrogen atoms.

To investigate hydrogen bonds in the BCN film, we took particular note of the FT-IR absorption bands at 2960 cm^{-1} due to the asymmetric stretching mode of C-H of the methyl group (CH_3). The FT-IR absorption band at 2960 cm^{-1} (C-H) was clearly seen for the BCNH film deposited by TMAB, as shown in the inset of Fig. 5. On the other hand, the BCN film by BCl_3 gas has no absorption band due to C-H bonds. It is thought that the BCNH film has methyl bonds in the film. Fig.5 shows the dependence of peak intensity of C-H absorption at 2960 cm^{-1} on RF plasma power. The intensity of C-H absorption of CH_3 increases with decreasing RF power. This indicates that the methyl bonds can easily remain in the BCNH at low RF power.

The interatomic binding energy of the C-H bond (416 kJ/mol) is larger than that of B-N(389 kJ/mol) and C-N bonds(292 kJ/mol). Thus, it is easy for C-H bonds to remain in the film as CH_3 at low RF power.

The I–V characteristics of the BCNH film were measured by the MIS structure. The leakage current of the films deposited in the RF plasma power range of 10~90 W was less than 1×10^{-10} A as shown in Fig.6. The resistivity of the films has high value more than 10^9 as shown in Table 1. This is sufficient for interlayer interconnections. The dielectric constant of the film was estimated by the accumulation capacitance of C–V characteristics using a MIS structure. Fig.7 shows the dielectric constant (k value) of as-deposited BCNH samples plotted as a function of the RF plasma power. The dielectric constant decreases with decreasing RF plasma power. A minimum dielectric constant of 1.8 was obtained at 10 W. It is thought that increasing the C-H bonds with low polarizability can realize a lower k-value. Moreover, it is possible that CH_3 bonds form nano space in the film and the terminations of dangling bonds by hydrogen achieve a stable k-value. Thus, BCNH film with a methyl group can easily keep a low-k value, due to the nano space.

Most porous low-k films have serious issues, such as weak strength (Young's modulus <5 Gpa). Young's modulus of BCNH film was measured by nano-indentation, and estimated to be more than 26 GPa sufficient for CMP processes, which require more than 10 GPa. BCNH film is a low-k material with suitable strength for interlayer materials.

978-1-4244-2784-0/09 $25.00 © 2009 IEEE

CONCLUSION

The k-value of BCNH film as-deposited by TMAB gas can be as low as 1.8 at 10 W of RF plasma power. The film has a sufficient Young's modulus (>26 GPa) for LSI interlayer interconnections. There is a possibility that the dielectric constant is decreased by maintaining a BCNH structure with high strength. It is thought that forming CH_3 bonds in the BCN film is an important approach to realizing lower k values. BCNH film can be readily applied as low-k material for next-generation technology devices.

ACKNOWLEDGEMENTS

We are indebted to ESCO Ltd. for helping with thermal desorption spectroscopy (TDS) measurements

REFERENCES

[1] S. Umeda, T. Yuki, T. Sugiyama and T.Sugino, Diamond and Relat. Mater. 13 (2004) 1135.
[2] S. Tokuyama, M. K. Mazumder, D. Watanabe, C. Kimura, H. Aoki and T. Sugino, Jpn. J. Appl. Phys. 42 (2008) 2492.

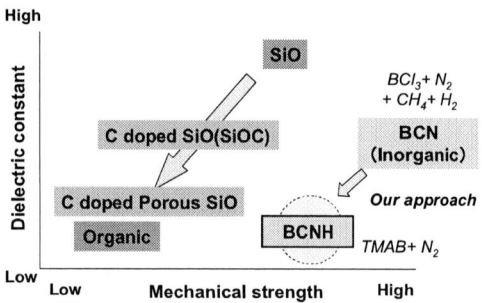

Fig. 1 Dielectric constant and mechanical strength of LowK films

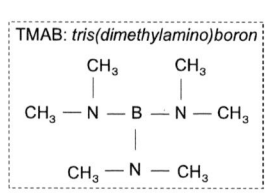

Fig.2 Chemical structure of TMAB: tris(dimethylamino)boron.

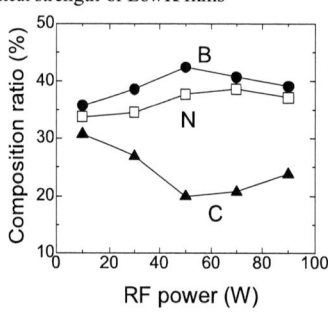

Fig. 3 Composition ratios of B, N and C in the film as a function of RF power.

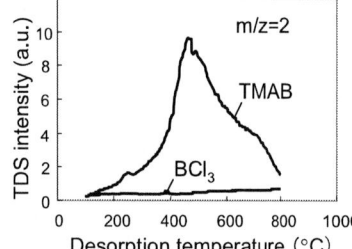

Fig.4 TDS spectra of hydrogen (m/z = 2) of the BCN film deposited by TMAB gas and BCl_3 gas with 10 W of RF power.

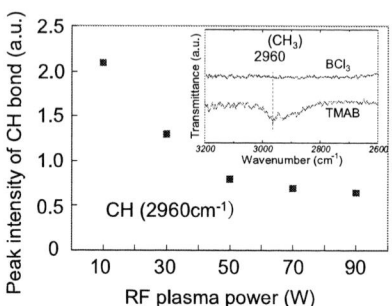

Fig. 5 Dependence of the peak intensity of C-H absorption at 2960 cm^{-1} on RF power.
(Inset) The FT-IR spectra around 2960 cm^{-1} for films using TMAB gas and BCl_3 gas.

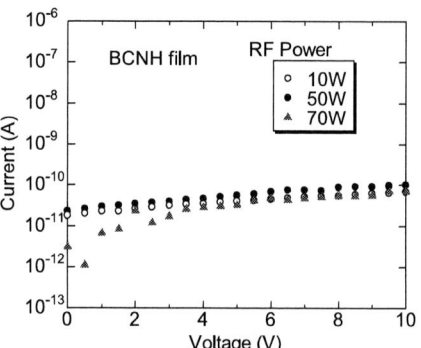

Fig. 6 I–V characteristics of the BCNH film for various RF power .

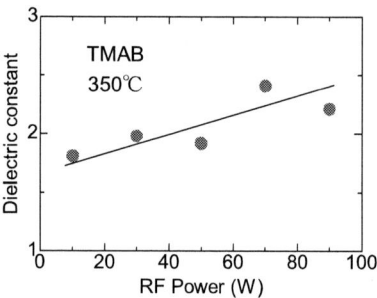

Fig. 7 Dielectric constant (k value) of as-deposited BCNH samples plotted as a function of the RF power.

Table. 1 Properties of BCNH films.

RF plasma power (W)	10 (W)	70 (W)
Resistivity (Ω cm)	6×10^{12}	1×10^{13}
Dielectric constant (k value)	1.8	2.1
Young modulus (GPa)	26 (GPa)	32 (GPa)

978-1-4244-2784-0/09 $25.00 © 2009 IEEE

Ge Shallow Junction Formation by As implantation and Flash Lamp Annealing

[1,2]Kosei Osada, [2,3]Tetuya Fukunaga, and [1,2,3]Kentaro Shibahara

[1]Research Institute for Nanodevices and Bio Systems, Hiroshima University
[2]Graduate School of Advanced Sciences of Matter, Hiroshima University
[3]Research Center for Nanodevices and Systems, Hiroshima University
1-4-2, Kagamiyama, Higashihiroshima, 739-8527 Japan
Phone +81-82-424-6267, FAX: +81-82-424-3499, e-mail: ksshiba@hiroshima-u.ac.jp

1. Introduction

Recently Ge is extensively investigated as an alternative material to Si because of its higher mobility and injection velocity [1]. Fabrication of 100-nm-scale Ge MOS devices [2,3] were already reported. However, research on doping was not so active that shallow junction formation process were not established. Different from Si, donors in Ge diffuse faster than acceptors. Therefore, n^+/p shallow junction is difficult to form. Previously, we have reported fabrication of n^+/p junctions with As and Xe^+ PAI process. The Xe^+ PAI was effective to retard diffusion [4]. However, diffusion during furnace annealing was still too large for sub-100 nm junctions.

In this paper, 20 nm junction formation is reported. This is the first report of ultra-shallow junction formation using arsenic. To minimize dopant diffusion, Xe lamp flash lamp annealing (FLA) which is one of the so-called ms annealing, was adopted. Arsenic was selected, as a dopant. Although, phosphorus was mainly used for shallow junction formation formerly [5-7], arsenic's high thermal-equilibrium solid solubility (about 2×10^{20} cm^{-3} at 780°C) [8] and heavy mass are suitable features for shallow junction formations.

2. Experimental Conditions

Ge (100) p-type substrates were dipped in diluted HF acid to remove a native oxide prior to ion implantation. For all samples, As^+ was implanted at 5 keV for 1×10^{15} cm^{-3}. Prior to the As^+ implantation, pre-amorphization implantation (PAI) was performed with Xe^+ or Ge^+ for some specimens. FLA duration was fixed to 1 ms. Substrates were not heated for FLA. Sheet resistance of an As doped layer was evaluated by the four point probe method. The probe head was designed for the measurements of shallow junction of fragile material like Ge.

3. Results and Discussions

Figure 1 shows thickness of an amorphous layer formed by Ge^+ or Xe^+ implantation. The thickness was evaluated mainly by spectroscopic ellipsometry and validated by XTEM for some. Figure 2 shows As depth profiles for various Xe^+ PAI doses. As expected, dopant ion channeling was well suppressed by the PAI. However, increase in PAI dose resulted in deeper profile. This anomalous result is attributable to bubble formation. The bubble is the huge precipitation of Xe [4], shown in Fig. 3(a). It is considered that the bubbles grow larger during annealing and break remaining craters seen in Fig. 3(b). By limiting Xe^+ dose up to 1×10^{14} cm^{-2}, PAI without bubbles is possible, as shown in Fig. 4(d). In the case of Ge^+ PAI, implantation dose for junction formation was 1×10^{15} cm^{-2}.

The maximum lamp energy for FLA was 22.5 J, because wafers broke into pieces by higher energy irradiation. Figure 5 shows As depth profiles before and after FLA at 22.5 J. Diffusion of As was so small that profiles for before and after FLA were almost identical to each other. The junction depth defined for 3×10^{18} cm^{-3} was about 20 nm for without PAI specimens. Figure 6 shows relationship between sheet resistance and FLA energy. Open circles represents the sheet resistance of specimens without PAI. The sheet resistance decreased as the FLA energy increased. To the contrary, the sheet resistance for Xe^+ PAI specimens hardly decreased. This difference is clearly explainable referring XTEM images in Fig. 4. The surface amorphous layer formed by As^+ implantation was completely recrystallized by FLA at 22.5 J, as shown in Fig 4(a) to 4(c). However, the amorphous layer formed by Xe^+ PAI almost remained even after FLA at 22.5 J. Similar result was obtained for the Ge^+ PAI case shown in Fig. 7. Improvement in sheet resistance by single FLA was slight and four-times annealing was necessary for clear reduction. The amorphous layer thicknesses before and after single FLA at 22.5J were 43 nm and 36 nm, respectively. Thus, retarding of solid phase epitaxy was clear for the PAI cases. Degradation in sheet resistance by Ge^+ PAI was also pointed out by Simonen et al. [9]. Vacancy is a predominant point defect concerning diffusion of atoms in Ge [10]. Excess interstitials formed by PAI would prevent recrystallization.

4. Conclusions

Shallow, about 20 nm, depth n^+/p junction of Ge was successfully fabricated by As^+ implantation and FLA. Since the junction depth was limited by implantation energy, much shallower junction would be fabricated by reducing the energy. High potential of arsenic as a dopant was clearly demonstrated, although FLA parameters were not optimized yet. Since SPE retardation was found in the specimens with PAI, other channeling suppression technique should be found.

Acknowledgements

Part of this work was supported by Grant-in-Aid for scientific research (C), 19560344, 2007 and "Interdisciplinary Research on Integration of Semiconductor and Biotechnology at Hiroshima University" based on "Creation of Innovation Centers for Advanced Interdisciplinary Research Areas" the Special Coordination Funds for Promoting Science and Technology, from the Ministry of Education, Culture, Sports, Science, and Technology of Japan.

References

[1] S. Takagi, Technical Digest of 2003 Symp. on VLSI Technology, pp. 115-116.
[2] T. Yamamoto et al., Technical Digest of IEDM 2007, pp.1041-1043.
[3] P. Zimmerman et al., Technical Digest of IEDM 2006, pp. 655-658.
[4] T. Fukunaga et al., Ext. Abst. of Int. Conf. on SSDM 2006, pp. 452-453.
[5] C.O. Chui et al., Appl. Phys. Lett. **83** (2003) pp. 3275-3277.
[6] A. Satta et al., Nucl. Instr. and Meth. B **257** (2007) pp. 157–160.
[7] S. Heo et al., Electrochemical and Solid-State Lett., 9 (2006) pp. G136-G137.
[8] F.A. Trumbore, Bell Syst. Tech. J, 39 (1960) pp. 205-233.
[9] E. Simoen et al., Mat. Sci. in Semiconductor Process. 9 (2006), pp. 634-639.
[10] J. Vanhellemont et al., J. Electrochemical Soc., **154** (2007) pp. H572-H583.

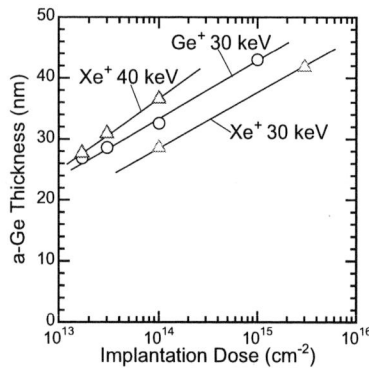

Fig. 1 Relationships between amorphized layer thickness and Ge^+ and Xe^+ implantation dose.

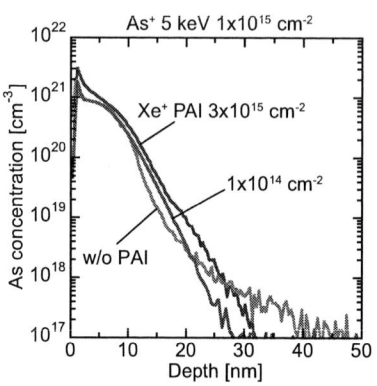

Fig. 2 As implanted SIMS depth profiles of arsenic. Xe^+ PAI was effective for channeling tail suppression, as usual. However, increase in Xe^+ dose resulted in a deeper profile.

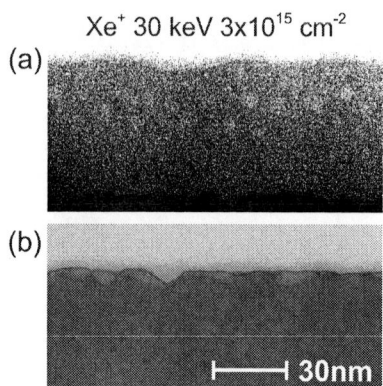

Fig. 3 XTEM images of Ge substrates (a) after Xe^+ implantation and (b) after annealing at 600°C for 15 s. High dose Xe^+ implantation formed Xe bubbles and craters were formed in such a specimen by the annealing.

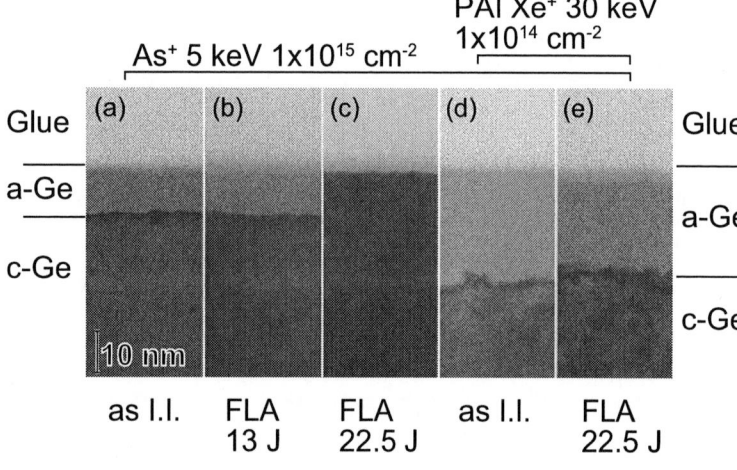

Fig. 4 Cross-sectional TEM images for As^+ and Xe^+ implanted Ge substrates. By reducing Xe^+ PAI dose to 1×10^{14} cm^{-2}, the Xe bubbles and the craters, shown in Fig. 2, disappeared, as shown in (d) and (e), respectively. Amorphous layer, in (a), formed by As^+ implantation for doping was completely recrystallized by FLA of 22.5 J, as shown in (c). However, amorphous layer in (e) formed by Xe^+ PAI almost remains after the same FLA. This result indicates retarding of solid phase regrowth in specimens with Xe^+ PAI.

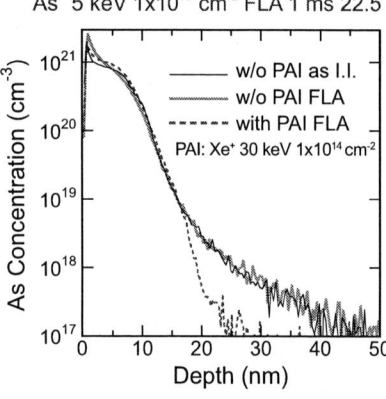

Fig. 5 Arsenic SIMS depth profiles before and after FLA. Diffusion by FLA was negligibly small.

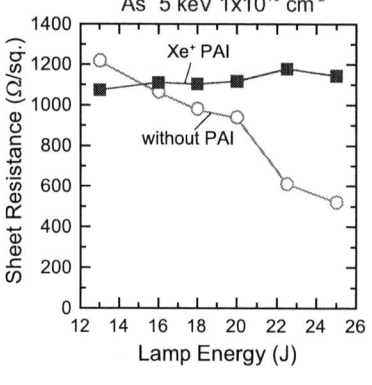

Fig. 6 Relationships between sheet resistance of As doped specimens and lamp irradiation energy. The specimens with Xe^+ PAI hardly showed reduction in the sheet resistance by As activation.

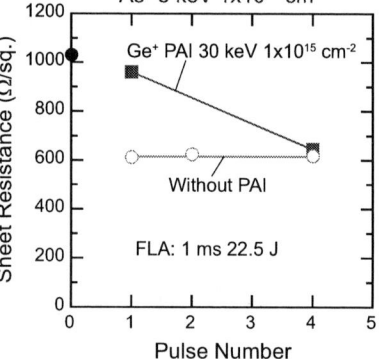

Fig. 7 Relationships between sheet resistance of As doped specimens and FLA pulse number. The specimens with Ge^+ PAI needed multiple irradiation for As activation.

Impact of Lithography Variations on Advanced CMOS Devices

J. Lorenz, C. Kampen, A. Burenkov, T. Fühner

Fraunhofer Institut für Integrierte Systeme und Bauelementetechnologie IISB

Schottkystrasse 10, 91058 Erlangen, Germany

ABSTRACT

Source and relevance of process variations are briefly discussed. A combination of own lithography and commercial TCAD simulation software is applied to assess the impact of some of the most relevant variations occurring in lithography on the electrical properties of three kinds of CMOS devices with 32 nm physical gate length.

INTRODUCTION

Process variations increasingly challenge the manufacturability of advanced devices and the yield of the circuits and systems produced. TCAD has the potential to make key contributions to minimize this problem, by assessing the impact of certain variations on the device and circuit in question and, in this way, providing the information necessary to decide about investment on process level or the adoption of a more variation-tolerant process flow, device architecture, or design on circuit or chip level. This presentation briefly classifies relevant sources of variations and discusses the impact of variations in two key parameters of lithography steps, namely focus and dose, on conventional and advanced device architectures.

CLASSIFICATION OF RELEVANT VARIATIONS

Processes vary between different wafers, different dies on the same wafer, or even within the die. Both inter-wafer and inter-die variations affect the electrical data of the devices or circuits and, in turn, the yield. Intra-die variations may affect even the overall functionality of a circuit e.g. via insufficient matching between different devices, and are, therefore, of primary interest to designers. Especially dopant fluctuations have been extensively discussed in the literature (e.g. [1]), using suitably extended device simulation programs. However, also several other intra-die process variations exist which have so far hardly been addressed w.r.t. their impact on device and circuit variability. In lithography, this includes variations of dose and defocus, some kinds of alignment errors, defects, lens aberrations, line edge/width roughness, and optical proximity effects. Lithography is one of the most important sources of process variability even on the intra-die level which is of great concern to designers, and is thus addressed in this paper. Other process-induced intra-die variations will be discussed elsewhere.

SIMULATION TOOLS

We used our Python-based simulation framework [2] to couple our rigorous lithography simulation tool *Dr.LiTHO* [3] to the *Sentaurus* TCAD suite of Synopsys [4], which was employed to simulate device properties and all process steps except for lithography. Our framework facilitates the organization and efficient conduction of the large amount of simulations, needed for this work, on a high performance PC-based computing cluster.

DEVICES INVESTIGATED AND SIMULATION SETUP

We investigated three types of NMOS and PMOS devices, each with a physical gate length of 32 nm, see Figures 1 to 3: Bulk, single-gate (SG) fully depleted SOI, and double-gate (DG) fully depleted

SOI MOSFETs. The SOI devices had a very lightly doped ($1x10^{15}cm^{-3}$) channel, a physical gate length of 32 nm, a body thickness of 10 nm for SG and 16 nm for DG, respectively, a 1.2 nm gate oxide and a buried oxide of 20 nm. Substrate doping was chosen as $10^{20}cm^{-3}$ for NMOS and $5x10^{19}cm^{-3}$ for PMOS, respectively, to achieve better DIBL control and therefore better stability versus process variations. Parameters for the bulk MOSFETs were similar except for a relatively lightly doped substrate of $10^{17}cm^{-3}$ and heavily doped pockets. Junction and gate work function engineering were employed to achieve similar threshold voltages of all three devices studied. More details on the simulation setup were presented in a paper limited to bulk and single gate SOI, which did not include the inverter results below [5]. Using Dr.LiTHO, the lithography process window analysis (allowing up to 10% CD variation) for the 193 nm water immersion process employed yielded a depth of focus of 52 nm and a "threshold latitude" of 8.5%. In order to study the impact of process variations not only under best conditions, the variation range has been increased to ±40 nm around best focus and to a threshold range of 0.25 – 0.4. A fixed etch bias of 13 nm was assumed.

RESULTS

Assuming random, normally distributed variations of defocus and a linear walk through the lithography threshold range within the process window given above, about 8600 variations of the physical gate length were simulated for each device architecture. Due to the physics of the lithography process, this results in an asymmetric General Extreme Values (GEV) distribution of the gate length (CD) of the transistors (Figure 4). Process and device simulation using *Sentaurus* then predicts only for the bulk MOSFET a symmetric threshold voltage distribution (Figure 5) which is due to the dominating role of the pockets. For both SOI devices, quite similar highly asymmetric GEV distributions result, see Figures 6 and 7. Figure 8 shows a strong short channel effect for both SOI devices, in contrast to the bulk transistor controlled by the pockets. This results in a large variability of both SOI devices in the I_{off} vs. I_{on} plot shown in Figure 9: Whereas their nominal I_{on}/I_{off} ratio is much better than for the bulk transistor, a penalty in terms of a higher leakage variation has to be paid. It is also at first glance surprising – but in view of the dopant distributions shown in Figure 3 easy to explain – that the DG device has a large variability and less good values of I_{on} for a given I_{off}. Finally, the performance of a static inverter, as shown in Figure 10, was studied for the bulk and the SG SOI device. Here, a penalty in terms of a larger uncertainty area for switching has to be paid for better performance of the SOI devices, see Figures 11 and 12.

CONCLUSIONS

An efficient MOSFET variability study has been performed using coupled lithography/process/device simulation. Besides the detailed results reported, there are three key messages: Real technologies with normal distributions of process parameters may lead to highly asymmetric parameter distributions at device level. A co-optimization of nominal device performance and device variability is of large importance, because these targets may compete against each other. Finally, not only nominal performance but also variability critically depends on the device architecture.

978-1-4244-2784-0/09 $25.00 © 2009 IEEE

REFERENCES

[1] A. Asenov et al., Simulation of intrinsic parameter fluctuations in decananometer and nanometer-scale mosfets, IEEE Trans. Electron Devices vol. 50. no. 9, 2003, p. 1837-1852

[2] T. Fühner et al., A simulation study on the impact of lithographic process variations on CMOS device performance, in Proc. SPIE, vol. 3, 2008, paper 692453

[3] T. Fühner et al., Dr.LiTHO: a development and research lithography simulator, in: Proc. SPIE, vol. 6520, 2007, paper 65203

[4] Sentaurus TCAD, Release z-2007.03-ed, Synopsys, Mountain View, CA, USA, 2007.

[5] C. Kampen et al., On the Stability of Fully Depleted SOI MOSFETs Under Lithography Process Variations, in Proc. ESSDERC 2008, pp. 194

ACKNOWLEDGEMENT

This research has in part been supported by the European Commission's Information Society Technologies Program, under PULL-NANO project contract No. IST-026828, and by the Fraunhofer Internal Programs under Grant No. MAVO 817 759, respectively.

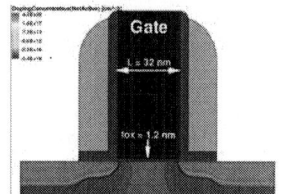

FIGURE 1. BULK MOSFET ARCHITECTURE

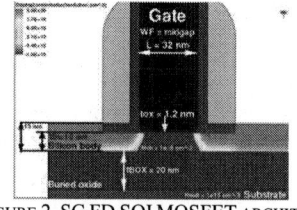

FIGURE 2. SG FD SOI MOSFET ARCHITECTURE

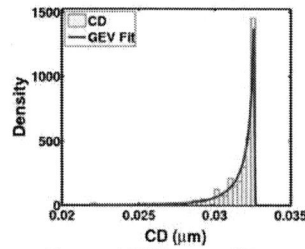

FIGURE 4. PDF OF THE CD AT THRESHOLD = 0.32: GENERALIZED EXTREME VALUES DISTRIBUTED

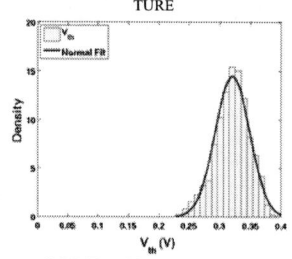

FIGURE 5. PDF OF V_{TH} OF THE BULK NMOS: NORMAL DISTRIBUTED

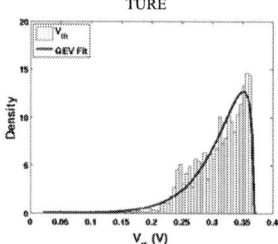

FIGURE 6. PDF OF V_{TH} OF THE SG SOI NMOS: GENERALIZED EXTREME VALUES DISTRIBUTED

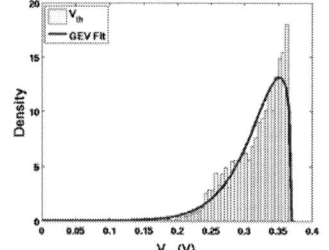

FIGURE 7. PDF OF V_{TH} OF THE DG SOI NMOS: GENERALIZED EXTREME VALUES DISTRIBUTED

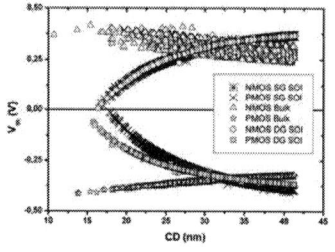

FIGURE 8. V_{TH}(CD) BEHAVIOUR OF ALL DEVICES OVER THE WHOLE EXPERIMENT

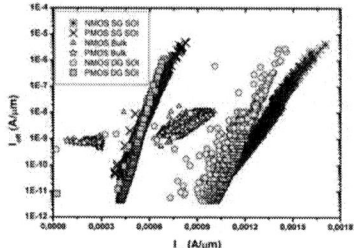

FIGURE 9. I_{ON}-I_{OFF} BEHAVIOUR OF THE ALL DEVICES OVER THE WHOLE EXPERIMENT: I_{ON} AND I_{OFF} OF THE DG WERE DIVIDED BY 2

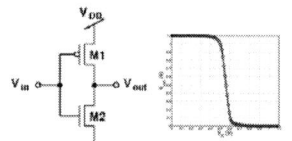

FIGURE 10. STATIC INVERTER SIMULATION CONCEPT

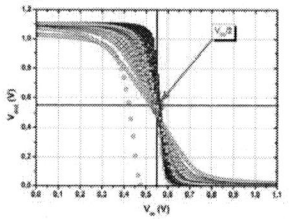

FIGURE 11. VOLTAGE TRANSFER CHARACTERISTIC (VTC) VARIATIONS OF THE BULK CMOS INVERTER: $V_{DD}/2 = 0.55$ V

FIGURE 12. VOLTAGE TRANSFER CHARACTERISTIC (VTC) VARIATIONS OF THE SG SOI CMOS INVERTER: $V_{DD}/2 = 0.55$ V

978-1-4244-2784-0/09 $25.00 © 2009 IEEE

Stress-Enhancement Technique in Narrowing NMOSFETs with Damascene-Gate Process and Tensile Liner

S. Mayuzumi, S. Yamakawa, Y. Tateshita, M. Tsukamoto, H. Wakabayashi, T. Ohno and N. Nagashima
Semiconductor Technology Development Division, Semiconductor Business Group, SONY Corporation
Atsugi Tec., 4-14-1 Asahi-cho, Atsugi, Kanagawa, 243-0014, Japan
Phone: +81-46-230-5508, Fax: +81-46-230-6556, E-mail: Satoru.Mayuzumi@jp.sony.com

ABSTRACT

Local channel stress behaviors induced by the combination of top-cut tensile SiN stress liner and damascene-gate (gate-last) process on the channel width for nFETs are investigated by using 3D stress simulations and demonstrations. It is found that the dummy-gate removal enhances high tensile channel stress along the gate length, especially at the edge of the channel beside the STI. Therefore, drivability enhancement is performed for damascene-gate nFETs with narrow channel width. High-drive current of 1430 uA/um at Ioff = 100 nA/um, Vdd = 1.0 V and the channel width of 0.3 um is achieved by the stress enhancement effects of the damascene-gate technology.

INTRODUCTION

High-performance metal/high-k gate MOSFETs with local channel-stress techniques using damascene-gate (gate-last) process have been recently reported [1-4]. Damascene-gate technology considerably enhances channel stress in shorter gate length (Lgate) for n- and pFETs and narrower channel width (Wch) for pFETs, because of dummy-gate removal [3, 5, 6]. Moreover, effects of the channel stress for damascene-gate pFETs have been precisely evaluated by using 3D stress simulation and UV-Raman spectroscopy [3, 6-8]. On the other hand, behaviors of the channel stress in narrow Wch for damascene-gate nFETs with top-cut tensile SiN stress liner (t-SL) haven't been studied yet. In this paper, effects of drivability enhancement on Wch are studied for damascene-gate nFETs with t-SL by using 3D stress simulations and demonstrations.

DEVICE FABRICATION AND SIMULATION

Damascene-gate process described in ref. [1] was carried out. Fig. 1 schematically shows cross-sectional structures of metal/high-k damascene-gate stacks for nFETs with t-SL on (100) substrate. 25-nm t-SL having the stress of 1.6-GPa was utilized. Metal/high-k gate stacks were formed with HfSi$_x$/HfO$_2$. For the references, devices with poly-Si/SiO$_2$ gate stacks were fabricated by using gate-1st process. 3D stress simulations were used to investigate channel-stress for damascene-gate nFETs after dummy-gate removal and gate-1st nFETs after top-cut t-SL formation [7].

RESULTS AND DISCUSSION

Stress Simulations

Fig. 2 shows 3D simulated lateral channel stress (Sxx) distributions for damascene-gate and gate-1st nFETs with 40-nm Lgate and 25-nm t-SL. A quarter of the channel area is shown. Variations of channel stress along the y- and z-axes on the Wch are much smaller than Sxx for both damascene-gate and gate-1st. Tensile Sxx at the edge of the channel beside the damascene gate is much higher than that for gate-1st. Fig. 3 shows simulated Sxx dependence on relative Wch position for damascene-gate and gate-1st, where the Wch is normalized, and the Wch center and edge correspond to 0 and 0.5, respectively. For damascene-gate, the tensile Sxx at the edge of the channel is almost the same for each Wch. However, the ratio of high tensile Sxx area to the Wch becomes large

as the Wch becomes narrow. On the contrary, gate-1st doesn't indicate the enhancement effect like damascene-gate. Calculated relative electron mobility dependence on relative Wch position for damascene-gate and gate-1st is shown in Fig. 4. Mobilities are estimated by using piezo-resistance coefficients [9]. The mobilities near the channel edge for damascene-gate are significantly improved more than those for gate-1st, as Wch becomes narrow. Damascene-gate concentrates the tensile Sxx at the channel edge by dummy-gate removal. However, gate-1st couldn't enhance the tensile Sxx because of reactive forces by the gate electrode (Fig. 5). Therefore, it is considered that drivability for narrow-Wch nFETs is remarkably enhanced by using damascene-gate process.

Characteristics

Fig. 6 shows normalized drain current dependence on Wch in linear and saturation regions for damascene-gate and gate-1st nFETs with 40-nm Lgate and 25-nm t-SL. Id enhancements in linear and saturation regions for damascene-gate are higher than those for gate-1st in narrower Wch. These results are supported by 3D stress simulation, as discussed in the previous sub-section. Ion dependence on Wch for damascene-gate nFETs with different Lgates is shown in Fig. 7. Ion is enhanced by not only narrower Wch but also shorter Lg. It has been only reported that damascene-gate nFETs with t-SL enhances the Sxx especially in shorter Lgate [1, 10]. Moreover, it is found that Ion enhancement becomes large in narrow Wch, as Lgate shortens. It is considered that the Sxx at the edge of the channel beside the STI is further enhanced by the effect on Lgate (Fig. 8). The combination of narrow Wch and short Lgate is significantly effective to enhance the channel stress by damascene-gate process even for nFETs. Fig. 9 shows Ion-Ioff characteristics on Wch. Drivability-enhancement effects on narrow Wch are demonstrated. Id-Vgs and Id-Vds characteristics of nFETs with 0.3-um Wch and 40-nm Lgate are shown in Fig. 10. High-drive current of 1430 uA/um at Vdd = 1.0 V and Ioff = 100 nA/um is obtained owing to the channel-stress enhancement by the combination of narrow Wch and short Lgate.

CONCLUSION

Drivability-enhancement effects of local channel stress on the channel width were investigated for damascene-gate nFETs with top-cut tensile stress liner by using 3D stress simulations and characteristics. Drivability enhancement in narrower channel width is achieved, because high tensile stress by dummy-gate removal is enhanced at the channel edge along the gate length. This technology must be suitable for the device scaling.

REFERENCES

[1] S. Mayuzumi et al., IEDM Tech. Dig., pp.293, (2007) [2] K. Mistry et al., IEDM Tech. Dig., pp.247 (2007) [3] S. Mayuzumi et al., VLSI Tech. Dig., pp.126 (2008) [4] C. Auth et al., VLSI Tech. Dig., pp.128 (2008) [5] J. Wang et al., VLSI Tech. Dig., pp.46 (2007) [6] S. Yamakawa et al., SISPAD, pp.109 (2008) [7] SELETE process simulator HySyProS. [8] A. Ogura et al., Jpn. J. Appl. Phys. 45, pp.3007 (2006) [9] S. E. Thompson et al., IEDM Tech. Dig., pp.681 (2006) [10] S. Yamakawa et al., ESSDERC, pp. 174 (2008)

978-1-4244-2784-0/09 $25.00 © 2009 IEEE

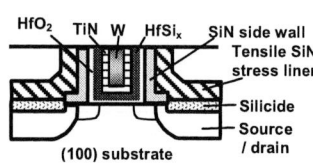

Fig. 1 Schematic diagram of damascene-gate nFET with metal/high-k gate stack and top-cut tensile SiN stress liner.

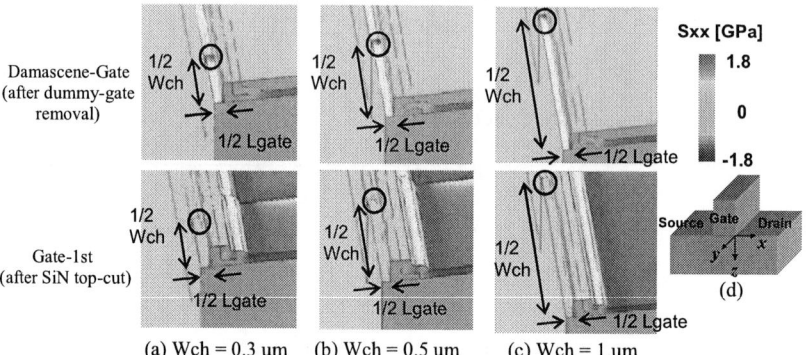

(a) Wch = 0.3 um   (b) Wch = 0.5 um   (c) Wch = 1 um

Fig. 2 Simulated lateral channel stress (Sxx) distributions for 40-nm Lgate nFETs with 25-nm t-SL. 1/4 of channel area is shown, respectively. (a) Wch = 0.3 um, (b) 0.5 um, (c) 1 um and (d) axial directions in stress simulation. Gate stacks for gate-1st are removed in these figures.

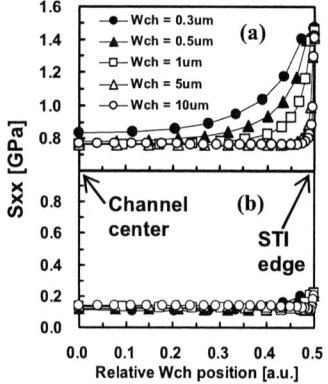

Fig. 3 Simulated Sxx dependence on relative Wch position at Lgate center for (a) damascene-gate and (b) gate-1st.

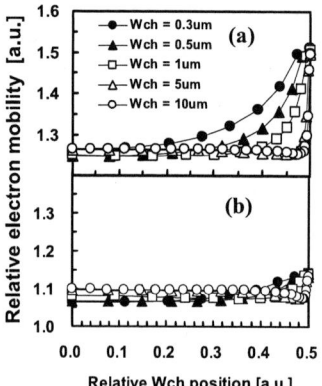

Fig. 4 Calculated relative electron mobility dependence on relative Wch position at Lgate center for (a) damascene-gate and (b) gate-1st.

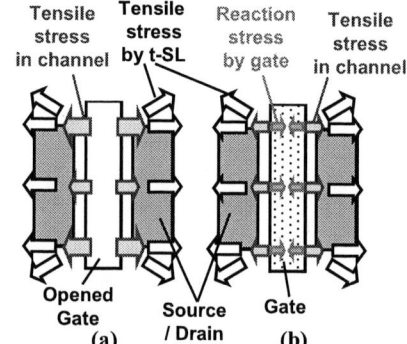

Fig. 5 Schematic diagrams of induced stress around channel for (a) damascene-gate and (b) gate-1st.

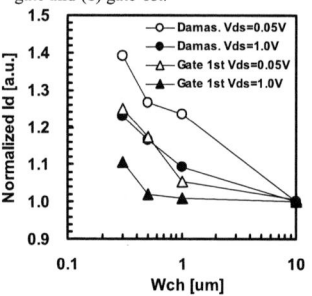

Fig. 6 Id dependence on Wch for damascene-gate and gate-1st nFETs with 25-nm t-SL.

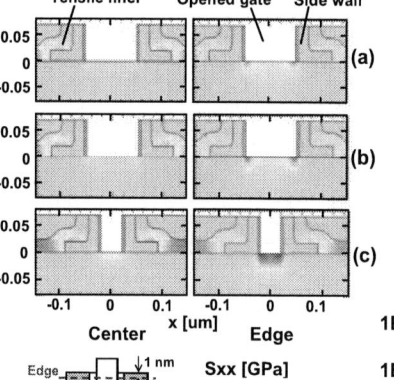

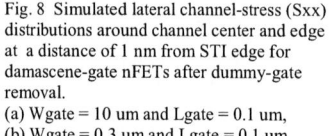

Fig. 8 Simulated lateral channel-stress (Sxx) distributions around channel center and edge at a distance of 1 nm from STI edge for damascene-gate nFETs after dummy-gate removal.
(a) Wgate = 10 um and Lgate = 0.1 um,
(b) Wgate = 0.3 um and Lgate = 0.1 um,
and (c) Wgate = 0.3 um and Lgate = 0.04 um.

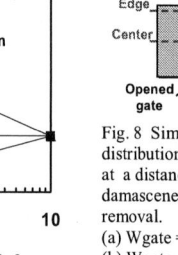

Fig. 7 Ion dependence on Wch for damascene-gate nFETs with three different Lgates.

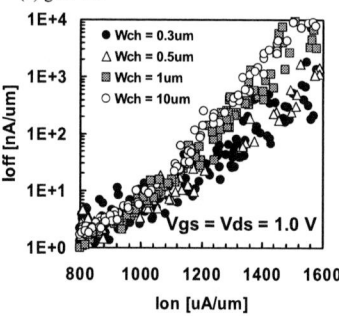

Fig. 9 Ion-Ioff characteristics of damascene-gate nFETs on Wch at Vgs = Vds =1.0V.

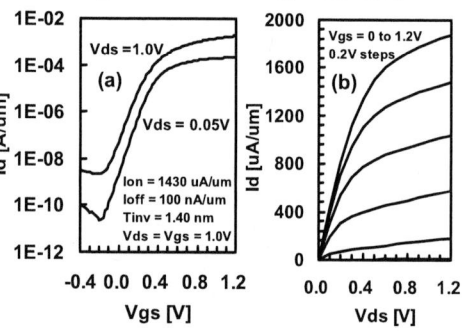

Fig. 10 Id-Vgs and Id-Vds characteristics of nFET with 0.3-um Wch and 40-nm Lgate. (a) Id-Vgs and (b) Id-Vds.

978-1-4244-2784-0/09 $25.00 © 2009 IEEE   18

Additive Mobility Enhancement and Off-State Current Reduction in SiGe Channel pMOSFETs with Optimized Si Cap and High-k Metal Gate Stacks

Jungwoo Oh, Prashant Majhi, and Raj Jammy
SEMATECH, 2706 Montopolis Drive, Austin TX 78741, USA

Raymond Joe, Anthony Dip, Takuya Sugawara, Yasushi Akasaka, Takanobu Kaitsuka,
Tsunetoshi Arikado, and Masayuki Tomoyasu
Tokyo Electron Ltd.

Abstract

We have demonstrated high mobility pMOSFETs on high quality epitaxial SiGe films selectively grown on Si (100) substrates. With a Si cap processed on SiGe channels, $HfSiO_2$ high-k gate dielectrics exhibited low C-V hysteresis (<10 mV), interface trap density ($7.5x10^{10}$), and gate leakage current ($\sim10^{-2}A/cm^2$ at an EOT of 13.4Å), which are comparable to gate stack on Si channels. The mobility enhancement afforded intrinsically by the SiGe channel (60%) is further increased by a Si cap (40%) process, resulting in a combined $\sim$100% enhancement over Si channels. The Si cap process also mitigates the low potential barrier issues of SiGe channels, which are major causes of the high off-state current of small bandgap energy SiGe pMOSFETs, by improving gate control over the channel.

Device Fabrication

Epitaxial SiGe (Ge=25%) channels and Si caps were selectively grown on shallow trench isolation formed Si (100) substrates. The Si cap thicknesses were varied (0~7nm). $HfSiO_2$ gate dielectrics were deposited by atomic layer deposition on SiGe channels after surface passivation, followed by sputtering of the metal gates.

Results and Discussion

Fig.1(a) shows TEMs of a 3nm Si cap deposited on a SiGe channel with a high-k $HfSiO_2$ dielectric/metal gate stack. After a high temperature spike anneal on the S/D, a Si cap layer on the SiGe channel is hardly distinguishable in Fig.1(b). EELS analysis in Fig.2, however, clearly detected the Si-rich layer at the interface. Subsequent electrical data also confirmed a Si cap layer and its impact on SiGe pMOSFETs. Fig.3 compares C-V curves of varying Si cap thicknesses (0~7nm). The C-V hysteresis at 100KHz was minimal (measured to be <10 mV). With a $HfSiO_2$ physical thickness of 40Å, the EOT was reduced from 14.1Å to 13.4~13.5Å after Si cap processing (Fig.4). EOT is further scaled down by thinning the $HfSiO_2$ gate dielectrics. Diffusion of Ge from the SiGe channel into the high-k dielectric was effectively suppressed by the Si cap, which lessened the formation of low-k interfacial oxide. A slightly lower EOT was obtained on thinner Si caps. Interface trap density (D_{it}) was measured using the charge pumping technique on SiGe pMOSFETs (Fig.5). Si cap processing significantly reduced the D_{it} from $1.1x10^{12}$ to $7.5x10^{10}$, which is comparable to $5.5x10^{10}$ for a Si channel. Fig.6 shows threshold voltage (V_{th}) dependency on Si cap for the SiGe channel. A 3nm Si cap shows a lower V_{th} than 5 or 7nm Si caps, when compared with the V_{th} of the Si channel. The SiGe channel without a Si cap shows the lowest V_{th}. Low V_{th} is attributed to the small bandgap energy of the SiGe channel (or valence band offset relative to the Si cap), which requires small surface potential for channel inversion. This behavior is in good agreement with the positive V_{th} reported on other Ge pMOSFETs [1-2]. C-V curves of SiGe MOSFETs in Fig.7 support varying V_{th} as a function of Ge content. Note flatband voltages are, however, relatively independent of Ge content. The high C-V stretch-out observed without the Si cap is due to high interface charge density. The slightly high EOT of the Si cap on SiGe channel (compared to Si channel) might be due to carrier confinement in the SiGe channel with the Si cap acting as an insulator. However, multi-V_{th}s, which are typically observed with Si cap on SiGe channel at a high surface field, are not observed in the C-V curves. Fig.8 compares the I_d-V_g curves of SiGe and Si channel pMOSFETs. High drain current is obtained in optimized Si caps on the SiGe channel when carriers transport on the SiGe channel and the Si cap serves as a passivation layer. Drain current is compromised when the Si cap becomes too thick for carriers to be transported through it. The off-state current of the SiGe channel is intrinsically higher than the Si channel because of a high subthreshold and substrate current, which are degraded by the low built-in potential of source-to-channel and S/D junctions (Fig.9). Si cap processing mitigates off-state current and short channel effects by improving the gate dielectric-channel interface quality and, therefore, providing better gate control over the channel. Fig.10 shows I_d-V_g for SiGe short channels. Fig.11 shows $\sim$100% enhanced drive current for SiGe pMOSFETs with an optimized Si cap. Mobility enhancement was confirmed in the transconductance and mobility results in Fig.12. SiGe channels without a Si cap exhibited $\sim$60% mobility enhancement over the Si channels. After applying the Si cap process, SiGe channel mobility was further enhanced by $\sim$40%. Therefore, a $\sim$100% mobility enhancement was achieved by the combined SiGe channel and Si cap processing.

Conclusions

Optimized Si caps reduced interface trap density from $1.1x10^{12}$ to $7.5x10^{10}$ and showed minimal C-V hysteresis (< 10 mV). Si cap processing effectively suppressed the off-state current of SiGe pMOSFETs by providing better gate control over the channel, while enhancing hole mobility by 40% in the SiGe channel, which resulted in a $\sim$100% mobility enhancement over Si channels after combining the SiGe channel and Si cap processing. Results obtained in this study advance the optimization of high mobility SiGe as an alternative channel material for future technology nodes.

References

[1] P.Zimmerman et al., IEDM, p.655, 2006.
[2] O.Weber et al., IEDM, p.143, 2005.

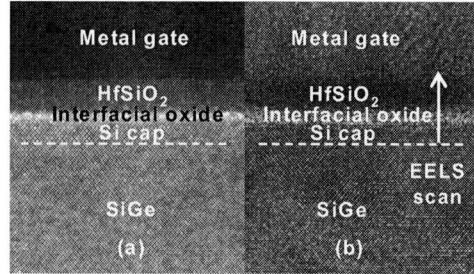

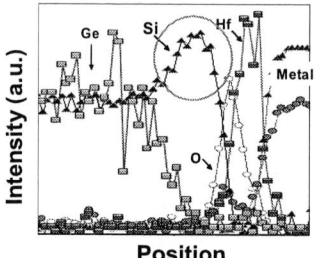

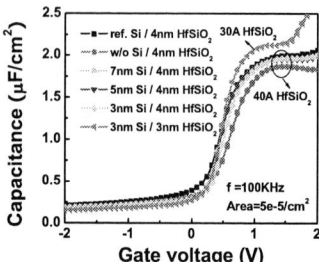

Fig.1 HR-XTEM of gate stack with Si cap, high-k $HfSiO_2$, and metal gates on epitaxial SiGe (Ge=25%) layers selectively grown on on Si (100) substrates. (a) Before and (b) after high temperature spike anneal for source/drain activation.

Fig.2 EELS analysis on the gate stack in direction shown in Fig.1(b) shows the Si-rich layer at the interface.

Fig.3 C-V of SiGe MOS capacitors showing scalable low EOT and minimal hysteresis (<10 mV) of 40Å $HfSiO_2$ on SiGe channels.

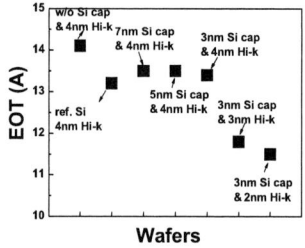

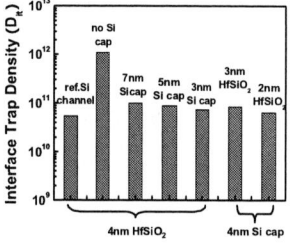

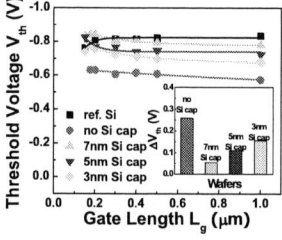

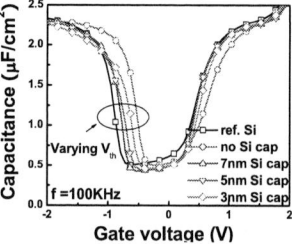

Fig.4 Scalable EOT on SiGe channels with varying Si cap and high-k dielectrics. EOT for Si cap is lower than without Si cap.

Fig. 5 Significant reduction in D_{it} from 1.1×10^{12} to 7.5×10^{10} after Si cap processing, providing better gate control over the SiGe channels.

Fig.6 V_{th} dependency on Si cap for the SiGe channel. Low V_{th} is attributed to the small bandgap energy of the SiGe channel.

Fig. 7 Varying V_{th} and relatively independent V_{fb} of Ge content. Slight C-V stretch-out without the Si cap due to high interface charge density.

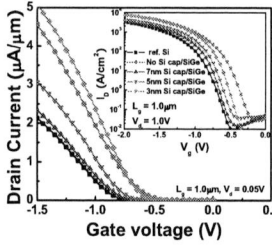

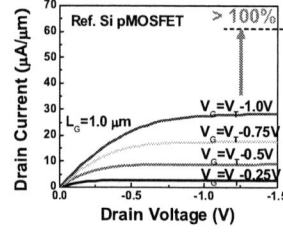

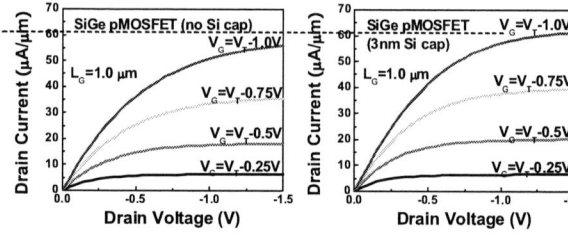

Fig.8 I_d-V_g of SiGe and Si channels. High drain current occurs in optimized Si caps when holes transport on SiGe channel and Si cap serves as a passivation layer.

Fig.11 Drive current enhancement afforded intrinsically by the SiGe channel is further increased by an optimized Si cap, resulting in a combined >100% enhancement over Si channels compared to reference Si channel at normalized gate voltages.

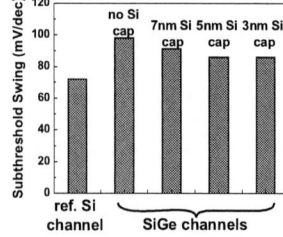

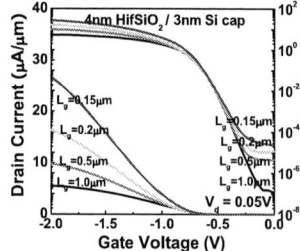

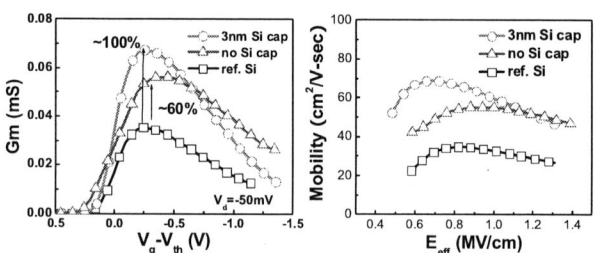

Fig.9 Intrinsically higher S.S. for SiGe than Si channels because of low built-in potential of source-to-channel and S/D junctions.

Fig.10 Si cap processing lessens short channel effects by improving interface quality and, therefore, providing better gate control over the channel.

Fig.12 SiGe channels without a Si cap exhibited ~60% mobility enhancement over the Si channels. After applying the Si cap process, SiGe channel mobility is further enhanced by ~40%. Therefore, a ~100% mobility enhancement was achieved by the combined SiGe channel and Si cap processing.

978-1-4244-2784-0/09 $25.00 © 2009 IEEE

Dopant and Thermal Interaction on SPE formed SiC for NMOS Performance Enhancement

P. W. Liu, T. F. Kuo, C. I. Li, Y. R. Wang, R. M. Huang, C. H. Tsai, C. T. Tsai, and G. H. Ma

United Microelectronics Corporation (UMC), ATD Division, No. 18, Nanke 2nd Rd., Tainan Science Park, Taiwan 741

e-mail: po_wei_liu@umc.com

ABSTRACT

The dopant and thermal interaction on solid phase epitaxy (SPE) formed SiC has been investigated. We have studied the impact on substitutional carbon concentration ($[C]_{sub}$) from various thermal steps including low temperature anneal, SiGe epitaxy thermal budget, RTP, and laser anneal (LSA). Regarding the integration scheme for implementing embedded SiC (eSiC) S/D on NMOS performance enhancement, both post-LDD and post-S/D schemes were studied. The higher $[C]_{sub}$ in post-LDD scheme was observed and the S/D dopants were found to enhance the carbon precipitation into interstitial with conventional RTP/LSA activation thermal processes. The phosphorous implant is also found to degrade $[C]_{sub}$ in comparison to As implant. The higher $[C]_{sub}$ and proximity to channel of formed eSiC in post-LDD scheme are beneficial to device performance. The fabricated eSiC S/D NMOS shows 31% mobility improvement and 7% current enhancement.

INTRODUCTION

In order to overcome the MOSFET device scaling difficulties, various strained-Si schemes have become essential components for 45nm and beyond CMOS technology. For example, the embedded SiGe (eSiGe) has been introduced in the S/D area to enhance PMOS performance [1]. Besides, the embedded SiC (eSiC) S/D stressors have gain interests for NMOS enhancement due to large tensile stress resulted from the smaller lattice constant of SiC. There are two different approaches to integrate the eSiC S/D for NMOS. One is to recess S/D region and grow SiC using selective epitaxy process [2]. Another approach was reported recently to use C ion implantation (I/I) and solid phase epitaxy (SPE) anneal to form SiC in the S/D [3]. The C I/I and SPE approach shows high substitutional carbon concentration ($[C]_{sub}$) and is easy to integrate into full CMOS process. However, the large sheet resistance (Rs) increase in doped SPE SiC will offset the performance gain obtained from the strain effect [3]. Moreover, the high temperature thermal processes will induce carbon atom precipitation out from substitutional sites and reduce channel stress as reported in other literatures [4]. Thus it is important to study the dopant and thermal interactions on SPE formed SiC, especially when integrated with high temperature rapid thermal processes (RTP) or mili-second anneal (such as the laser anneal, LSA).

STUDY OF C I/I AND THERMAL CONDITIONS

We have firstly studied the interaction of carbon and thermal conditions on blanket wafers with pre-amorphization implant (PAI) and carbon implant only. As shown in the X-ray diffraction (XRD) measurement (Fig. 1), the low temperature SPE anneal formed SiC shows 1.1% $[C]_{sub}$. High quality SiC is successfully formed and resulted in the same $[C]_{sub}$ using the SiGe epitaxy thermal budget as the SPE anneal (processed using the same thermal budget only, no SiGe epi-layer grown). This implies that we can use the SiGe epitaxy thermal to re-crystallize the SiC when integrating eSiC S/D for NMOS and eSiGe S/D for PMOS. In Fig.2, the $[C]_{sub}$ with different thermal steps were investigated. The RTP will degrade $[C]_{sub}$ while adding LSA may slightly improve $[C]_{sub}$ compared to the SPE only sample. The LSA only sample shows the largest $[C]_{sub}$.

INTERACTION BETWEEN DOPANTS AND CARBON

In real device, there will be dopants in the LDD and S/D area. Thus the interaction between dopants and implanted carbon will be very important. As shown in Fig. 3, the C I/I and SPE steps can be inserted either after NMOS LDD I/I (post-LDD) or after S/D I/I

(post-S/D). The optimization of integration schemes for SPE formed eSiC S/D NMOS was also studied. Table I shows the split table for samples used to study the combination effects of NMOS LDD & S/D dopants vs. RTP/LSA thermal processes. Both post-LDD and post-S/D schemes were studied. For the post-LDD scheme, the RTP/LSA will induce carbon precipitation and resulted in lower $[C]_{sub}$ as shown in Fig. 4. If the samples were implanted with S/D dopants, the final $[C]_{sub}$ tends to be lower (Fig. 5). The thermal processes such as the RTP and LSA deteriorate the substitutional C incorporation, especially with LDD or S/D dopants. It is worth noted that the sample with LDD implant shows 1.1% $[C]_{sub}$ and is comparable to the sample with PAI & C I/I only. Which means that the interference of LDD implant on the C incorporation is smaller. Since the LDD junction is shallow (<200A), the implanted C will be below LDD junction. The majority of implanted C is located in the region with low LDD/pocket dopants interference. For the post-S/D scheme (Fig. 6), the $[C]_{sub}$ is lower than post-LDD scheme even without RTP/LSA. And the RTP/LSA processes will degrade $[C]_{sub}$, too. Furthermore, the Rs will be too high for wafers with SPE anneal only. Except the implantation energy, the major difference between LDD and S/D implantation is the dopant species used. Phosphorus is used in S/D implantation and not in the LDD implantation. The blanket wafer results (Fig. 7) clearly indicates the phosphorus implant will degrade $[C]_{sub}$ much more than arsenic implant. It is also reported that SiC layers with P dopants will exhibit an enhanced substitutional-to-interstitial diffusivity of C during thermal processes [5-6]. As shown in Fig. 8, if post-LDD scheme is applied, the final $[C]_{sub}$ will be higher in the region under LDD junction and the tensile stress is higher. If C is implanted using post-S/D scheme, the deep S/D dopants will reduce the final $[C]_{sub}$ after RTP/LSA processes and result in lower tensile stress (Fig. 9). Besides, the formed SiC will be closer to channel in the post-LDD scheme. The above results clearly indicate the benefit of post-LDD scheme for SPE formed eSiC S/D NMOS.

Device Electrical Results

We have fabricated the SPE formed eSiC S/D NMOS. In order to reduce the sheet resistance in S/D area, RTP and LSA were applied to the devices. Short channel electron mobility enhancement is 31% from the slope reduction of Ron-Lpoly in Fig. 10. The eSiC S/D NMOS shows 7% current enhancement compared to control devices in Fig. 11.

CONCLUSION

High performance SPE formed eSiC S/D NMOS was successfully fabricated. The interaction of dopants and thermal effect on SPE formed SiC was investigated. The S/D dopants which contains P will reduce final $[C]_{sub}$ incorporation with RTP/LSA thermal processes. By introducing the post-LDD C I/I and SPE anneal scheme, the resulted $[C]_{sub}$ will be higher. The higher $[C]_{sub}$ and proximity to channel of post-LDD eSiC S/D NMOS will result in higher channel stress. The optimized eSiC S/D NMOS shows 31% mobility improvement and 7% current enhancement.

REFERENCES

[1] T. Ghani et al., *IEDM Tech. Dig.*, p. 978, 2003 [2] K. W. Ang et al., *IEDM Tech. Dig.*, p. 1069, 2004 [3] Y. Liu et al., *Symp. VLSI Tech.*, p. 44, 2007 [4] Y. J. Kim et. al., *Jpn. J. Appl. Phys.*, p. 773, 2001 [5] P. Grudowski et al., *SOI Conf.Proc.*, p. 17, 2007 [6] L. A. Ladd, et al., *MRS. Proc.*, v59, p. 419, 1985

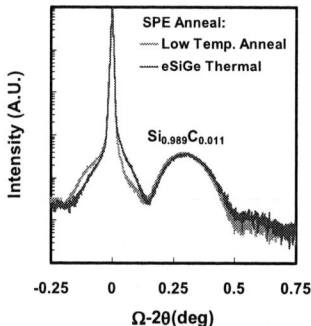

Fig. 1 The XRD rocking curves of the SPE SiC films. Both samples were implanted with PAI and C only. The sample with eSiGe thermal shows comparable $[C]_{sub}$ vs. low temperature anneal.

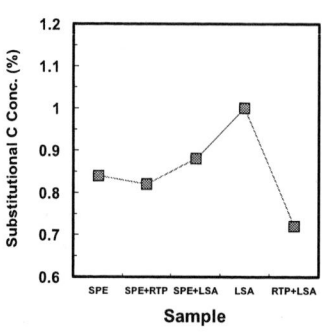

Fig. 2 The substitutional C concentration $[C]_{sub}$ vs. different thermal steps. All samples were implanted with PAI & C only. The sample with laser anneal (LSA) shows largest $[C]_{sub}$.

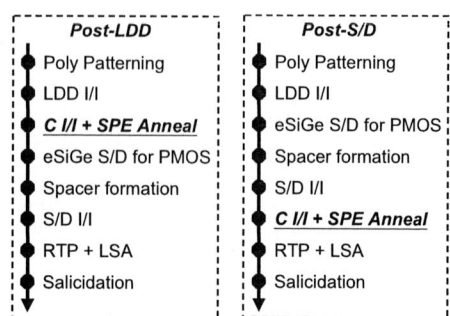

Fig. 3 Process flow illustration for post-LDD **(left)** and post-S/D **(right)** eSiC S/D NMOS integration schemes. For post-LDD scheme, the eSiGe thermal budget can be used as the SPE anneal to simplify process.

Post-LDD C I/I and SPE anneal

Sample	LDD I/I	C I/I+SPE	S/D I/I	RTP	LSA
A	O	O			
B	O	O		O	
C	O	O			O
D	O	O	O	O	
E	O	O	O	O	O

Post-S/D C I/I and SPE anneal

Sample	LDD I/I	S/D I/I	C I/I+SPE	RTP	LSA
F	O	O	O		
G	O	O	O	O	
H	O	O	O	O	O

Table I. Split table to study the combination effects of LDD & S/D dopants and RTP/LSA. Both post-LDD **(top)** and post-S/D **(bottom)** schemes were studied.

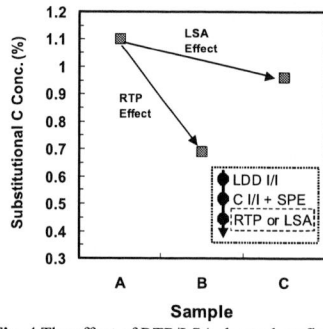

Fig. 4 The effect of RTP/LSA thermal on final $[C]_{sub}$ for post-LDD scheme. The RTP thermal will reduce $[C]_{sub}$ more than LSA thermal.

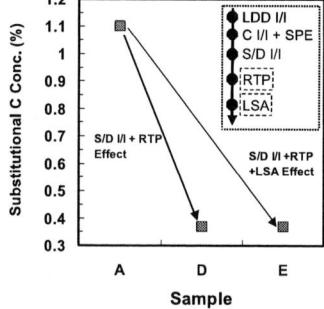

Fig. 5 The effect of S/D dopants and RTP/LSA on final $[C]_{sub}$ for post-LDD scheme. It is found that the S/D dopants will reduce final $[C]_{sub}$ after RTP/LSA.

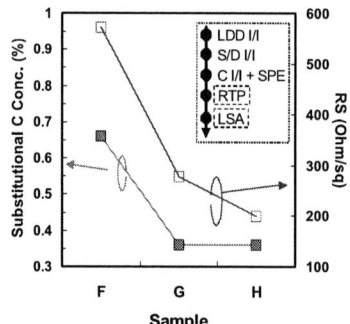

Fig. 6 The effect of RTP/LSA on $[C]_{sub}$ for post-S/D scheme. It is found that RTP will reduce $[C]_{sub}$. The sheet resistance Rs is also higher for SiC film with SPE only.

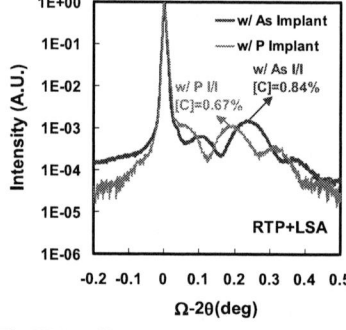

Fig. 7 The XRD rocking curves of SiC wafers with As and P implant. The wafer with P implant shows reduced $[C]_{sub}$ compared with As implanted wafer. The P implant will degrade C incorporation.

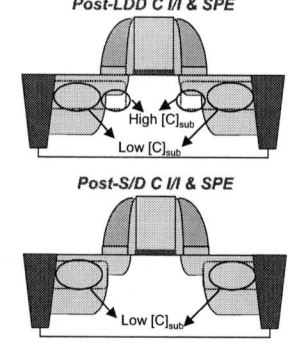

Fig. 8 Schematic illustration of final $[C]_{sub}$ distribution. The final $[C]_{sub}$ will be higher in the region under LDD due to less dopant interference.

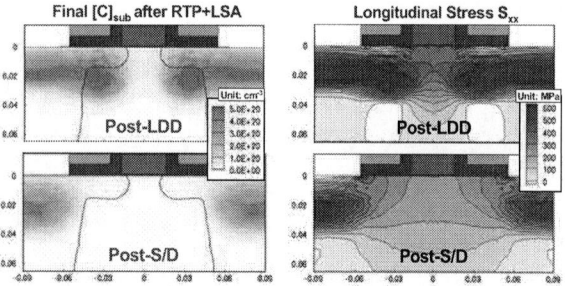

Fig. 9 (left) The simulated final $[C]_{sub}$ after full device integration process (including thermal activation processes). The $[C]_{sub}$ will be higher in post-LDD scheme due to lower dopant interference in the area below LDD junction. **(right)** The simulated longitudinal stress S_{xx} according to the simulated final $[C]_{sub}$.

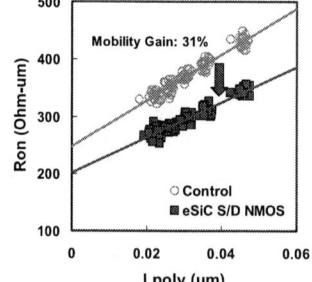

Fig. 10 NMOS resistance Ron vs. Lpoly plots. The eSiC S/D NMOS shows 31% mobility improvement as observed in the slope reduction.

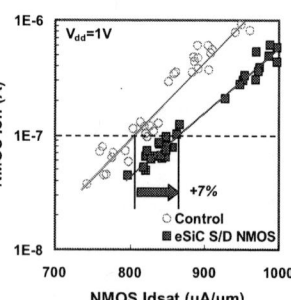

Fig. 11 NMOS Idsat-Ioff performance comparison. The eSiC S/D can improve NMOS drive current by 7%.

978-1-4244-2784-0/09 $25.00 © 2009 IEEE

Successful Integration Scheme of Cost Effective Dual Embedded Stressor Featuring Carbon Implant and Solid Phase Epitaxy for High Performance CMOS

M. Nishikawa, K. Okabe, K. Ikeda, N. Tamura, H. Maekawa, M. Umeyama, H. Kurata, M. Kase and K. Hashimoto

Fujitsu Microelectronics Ltd., 1500 Mizono, Tado-cho, Kuwana, Mie, 511-0192, Japan

Phone: +81-594-24-2294, Fax: +81-594-24-5583, Email: m_nishikawa@jp.fujitsu.com

Abstract

We have developed a device integration scheme for embedded silicon carbon (Si:C) SD structures induced by the solid phase epitaxy (SPE) technique. Our integration scheme comprises a combination of three key processes: carbon ion implantation (I/I) with Ge pre-amorphization implantation (PAI), sRTA and LSA. The guideline of our scheme is as follows. First, carbon I/I with Ge PAI plays large roll in this scheme since we can independently control both damage and stressor. Second, Ge PAI prior to carbon I/I is also performed to realize a steep carbon profile. Third, the embedded Si:C is required to be positioned beneath the Rp of n+dopant to maximally utilize the low resistance deep SD I/I region. Finally, optimizing thermal budget enables us to suppress both carbon clustering and residual defects induced by Ge PAI without a degradation of Vth-rolloff characteristics and a strain relaxation in embedded SiGe (eSiGe) in PMOSFETs. By using this scheme, we have controlled both parasitic resistance and junction leakage current simultaneously. In addition, UV-Raman spectroscopy and HR-XRD clarified the achievement of more than 1 at% effective substitutional carbon concentration by this scheme. Consequently, a 5.1% improvement in Ion of NMOSFETs for Ioff = 100 nA/μm at Vd = 1.0 V and Ion = 1154 μA/μm was obtained. For PMOSFETs, thanks to an optimized annealing process, strain relaxation in eSiGe was avoided, and thus Ion = 818 μA/μm for Ioff = 100 nA/μm at Vdd = 1.0 V, was obtained. We have successfully demonstrated the CMOS integration with a cost-effective "dual" embedded stressor.

Introduction

Recently, using embedded Si:C as a novel stressor for NMOSFETs has been widely investigated by many groups [1-11]. So far, two major approaches toward forming Si:C to induce tensile stress into NMOSFET channels have been reported. One is combining the SPE technique with carbon I/I [1-6]. The other is a recess etch and selective epitaxial growth approach [7-11]. The latter appears to be more attractive, since it enables the stressor shape to be designed flexibly to maximize channel stress. This approach utilizes Si:C (C_{sub} = 1-2 at%) which generates a high strain to improve the electron mobility for NMOSFETs. With this approach, however, it is difficult to selectively grow Si:C with C concentration > 1 at% due to the extremely low solid solubility of C in Si [12]. In contrast, the former approach, i.e., combining the SPE technique with carbon I/I, is advantageous in two ways. First, carbon I/I is a lower-cost process than epitaxial growth since no new facilities or surface treatments (e.g. embedded SiGe technique) are required. Second, using conventional deep SD I/I process enables us easily to selectively implement the Si:C stressor into SD regions of NMOSFETs.

This paper describes our development of a device integration scheme for combining the SPE technique with carbon I/I. By optimizing stressor positions to deep I/I region and thermal budget, we have successfully integrated dual embedded stressor into a high performance CMOS technology.

Comprehensive Investigation of SPE Growth of Si:C

We used UV-Raman spectroscopy to investigate both Raman shift corresponding to Si-Si bonds and the existence of Raman shift induced by Si-C bonds. In general, Raman spectroscopy can bring out an in-plane strain. Since C I/I with Ge PAI seems to be an advantageous way to independently control both damage (Ge PAI) and stressor (carbon), we investigated Si:C induced by SPE with carbon I/I and Ge PAI. Fig. 1 (a) shows the intensity of Raman shift by C I/I + Ge PAI with various annealing schemes. The concentration of C in epi-Si:C was found to be 1 at% which was cross checked by SIMS analysis. By using both laser spike annealing (LSA) and spike rapid thermal annealing (sRTA), we have successfully obtained as large Raman peak shift to a shorter wavelength (from 520 cm^{-1}) as that by epi-Si:C (Fig. 1 (b)) This means that Si-Si bonds were expanded by substititional carbon. In addition, a stronger Raman peak shift corresponding to Si-C bonds was also clearly observed (Fig. 1 (c)). Moreover, synchrotron orbital radiation (SOR) HR-XRD Si-C(004) scans clarified that the out-of-plane Si lattice constant of a sample with C I/I was smaller than that of epi-Si:C, as shown in Fig. 2. This result indicated that optimizing annealing process increases the amount of effective substitutional carbon. These physical analyses show that our embedded Si:C makes the out-of-plane Si lattice constant become smaller and therefore allows the Si horizontal lattice to expand in the channel direction to enhance the electron mobility.

Figure 3 shows the cross-sectional TEM micrographs of our NMOSFETs (a) with Si:C induced by carbon I/I and LSA and (b) w/o SPE technique (conventional flow). It indicates that excellent crystalline quality was obtained. This is important since it is essential to suppress the defect formation in the case like eSiGe [13] to form high quality Si:C in the SD region for stressor application.

Concept of Device Integration

It seems reasonable to assume that from the perspective of device characteristics, two key factors are required to integrate Si:C induced by SPE. First, the "stressor" position should be designed so as to reduce parasitic resistance. It has been pointed out that carbon clustering in the SD region are associated with parasitic resistance [5, 11]. High temperature activation annealing is an effective means of suppressing carbon clustering. However, the diffusion of SD dopants induced by high temperature annealing degrades Vth-rolloff characteristics. Second, high quality crystal regrowth should be achieved to suppress junction leakage current even with a lower thermal budget. This would enable shallow junction abruptness to be maintained. In order to resolve this issue, it is necessary to optimize the overlap between the Si:C stressor and the deep SD region since the rate of crystal restoration is much larger in n-type Si. On the other hands, the embedded Si:C is required to be positioned beneath the Rp of n+dopant in order to maximally utilize the low resistance region by deep SD I/I. Therefore, we optimized C dosage and implantation energy with Ge PAI which enables us to realize a steep carbon profile and a high concentration in various thermal budgets. To overcome the second issue, both PAI conditions and annealing temperature are required to optimize in SPE process, i.e. minimized damage even for lower activation anneal temperature is very advantageous from the viewpoint of a crystal regrowth. However, the strain relaxation in eSiGe also limits the maximum annealing temperature that can be attained.

In this work, we used three types of process sequences to form the Si:C induced by SPE in SD structures, as shown in Fig. 4. Our SPE technique comprises a combination of three key processes: sRTA, LSA (no melt temperature region) and carbon I/I (Ge PAI + C). These processes are easy to incorporate into a conventional CMOS flow (Flow A). Figure 5 shows the feature of 2D distribution of key I/I processes such as Ge PAI, C I/I and deep I/I. C I/I with Ge PAI is used in Flow B and Flow C since we can independently control both damage (Ge PAI energy and dosage) and stressor (carbon I/I energy and dosage). We can understand that amorphous thickness and interface roughness control are important. In addition, Ge PAI suppresses the channeling tails of both n+dopant and carbon, thus it enables us easily to optimize the overlap between Si:C stressor and deep SD region.

Device Characteristics

In Flow B, we speculate that C I/I after sRTA appears to have advantages in minimizing carbon out-diffusion by MSA, which can induce higher stress than the case prior to sRTA. However, both Ion-Ioff (Fig.6 (a)) and Idlin-Ioff (Fig.6 (b)) characteristics were severely degraded by heavy carbon doping in Flow B. In addition, the Vth-rolloff curve also deteriorated more than that of control device (Fig 7). These results are caused by two factors. The first is an increase in parasitic resistance due to carbon clustering in the SD region with low thermal budget. Indeed, significant degradation (28%) of Rpara (extracted from total device resistance) is obtained in this case due to clustering or interstitial carbon in SD region (Fig. 8: inset). The other factor is the residual defect induced by Ge PAI that results in increased off-state current, thereby causing Vth-rolloff to deteriorated. Junction leakage current increased significantly more than it did in the control device (Fig. 13). Furthermore, SD silicide sheet resistance of NMOSFETs also increased due to an amorphous region existing the beneath of the silicide. This indicates that using only LSA after C I/I is not enough to regrow the Si crystals in our temperature range (no-melt region) that are compatible with PMOSFETs.

On the other hand, Device C appears to include a more sufficient thermal budget than that of Flow B. Indeed, we found that using optimized carbon dose and energy in Flow C can actually improve both the Ion and the Idlin of NMOSFETs by 5.1% and 7.1% respectively at Ioff = 100 nA/μm resulting from the high quality stressor, as shown in Fig. 11(a) and (b). The Vth-rolloff characteristics between control and Device C are very comparable (Fig. 12). In addition, significant improvement was obtained in Id characteristics over the device with Si:C SD, in particular, linear current characteristics in the Id-Vd characteristic of 35-nm gate length devices (Fig. 13). In order to confirm the mechanism validity, we measured effective mobility as a function of surface carrier concentration (Fig. 11). Owing to the optimized thermal budget and carbon conditions, a 27.7% electron mobility improvement was successfully achieved in a 140-nm gate length device. Moreover, we measured Rtotal as a function of gate lengths (Figure 8). The shallow slopes of total device resistance for optimized Si:C indicate enhanced mobility. Hence the mobility of Device C is improved by 13%. In the inset of Fig. 8, Rpara of device C is almost comparable to that of control device due to the optimized annealing process that includes a sufficient thermal budget. As for the residual defects, junction leakage current is controlled to be comparable to control device. And SD silicide sheet resistance is almost same as control device due to high quality crystal regrowth (no residual amorphous region) . From the viewpoint of deep SD profile design, Ge PAI suppressed the channeling tails of both n+dopant and carbon: thus it enabled us to keep high carbon concentration in Si:C stressor, and to realize abrupt n+dopant profile with low sheet resistance. For PMOSFETs, no degradation in device performance was seen due to the optimized annealing process which is compatible with the suppression of eSiGe relaxation. Hence, Ion = 818 μA/μm for Ioff = 100 nA/μm at Vdd = 1.0V was obtained (Fig. 15). Finally, with optimized I/I conditions and thermal budget, we were able to successfully integrate this technique into CMOS process.

Conclusion

We have demonstrated a device integration scheme for Si:C induced by the SPE technique. Physical analyses clarified that this scheme produces enough substitutional carbon (> 1%) in SD region. By using the optimized all I/I conditions and annealing process, we were able to control both parasitic resistance and residual defects simultaneously. Consequently, we successfully achieved a 5.1 % improvement in Ion for Ioff = 100 nA/μm at Vd = 1.0 V and Ion = 1154 μA/μm. For PMOSFETs, thanks to our optimized annealing process which is compatible with the eSiGe process, we obtained Ion = 818 μA/μm for Ioff = 100 nA/μm at Vdd = 1.0 V.

Reference

[1] Y. Liu et al., VLSI Tech., pp. 44, 2007. [2] G. -H. Wang et al., VLSI Tech., pp. 207, 2008. [3] S. -M. Koh et al., et al., Ext. Abst. SSDM, pp. 872, 2008. [4] M. Tanjyo et al., Ext. Abst. IWJT, pp., 2008. [5] H. Maynard et al., RTP conf. proc., pp.147, 2008. [6] Y. Cho et al., IEDM Tech. Dig., pp. 959, 2007 [7] Y. -C. Yeo, Semicond. Sci. Technol. 22, pp. S177, 2007. [8] K. W. Ang et al., IEDM Tech. Dig., pp. 503, 2005. [9] Z. Ren et al., VLSI Tech., pp. 172, 2008. [10] R. T. Lee et al., IEDM Tech. Dig., pp. 685, 2007. [11] H. Itokawa et al., Appl. Surf. Sci., **254**, pp. 6135, 2008. [12] A. C. Mocuta et al., J. Appl. Phys., **85**, pp. 1240, 1999. [13] M. Nishikawa et al., Ext. Abst. SSDM, pp. 262, 2008.

978-1-4244-2784-0/09 $25.00

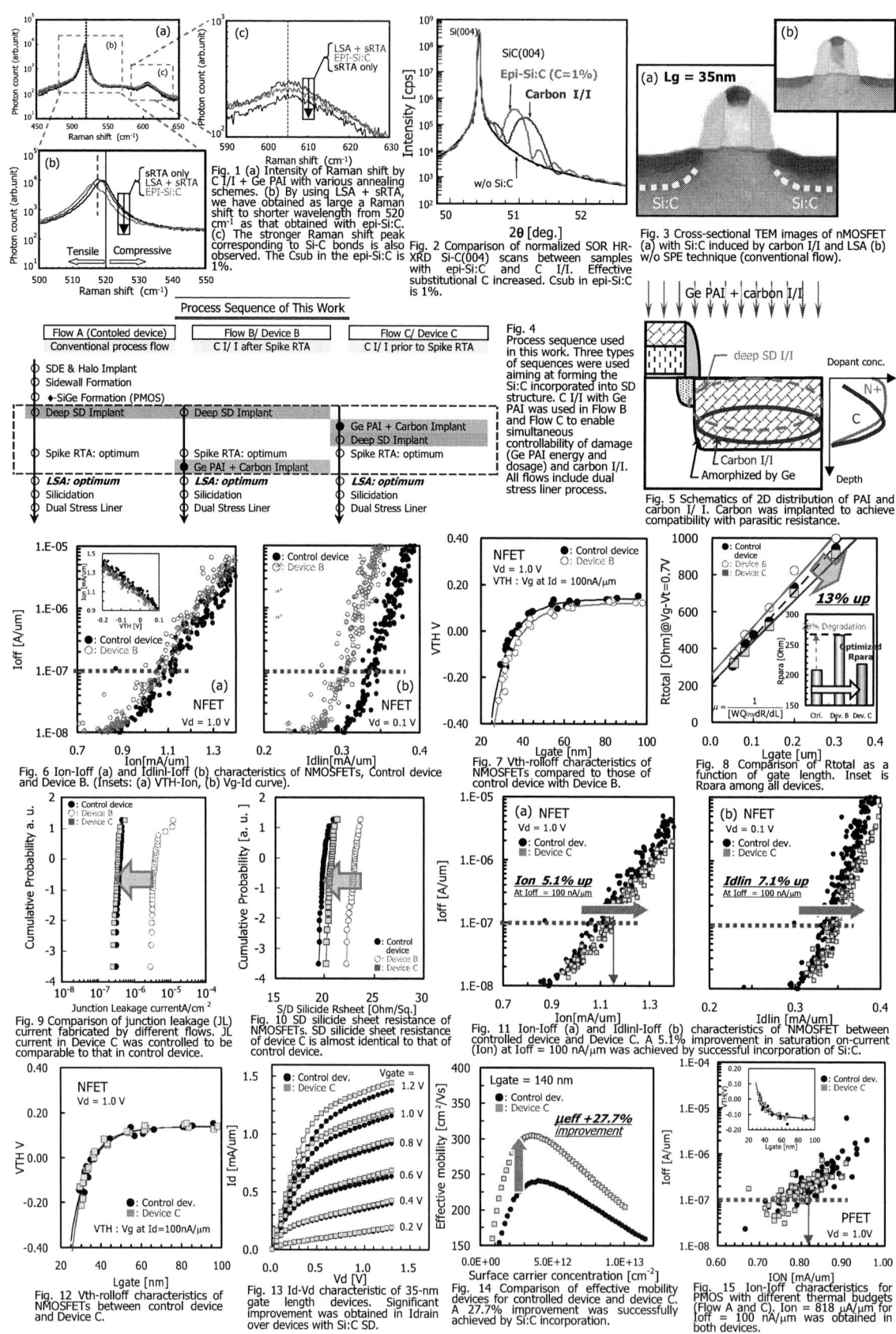

Fig. 1 (a) Intensity of Raman shift by C I/I + Ge PAI with various annealing schemes. (b) By using LSA + sRTA, we have obtained as large a Raman shift to shorter wavelength from 520 cm^{-1} as that obtained with epi-Si:C. (c) The stronger Raman shift peak corresponding to Si-C bonds is also observed. The Csub in the epi-Si:C is 1%.

Fig. 2 Comparison of normalized SOR HR-XRD Si-C(004) scans between samples with epi-Si:C and C I/I. Effective substitutional C increased. Csub in epi-Si:C is 1%.

Fig. 3 Cross-sectional TEM images of nMOSFET (a) with Si:C induced by carbon I/I and LSA (b) w/o SPE technique (conventional flow).

Fig. 4 Process sequence used in this work. Three types of sequences were used aiming at forming the Si:C incorporated into SD structure. C I/I with Ge PAI was used in Flow B and Flow C to enable simultaneous controllability of damage (Ge PAI energy and dosage) and carbon I/I. All flows include dual stress liner process.

Fig. 5 Schematics of 2D distribution of PAI and carbon I/I. Carbon was implanted to achieve compatibility with parasitic resistance.

Fig. 6 Ion-Ioff (a) and Idlinl-Ioff (b) characteristics of NMOSFETs, Control device and Device B. (Insets: (a) VTH-Ion, (b) Vg-Id curve).

Fig. 7 Vth-rolloff characteristics of NMOSFETs compared to those of control device with Device B.

Fig. 8 Comparison of Rtotal as a function of gate length. Inset is Rpara among all devices.

Fig. 9 Comparison of junction leakage (JL) current fabricated by different flows. JL current in Device C was controlled to be comparable to that in control device.

Fig. 10 SD silicide sheet resistance of NMOSFETs. SD silicide sheet resistance of device C is almost identical to that of control device.

Fig. 11 Ion-Ioff (a) and Idlinl-Ioff (b) characteristics of NMOSFET between controlled device and Device C. A 5.1% improvement in saturation on-current (Ion) at Ioff = 100 nA/μm was achieved by successful incorporation of Si:C.

Fig. 12 Vth-rolloff characteristics of NMOSFETs between control device and Device C.

Fig. 13 Id-Vd characteristic of 35-nm gate length devices. Significant improvement was obtained in Idrain over devices with Si:C SD.

Fig. 14 Comparison of effective mobility devices for controlled device and device C. A 27.7% improvement was successfully achieved by Si:C incorporation.

Fig. 15 Ion-Ioff characteristics for PMOS with different thermal budgets (Flow A and C). Ion = 818 μA/μm for Ioff = 100 nA/μm was obtained in both devices.

978-1-4244-2784-0/09 $25.00 © 2009 IEEE

An Investigation about the Limitation of Strained-Si Technology

M. H. Liao, Lingyen Yeh, J. C. Lu, M. H. Yu, L. T. Wang, J. Wu, P.-R. Jeng, T.-L. Lee, and Simon Jang

Research & Development, Taiwan Semiconductor Manufacturing Co. Ltd., Hsinchu, Taiwan, R. O. C.

Tel: 886-3-5636688 ext. 7125321 Fax: 886-3-5637000 E-mail: mhliao@tsmc.com

Strained-Si technology is the Holy Grail for present semiconductor industry and is used extensively to boost the device performance, recently. However, the limitation of strained-Si technology has greatly perplexed us and need to investigate in detail. In this work, the low temperature ballistic measurement enables us to discriminate the origin of mobility enhancement under stress from the reduction of effective mass and/or the influence of different scattering mechanisms. It is found that the electron mobility enhancement under stress will become less sensitive when the gate length of device reaches ~100 nm. The real mechanism of this phenomenon have be proved to the characteristic of device ballistic transport and the optimal stress design developed in this work can further extend the limitation of Strained-Si technology to the smaller gate length region (technology node) (Fig. 1).

EXPERIMENT AND DISCUSSION

Fig. 1. shows the drain-current enhancement as the function of the gate length with stressor in the modern CMOS fabrication. It can be found that the Ion improvement with stress becomes less sensitive when the device gate length is smaller than 100 nm (Quasi-Ballistic region). The corresponding Idsat v.s. Vg transistor characteristics are also shown in Fig. 2. This phenomenon limits seriously the development and the practicability of Strained-Si technology in the short gate length region (next technology node). To understand the origin of this phenomenon, 3D stress simulation/modeling with the full process flow such as active area and isolation process [2], surface roughness scattering, coulomb (neutral) scattering, and ballistic measurement [3] at different temperature is given in this work. Fig. 3 shows the 3D stress distribution in the device through the full process. As the channel is smaller than 100 nm, the stress along the channel and out of plane direction becomes more tensile and compressive stress, respectively. These stresses are both good for the electron transport, based on full band mobility calculation, and cannot explain the experiment observed phenomenon in Fig. 1. Fig. 4 - Fig. 5 show the Exp. mobility data at different electric field and the simulated electron mobility, considering the surface roughness-limited and phonon-limited scattering mechanism. At the higher operation electric field in the device of short gate length, the more mobility enhancement with stress is observed (Fig. 4 & 5), due to the electron population. It still cannot explain ideally why the Ion improvement becomes less sensitive with stress in the short channel region. Ref. 4 & 5 demonstrate that the less Ion improvement in strained Si device with the short gate length is also not resulted from the coulomb (neutral)-limited scattering, provided in other works. The real origin and mechanism of the experimental phenomenon should be due to the different electron transport characteristic in the Drift-Diffusion Region and Ballistic Region, which will be provided/investigated in next paragraph in detail.

As we known the electron transport characteristics in the two operation regions such as Drift-Diffusion Region and Ballistic Region, which are suitable for long channel and short channel device, respectively. The drain current enhancement is almost proportional to mobility enhancement. The mobility enhancement under stress has the inverse proportion with the reduction of the effective mass in the Drift-Diffusion Region (Exp. data in Fig. 6), while the relationship in the Ballistic Region has became square root. Fig. 6 shows the first observation on the change of effective mass under stress with

different gate length devices by the Ballistic measurement [3]. The less relationship between mobility enhancement and the reduction of effective mass (square root) under stress in the short channel device (Ballistic region) is the **main reason** why the Ion improvement becomes less sensitive. **By the optimum stress design further proposed in this work, the "turning point" in Fig. 1 can be "pushed" to the shorter gate length region. It means that the limitation of Straied-Si Technology can be extended/used in the next technology node.** Fig. 7 shows the extracted Ballisticy. The optimum stress design indeed can reduce the electron effective mass and increase the Ballisticy (Fig. 7). Note that the Ballisticy reported in this work is **High-Record**. Fig. 8 shows the Ion enhancement ratio as a function of gate length and two theoretical calculation models. The Exp. data agrees well with the theoretical model and explain successfully the phenomenon observed in Fig. 1. The change of effective mass resulted from the 3D stress characteristic play the main role on the mobility/Ion enhancement ratio in the ballistic region. Therefore, the 3D stress characteristic with different device structure has been simulated/designed and the change of effective mass under different device design has also been calculated by full band model further (Fig. 8). It is found that as the gate width decrease even on the same gate length of 30 nm, the effective mass will further decrease and result in more device performance enhancement. Thus, the mutli-channel device with symmetric W and L is proposed to enhance the device performance in the circuit design for n FET. **The limitation of Strained-Si technology can successfully be boosted by optimum stress design and device structure (Fig. 9).**

Besides the On state characteristics, the change of Vt (Fig. 10) and the Off-state characteristics (Fig. 11) are also been studied. The stress characteristic from symmetry structure (W~=L) will give more valence band splitting. It will increase the hole-tunneling barrier (Fig. 11) and reduce the leakage current. Fig. 12 summarizes the results observed in this work. The optimum stress design and device structure has the higher Ion current and lower leakage current, due to the reduction of effective mass and the increase of hole tunneling energy barrier, respectively.

We investigate the limitation of Strained Si technology and explain why the Ion improvement with stress becomes less sensitive as the device gate length scales down in the next technology node. The high-record low temperature Ballistic measurement shows that the change of effective mass under stress is the main key factor on the device performance booster when the device enters ballistic region. The full 3D stress characteristic has been simulated and the effective mass with stress characteristic has been calculated by full band model. **With the optimum device design/structure design, the limitation of Strained-Si technology has successfully boosted in the smaller device gate length (technology node) and more Ion (due to the reduction of effective mass)-Ioff (due to the larger hole tunneling barrier) has been achieved.**

References:
[1] F. Payet, et al., TED, p. 105, 2008.
[2] M. H. Liao., et al. EDL, p. 402, 2008.
[3] M. H. Liao., et al. APL, p. 063506, 2008.
[4] O. Bonno., et al., VLSI, p. 134, 2007.
[5] A. Cros., et al. IEDM, 2007.

978-1-4244-2784-0/09 $25.00 © 2009 IEEE

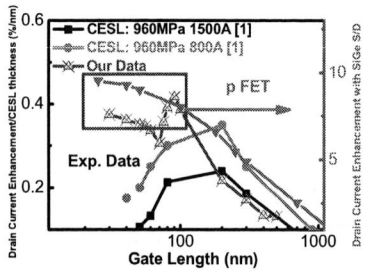

Fig. 1. Drain-current enhancement as a function of the gate length. When the gate length (L) is smaller than 100 nm, the Ion improvement in n FET with stress becomes less sensitive in the channel. By the device optimum design, the "turning point" can be "pushed" to smaller gate length region (technology node). For p FET, the strained-Si benefit can continue to the next generation, due to the change of hole injection velocity (mass) under stress.

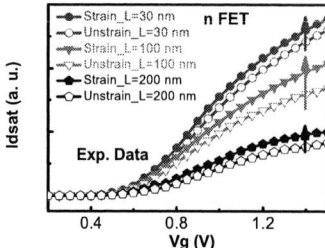

Fig. 2. Typical transistor characteristics of strained-N-MOSFET with different gate length. The maximum device improvement with stress is observed in L=100 nm device. The Ion improvement becomes less sensitive to stress when the device gate length is smaller than 100 nm.

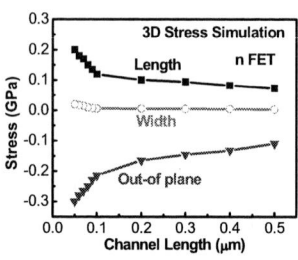

Fig. 3. The stress components as a function of gate length. The stress is the average value between channel center and edge. The smaller gate length device has the larger tensile stress and compressive stress along the gate length and out of plane direction, which is good for electron transport. It cannot explain why the Ion improvement becomes less sensitive with stress in the short channel device. (Fig. 1).

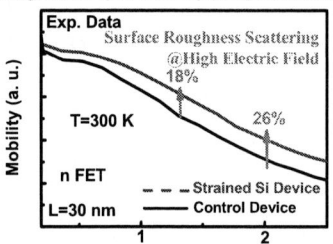

Fig. 4. High field mobility Exp. Data for control Si and strained Si device. There is more mobility enhancement at high electric field, due to the reduction of surface roughness scattering with stress. The mobility Exp. Data cannot explain why the Ion improvement becomes less sensitive with stress in the short channel device (Fig. 1).

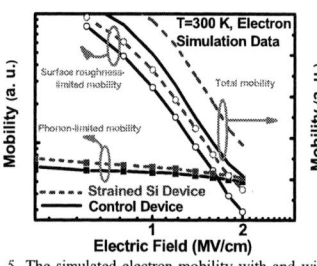

Fig. 5. The simulated electron mobility with and without stress as a function of electric field, considering the surface roughness-limited and phonon-limited scattering mechanism. The simulation data shows that the strained Si device has more mobility enhancement at high electric field, which agrees well with the Exp. Data (Fig. 4).

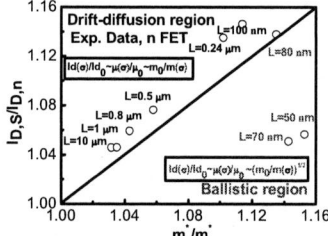

Fig. 6. The comparison between Ion current enhancement and the reduction of effective mass under stress. In the Drift-diffusion region (Long channel device), the Ion improvement has the inverse proportion with the change of effective mass. This relationship is discordant in the Ballistic region (Short channel device: L<100 nm). It is the main reason why the Ion improvement becomes less sensitive with stress in the short channel device (Fig. 1).

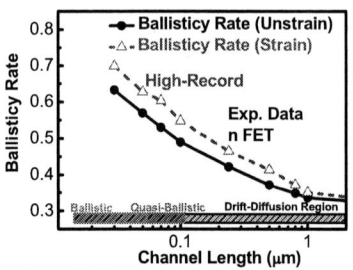

Fig. 7. The Extracted Ballisticy Rate on the control and strained-Si device. The strained Si device has the higher ballisticy rate than the control Si device in the short channel region, due to the reduction of the effective mass. When the gate length is smaller than 100 nm, the ballistic model is adopted. High-Record Ballisticy is reported in this work.

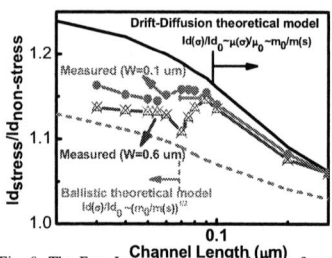

Fig. 8. The Exp. Ion enhancement ratio as a function of gate length. When the gate length is smaller than 100 nm, the electron transport characteristic is Ballistic and the Ion improvement becomes less sensitive with stress in this region. The Exp. Data agrees well with the theoretical model. Moreover, the optimum device structure design leads to biaxial-like stress and further reduce the electron effective mass to enhance the Ion sensitivity in the short channel device.

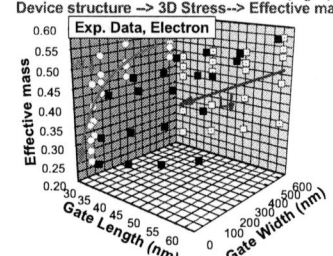

Fig. 9. The extracted effective mass in different device structure. The different device structure (W/L) will result in the different stress distribution. It will influence the effective mass and current enhancement ratio in Drift-Diffusion Region and Ballistic Region. The smaller gate width device has the lower electron effective mass, even on the same gate length. The Ion improvement will be more sensitive with stress, if the effective mass is lower. It is the very important characteristic in the Ballistic Region (Short Channel Device).

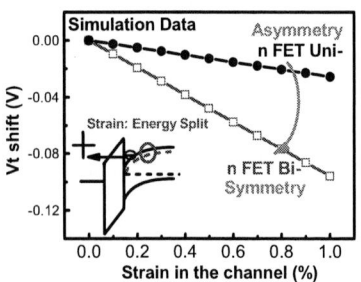

Fig. 10. The shift of Vt as the function of stress. The biaxial-like stress resulted from the symmetry structure leads to more Vt shift, due to the energy band splitting.

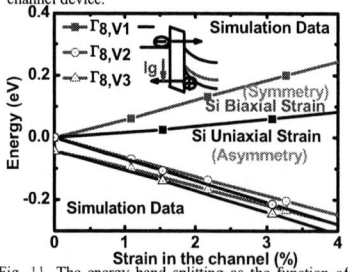

Fig. 11. The energy band splitting as the function of stress under the accumulation region. Besides the better Ballistic performance and higher Ion sensitivity with stress, the symmetry structure leads to biaxial-like stress distribution and further reduce the Ioff state current (The barrier height for hole tunneling becomes higher).

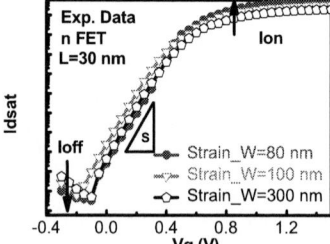

Fig. 12. The Id-Vg characteristic on L=30 nm n FET devices with different gate width. The summary in this work is that we provide the high stress sensitivity, high Ion/Ballisticy, and low leakage current with optimum device design.

978-1-4244-2784-0/09 $25.00 © 2009 IEEE

Characterization of Poly-silicon Emitter BJTs as Access Devices for Phase Change Memory

B. Rajendran[†], M. Breitwisch[†], R. Cheek[†], M-H. Lee[°],Y-H. Shih[°], H-L. Lung[°] and C. Lam[†]

IBM Macronix PCRAM Joint Project

[†]IBM T.J. Watson Research Center, 1101 Kitchawan Road, Yorktown Heights, NY, 10598, USA
[°]Macronix International Co., Ltd., 16, Li-Hsin Road, Science Park, Hsin-chu, Taiwan, R.O.C.
Tel: +1-914-945-1809, email: brajend@us.ibm.com

Abstract

We demonstrate poly-Silicon emitter vertical PNP Bipolar Junction Transistors (BJTs) that could be used as access devices for Phase Change Memory. The device arrays fabricated using a 180nm BiCMOS process exhibit current drive capability in excess of $10mA/\mu m^2$, On-Off ratio greater than six orders of magnitude and excellent cross-talk immunity. Our process integration scheme could be extended to enable a high-density Phase Change Memory technology.

Keywords: PCRAM, poly-emitter BJTs, NV memory

Introduction

Phase change memory (PCM) [1] requires an access device that can satisfy its high programming current requirements with minimum foot print, low leakage in the OFF state, and minimal cross-talk between adjacent devices. Since the drive capability of FET devices are smaller compared to bipolar devices, FET based PCM cells [2] are relatively larger in size ($16–25F^2$) compared to a $\sim 12F^2$ BJT [3] or $\sim 6F^2$ diode [4] based cell. However, minority carrier coupling between adjacent devices in the diode array and the low-gain BJT array causes cross-talk and their process integration with conventional CMOS peripheral circuitry is very challenging.

Moreover, the diode and low-gain BJT based array also suffers from significant variability in the operating conditions, as a function of the distance from the WL contact, due to the non-negligible resistance of the buried semiconductor WL. Earlier demonstration of a diode driven PCM array [3] utilized a WL strapping contact for every eight devices, and the low gain BJT driven PCM array [4] used a strapping contact after every device to minimize this variability, which reduces the array utilization efficiency. Here, we present a proof-of-concept demonstration of a high gain poly-silicon emitter based vertical PNP BJT array suitable for PCM access devices that could potentially lead to higher array utilization efficiency.

Process Technology

We demonstrate a 10×10 array of vertical PNP BJT devices, with poly-silicon emitter, fabricated utilizing a 180 nm BiCMOS process technology. The process steps are shown in Figure 1. The devices on the same word-line share a common base, with a single Base contact at one end of each word-line. The entire array has a single buried Collector, with a ring collector contact (Figure 2). The cross-sectional TEM image of a string of devices in a word-line is shown in Figure 3. The foot print of the fabricated devices were determined by the emitter-base contact area of 1×0.24 μm, and scaling this area could potentially allow the fabrication of high density ($6F^2$) access device arrays for PCM (Figure 4).

Array and Device Characterization

The biasing scheme for read and write operation of a single device from the array is shown in Figure 5; in order to characterize the poly-emitter BJTS, we used the following biasing scheme: the Collector contact is held at ground, and we sweep the selected BL (Emitter) bias from 0 to 3V, as the selected WL (Base) bias is stepped from 0 to 3 V. The unselected BLs and WLs remain held at 0V and 3V respectively (Figure 6).

A typical measurement result for a selected device in the array is shown in Figure 7. The WL bias of the selected device was 1V, the unselected BLs and WLs were held at 0V and 3V respectively. The device turns on at a Base-Emitter forward bias of approximately 0.6V, and for relatively small biases, the measured BL current is ~100 times larger than the WL current, indicative of the gain of the BJT. Also note that at sufficiently high bias condition, the devices have a drive current capacity in excess of 10^6 A/cm^2. The maximum reverse current leakage of a single device (measured at V_{BL}=0V, V_{WL}=3V, and equal to $1/10^{th}$ of the BL current), is found to be in the order of few 10s of nA (Figure 8), indicating that the poly-emitter process maintains excellent interface quality. Silicidation of the WL region between the devices limits minority carrier coupling and hence, the cross-talk between adjacent devices; the observed current in the unselected neighboring cells on the same WL is found to be smaller by ~6 orders of magnitude (Figure 9).

The variability in the operating conditions of a BJT based array as a function of the distance from the WL contact is minimized if it is operated in its high-gain mode – the current that flows in the base region of the selected WL (base current) is ~ β times smaller (where β is the current gain of the BJT) compared to the programming current that flows through the selected BL (emitter current). With a smaller WL current, the parasitic voltage drop along the WL is negligible, resulting in effectively same bias across the devices on the WL, independent of the distance from the WL contact. Our experimental BJT array clearly demonstrates this; the drive current of the devices decreases as a function of the distance from the WL contact when operated at the low gain mode, while this variation is minimized in the high gain mode (Figure 10).

One of the challenges with using a BJT based device for the PCM array is that at the high currents, the BJT devices suffer from a lower gain. This is due to the fact that at larger current density, the gain drops due to the base push-out Effect, as can be seen from the fact that the device current is independent of the bias on the collector (Figure 11). Further device optimization is required to mitigate this problem.

Conclusions

Poly-silicon emitter vertical PNP device technology is demonstrated to be a viable candidate for use as access device for high density Phase Change Memory. Our demonstration shows that it is possible to build arrays with high drive current, low leakage and excellent cross-talk immunity.

Acknowledgments

Expert processing support from the IBM Essex Junction facility and valuable discussions with W. Gallagher, T. H. Ning, T. C. Chen and R. Liu are gratefully acknowledged.

References

[1] S. Lai, *IEDM Tech. Digest* (2003).
[2] S. Kang, et al, *IEEE Journal of Solid-State Circuits 42* (2007)
[3] F Pellizzer et al, *Symposium on VLSI Technology* (2006)
[4] J. H. Oh et al, *IEDM Technical Digest* (2006)

978-1-4244-2784-0/09 $25.00 © 2009 IEEE

1. STI
2. Collector implant
3. Gate Stack Deposition & patterning (clears the BJT area) for CMOS
4. Base implants
5. Spacer process
6. Emitter opening & poly-Si emitter deposition for BJT
7. Emitter isolation patterning
8. S/D implants & Activation
9. Silicidation
10. Contact formation

Fig. 1: BiCMOS process steps to fabricate the PNP poly-emitter BJT devices.

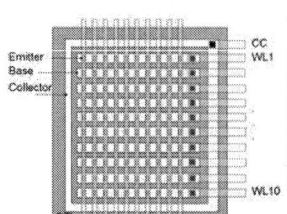

Fig. 2: Layout of a 10×10 VPNP BJT array. The devices on each WL share a common base, and the array has a single buried Collector (ring contact). Metal BLs contact Emitters in the vertical direction.

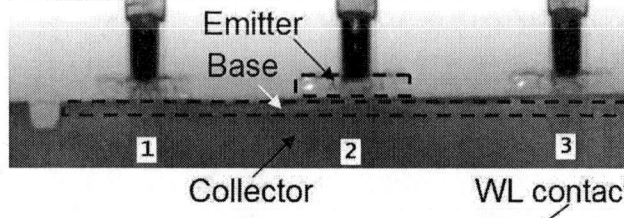

Fig. 3: a) Cross-sectional TEM images of devices on BL1, BL2 and BL3 And b) devices on BL8, BL9 and BL10, along with the WL contact.

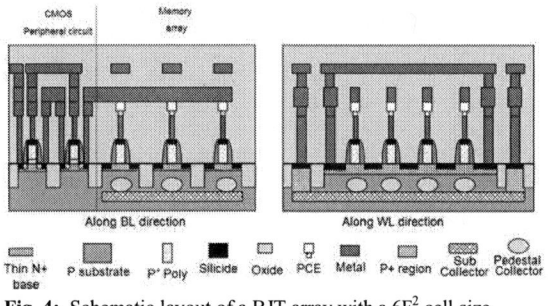

Fig. 4: Schematic layout of a BJT array with a 6F^2 cell size.

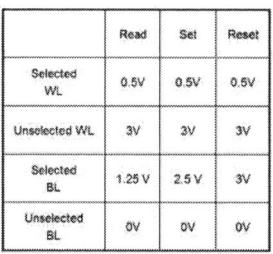

	Read	Set	Reset
Selected WL	0.5V	0.5V	0.5V
Unselected WL	3V	3V	3V
Selected BL	1.25 V	2.5 V	3V
Unselected BL	0V	0V	0V

Fig.5: Array biasing scheme for read and write (RESET and SET) operation of PCM elements.

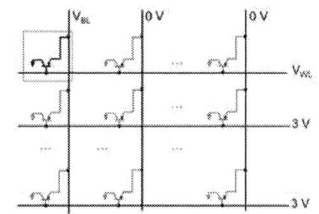

Fig. 6: The biasing scheme used to characterize the array. The unselected BLs are held at 0 V and unselected WLs are held at 3V.

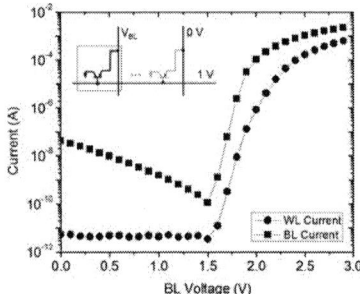

Fig. 7: The transfer characteristics for a typical device in the array. The WL bias of the selected device was 1V, the unselected BLs and WLs were held at 0V and 3V respectively.

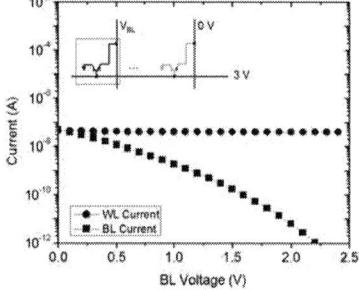

Fig. 8: Leakage characterization of the devices in the array. The WL bias of the selected device was 3 V, the unselected BLs and WLs were held at 0V and 3V respectively. At V$_{BL}$=0V, the measured BL current is the leakage of 10 devices.

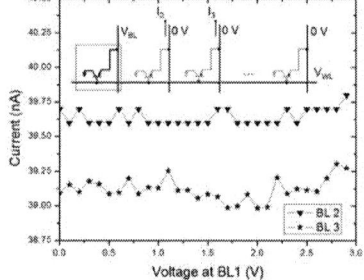

Fig. 9: Cross-talk characterization of the array. The current on two neighboring unselected BLs are measured as a function of the bias voltage at the selected BL.

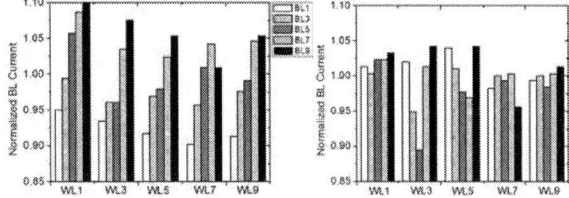

Fig. 10: At small current gain (β~2, left), identical bias conditions leads to larger current on the device closer to the WL contact (BL 9) compared to the device farthest from the WL contact (BL1). At high current gain (β~100, right), there is no such consistent variation in the drive current characteristics.

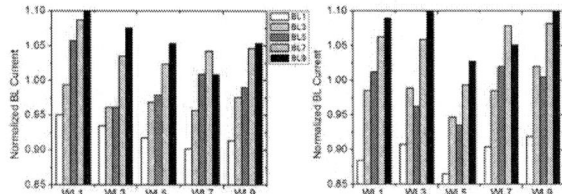

Fig.11: The collector bias has very little impact at high current density (left V$_C$=0V), (right V$_C$=−1V), indicating the base push-out effect.

978-1-4244-2784-0/09 $25.00 © 2009 IEEE   28

Low Current and Voltage Resistive Switching Memory Device Using Novel Cu/Ta$_2$O$_5$/W Structure

S. Z. Rahaman[1], S. Maikap[1,*], C.-H. Lin[2], T.-Y. Wu[2], Y. S. Chen[2], P.-J. Tzeng[2], F. Chen[2], C. S. Lai[1], M.-J. Kao[2], and M.-J. Tsai[2]

[1]Department of Electronic Engineering, Chang Gung University, Tao-Yuan, Taiwan
[2]Electronic and Opto-Electronic Research Laboratories, Industrial Technology Research Institute, Hsinchu, Taiwan
*Corresponding author: Tel: 886-3-2118800 ext. 5785 Fax: 886-3-2118507 E-mail: sidhu@mail.cgu.edu.tw

ABSTRACT

Low current/voltage (~10 nA/1.0V) resistive switching memory device in a Cu/Ta$_2$O$_5$/W structure has been proposed. The low resistance state (R_{Low}) of the memory device decreases with increasing the programming current from 10 nA to 1mA, which can be useful for multi-level of data storage. This resistive memory devices have stable threshold voltage, good resistance ratio (R_{High}/R_{Low}) of 5.3×10^7, good endurance of $>10^3$ cycles, and excellent retention (>11 hours) with resistance ratio of $> 9 \times 10^3$ can be useful in future non-volatile memory applications.

INTRODUCTION

Now a day the resistive switching memory device is a promising candidate due to its highly non-volatility, scalability potential, low power consumption and reliability aspects. The resistive memory devices using oxide materials such as HfO$_2$, NiO, CuO, Cr doped SrTiO$_3$, etc. have been reported by several groups. Recently, resistive switching memory devices using Ag doped GeSe or GeS solid electrolyte have been also reported [1-2]. Sakamoto et al. reported the resistive memory device in a Cu/Ta$_2$O$_5$/Pt structure with a current compliance of 100 μA [3]. However, the Pt as a bottom electrode can not be used fairly in the CMOS technology. In our previous study, we have reported the resistive switching memory device in a Cu/Ta$_2$O$_5$/TiN structure using shadow mask [4]. The resistive memory device using Cu/Ta$_2$O$_5$/W structure is not reported yet. In this study, the low current/voltage (~10 nA/1.0V) resistive switching memory device with a small via of 0.2 μm in a Cu/Ta$_2$O$_5$/W structure has been investigated for the first time. Furthermore, both Cu and W electrodes are more compatible with the CMOS technology.

EXPERIMENT

At first the tungsten (W) metal as a bottom electrode was deposited by sputtering on Si-substrate. After that the SiO$_2$ layer with a thickness of 150 nm was deposited. Then the different sizes of via (0.2 μm to 8 μm) were fabricated by using optical lithography and etching process. To perform lift-off process, photo-resist was used. Then high-κ Ta$_2$O$_5$ as a solid electrolyte material with the thicknesses of 10 nm to 35 nm were deposited by sputtering using pure Ta$_2$O$_5$ target. The sputtering power was maintained at 50 W in Ar (25 SCCM) ambient and the deposition time was 10 min to 35 min. After that Cu as a top electrode metal with a thickness of 100 nm was deposited by thermal evaporator. The lift-off process was done to fabricate the memory device. The schematic structure of our fabricated two terminal memory device is shown in Fig. 1. Electrical characteristics were performed using HP4156C semiconductor measurement analyzer.

RESULTS AND DISCUSSION

The Ta$_2$O$_5$ solid electrolyte film is confirmed by x-ray photoelectron spectroscopy as shown in Fig.2. The peak binding energies of the Ta4$f_{7/2}$ and Ta4$f_{5/2}$ core level spectra are found to be 26.3 eV and 28.2 eV, respectively, which correspond to the Ta$_2$O$_5$ film. The other binding energies of the Ta4$f_{7/2}$ and Ta4$f_{5/2}$ doublets are found to be 21.7 eV and 23.6 eV, respectively, which correspond to the Ta metal. Fig. 3 shows the current-voltage (I-V) hysteresis characteristics of the resistive memory device in a Cu/Ta$_2$O$_5$/W structure. The I-V hysteresis loop can be explained as follows. Initially, the device was in the OFF-state (arrow 1). By applying a positive voltage on the top electrode, there is an instantaneous switching from the high resistance (low current: R_{High}) state to the low resistance (high current: R_{Low}) state (arrow 2) at a threshold voltage (V_{Th}) of 0.6 V. To form the conducting path of Cu chain into the solid electrolytes, the applied bias should be larger than the V_{Th}. The memory device can continue the low resistance state (ON-state: R_{Low}) until the negative voltage of -0.2 V (arrows 3-6). The resistive memory device is going to be the high resistance state (R_{High}) if the applying voltage is more negative (<-0.2V) as shown in arrow 7. It will continue the high resistance state as shown in arrows 8 & 9. Corresponding resistance-voltage (R-V) hysteresis characteristic curve of Fig. 3 is shown in Fig. 4. The high resistance ratio (R_{High}/R_{Low}) of $\sim 5.3 \times 10^7$ has been observed (Fig. 4). Fig. 5 is the schematic illustration of resistive switching mechanism which can be controlled by applying external bias. The switching mechanism is based on the formation and dissolution of conductive Cu chain into the Ta$_2$O$_5$ solid electrolyte in a Cu/Ta$_2$O$_5$/W structure. By applying the positive bias on the top electrode, the Cu ions diffuse into the Ta$_2$O$_5$ solid electrolyte layer. Therefore there is the formation of metallic conductive path between Cu and W electrodes. Beyond the V_{Th}, the high resistance state will change to the low resistance state due to the metallic conductive bridge. This is called the ON-state [Fig. 5(a)]. This conductive Cu chain can be broken by applying opposite polarity. Therefore there is the changing from low resistance state to the high resistance state. This is called the OFF-state [Fig. 5(b)]. This formation and removal of the metallic pathway are due to the electro-migration and oxidation, respectively. Due to the formation of Cu chain into the Ta$_2$O$_5$ film, the ON-state resistance (R_{Low}) decreases with increasing the programming current from 10 nA to 1000 μA (Fig. 6), which can be applicable for multi-level data storage (MLC). There is the variation (slightly) of threshold voltage (V_{Th}) with programming current from 10 nA to 1000 μA and different device size from 0.2 μm to 8 μm (Figs. 7 & 8). It is expected that the ON-state resistance is independent on the device size and the thickness of Ta$_2$O$_5$ film (Figs.6-9). The R_{High} decreases with increasing the device size (Fig. 8), due to higher leakage current through the Ta$_2$O$_5$ film. The R_{high} can be increased (slightly) with increasing the thickness of Ta$_2$O$_5$ film (Fig. 9). Good endurance characteristics are observed $>10^3$ cycles (Fig. 10). Both unstable states can be improved by using series transistor on the memory element. The excellent data retention characteristics up to 11 hours is observed (Fig. 11) for a low programming current of 100 μA, due to the Cu chain formation into the Ta$_2$O$_5$ solid electrolyte.

CONCLUSION

Novel nonvolatile resistive switching memory device using Ta$_2$O$_5$ as a solid electrolyte in a Cu/Ta$_2$O$_5$/W structure with a very low current (~10 nA) operation has been proposed. Due to the R_{Low} variation with programming current, high R_{High}/R_{Low} of 9×10^3, good endurance (>10^3 cycles) and excellent retention (>11 hours), etc., this resistive memory device can be useful in future nanoscale nonvolatile memory applications.

ACKNOWLEDGEMENT

The Chang Gung University group is grateful to EOL/ITRI, Hsinchu, Taiwan, to support this work. Thanks a lot to National Science Council (NSC), Taiwan, under contract no. 97-2221-E-182-051-MY3 to support also this project.

REFERENCES

[1] M. N. Kozicki et. al., IEEE Trans. Nanotechnol., 4, 331 (2005). [2] M. N. Kozicki et. al., Symp. NVMTS05 1541405, p. 83, (2005). [3] T. Sakamoto et. al., Appl. Phys. Lett. 91, 092110 (2007). [4] Y. –R. Tsai et. al., SSDM-2008, p. 458-459.

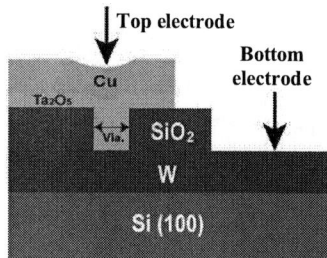

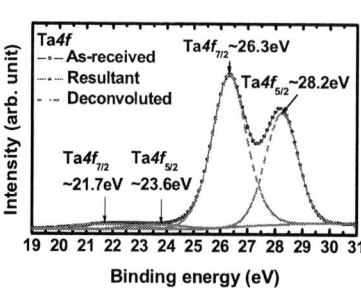

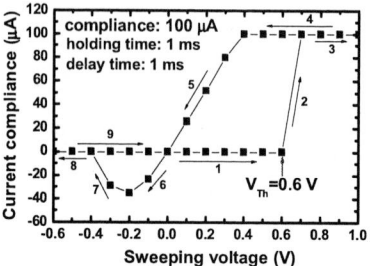

Fig. 1 Schematic of the resistive memory device using novel $Cu/Ta_2O_5/W$ structure. The thicknesses of Ta_2O_5 are 10-35 nm. The size (via) of memory device is 0.2-8 μm.

Fig. 2 X-ray photoelectron spectra of $Ta4f$ core level. The deconvoluted peaks show the Ta-rich Ta_2O_5 films.

Fig. 3 Current-voltage (I-V) hysteresis characteristics of $Cu/Ta_2O_5/W$ memory device with a size of 0.2 μm. The thickness of Ta_2O_5 film is 10 nm.

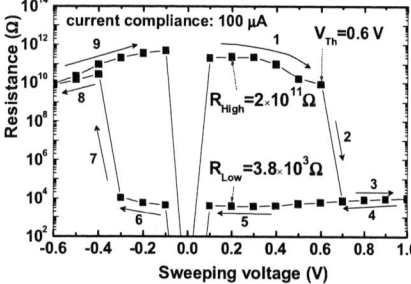

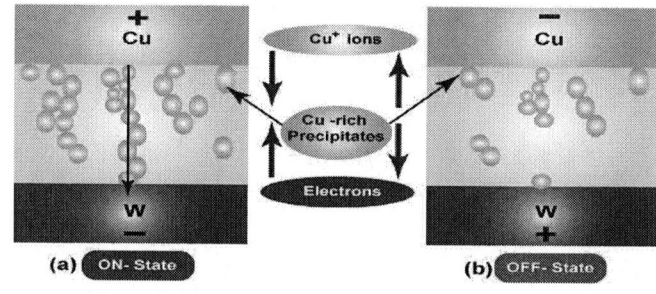

Fig. 4 R-V hysteresis characteristics have been calculated from the Fig. 3. A high resistance ratio (R_{High}/R_{Low}) of 5.3×10^7 is observed.

Fig. 5 Schematic illustrations of (a) Cu chain formation in the Ta_2O_5 solid electrolyte under positive bias (>V_{Th}) on the top electrode (R_{Low} state) and (b) Cu chain broken in the Ta_2O_5 solid electrolyte under negative bias on the top electrode (R_{High} state).

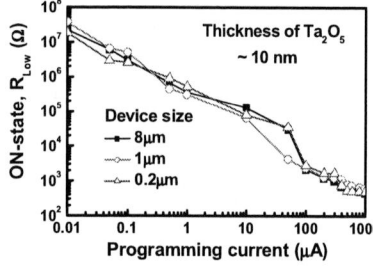

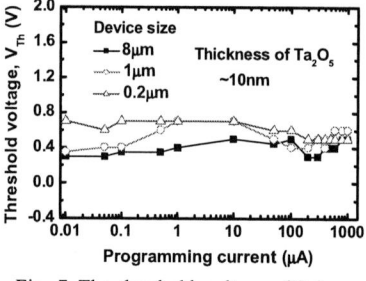

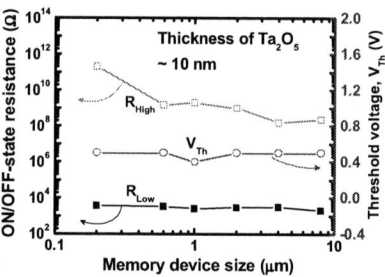

Fig. 6 Variation of R_{Low} with programming current from 10 nA to 1000 μA for 0.2 μm, 1 μm and 8 μm devices. R_{Low} decreases with increasing the programming current, which can be useful for multi-level of data storage (MLC). The resistive memory device shows a smallest programming current of 10 nA.

Fig. 7 The threshold voltages (V_{Th}) are almost the same with different sizes (0.2 μm to 8 μm) and programming currents (10 nA to 1mA) of the resistive memory devices. It is important to design the memory circuit.

Fig. 8 Variations of ON- (R_{Low}) and OFF-states (R_{High}) resistance with different memory device size from 0.2μm to 8 μm are shown. R_{Low} state is almost the same but RHigh decreases with increasing device size. V_{Th} is almost constant with device size under 100 μA programming current.

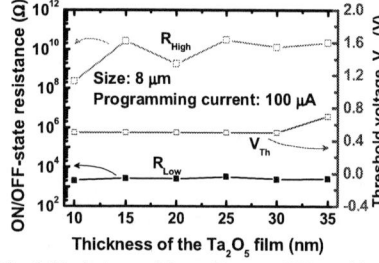

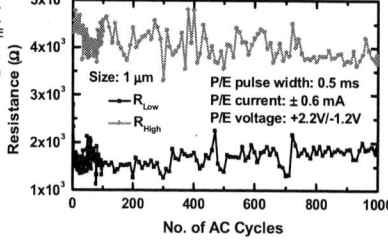

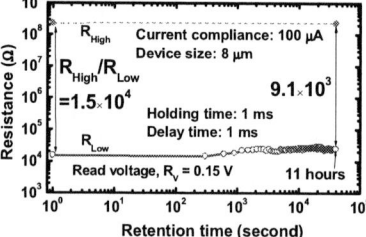

Fig. 9 Variations of R_{Low}, R_{High} and V_{Th} with different thicknesses of the Ta_2O_5 solid electrolyte films. R_{Low} states and V_{Th} are almost the same but there is slight variation of R_{High} and V_{th} with the thickness of Ta_2O_5 films.

Fig. 10 Endurance characteristics of E-Gun deposited Ta_2O_5 film with a thickness of 15 nm. The memory device can have endurance >10^3 cycles. The resistance states are unstable due to un-optimized programming/erasing (P/E) conditions and no transistor used in the series of memory element.

Fig. 11 Excellent retention characteristics of Ta_2O_5 film with a thickness of 10 nm in the Cu Ta_2O_5/W memory structure are observed at room temperature. The high resistance ratio (R_{High}/R_{Low}) of 9.1×10^3 is observed after 11 hours of retention.

978-1-4244-2784-0/09 $25.00 © 2009 IEEE

A Novel Multi - Nitridation ONO Interpoly Dielectric (MN-ONO) for Highly Reliable and High Performance NAND Flash Memory

C. H. Liu, Y. M. Lin, *Y. Sakamoto, R. J. Yang, D. Y. Yin, P. J. Chiang, H. C. Wei, C. Y. Ho. S. H. Chen, H. P. Hwang, C. H. Hung, S. Pittikoun and S. Aritome.

Powerchip Semiconductor Corp. No. 12, Li-Hsin Rd. 1, Hsinchu, Taiwan, R.O.C.   *Vantel Corp. Japan

Email:liuch@psc.com.tw

ABSTRACT

Multi-Nitridation ONO has been demonstrated for the first time. Significant improvement are obtained in NAND Flash performance and reliability. (1) 1V program voltage reduction owing to 10A EOT (equivalant oxide thickness) reduction (2) More than 20% tighter cell Vt distribution width can be achieved from ONO bird's beak free due to supressing encroachment of gate re-oxidation by Floating Gate (FG) / top oxide nitridation. And also, (3) good data retention can be realized by applying plasma oxidation on bottom oxide to suppress the trap assisted charge loss. MN-ONO is a promising technology for high density NAND Flash beyond 40nm generation.

INTRODUCTION

As the NAND Flash dimension scales down to 45nm and beyond, the gate coupling ratio (GCR) will be degraded [1]. The EOT of the Inter-poly dielectrics (IPD) must be scaled to around 13nm or thinner to meet GCR and fast operation level [2]. However, it is difficult to scale down ONO EOT due to reliability issues [3] and worse bird's beak encroachment from the poly gate re-oxidation as the gate dimension shrinks. High-K IPD to reduce EOT has paid much attention. However, it usually suffers data retention fail issue [4].

In this work, Multi-Nitridation ONO (MN-ONO) has been successfully demonstrated on a 70nm NAND Flash, with thinnest EOT around 11.6nm and better cell performance.

EXPERIMENTS

The devices were fabricated by 70nm node NAND Flash process with tunnel oxide 80A and periphery gate oxide 410A on the p-type epitaxy Si substrate. After self-aligned STI and Floating Gate patterning, IPD is formed with nitridation by radical plasma on floating gate and top oxide. The WSix gate, inter-layer dielectric contact and back end of line were fabricated by conventional process.

RESULTS AND DISCUSSION

Cell program/erase voltage and EOT reduction of MN-ONO are summarized in Figure 1. Compared with conventional ONO (Sample A), MN-ONO with FG /top oxide nitridation can achieve 10~14A EOT reduction, 0.8~1.1V cell program voltage reduction and 0.4~0.6V cell erase voltage reduction. FG nitridation and top oxide nitridation contribute 4A and 10A EOT reduction, respectively. The plasma O_2 densification after bottom oxide increase EOT 3~4A, which should be due to FG nitridation film re-oxidized during O_2 densification process. However, the EOT of the MN-ONO with O_2 densification is still much thinner than conventional ONO.

The 1-pulse-programmed cell Vth distributions are shown in Figure 2. MN-ONO show more than 20% tighter Vth distribution than conventional ONO. Either FG nitridation or top oxide nitridation process can also have 18% and 13% tighter distribution than conventional ONO. The TEM cross-section images of ONO and MN-ONO are shown in Figure 3. There is no bird's beak encroachment in MN-ONO case, but 9~12A thicker offset (Bird's beak) is observed in conventional ONO. From these data, the tighter Vth distribution of MN-ONO sample seems to be obtained by suppressing bird's beak encroachment by FG nitridation and top oxide nitridation.

Data retention characteristics of MN-ONO (sample D) in two data patterns are shown in Figure 4. The serious Vt shift is found in the BL - BL direction path rather than the WL - WL direction. It indicates that the charge loss is related to IPD process rather than gates etch process. The fail bit maps of the retention failure pattern are shown in Figure 5 (a), including one bit fail and two bits fail (1 -bit-high-Vt / 1-bit-low-Vt-bit). The schematic plot of possible leakage current path is shown in Figure 5 (b). A novel model is proposed to explain this charge loss phenomenon: Trapped electrons in bottom nitride are emitted through bottom oxide to internitride layer, then these electrons go through to adjacent cell by hopping conduction mechanism.

In order to suppress BL-BL direction charge loss, O_2 plasma densification process is applied to bottom oxide. The TEM results of the MN-ONO with /without O_2 densification after bottom oxide are shown in Figure 6. The bottom oxide with O_2 densification get 8A thicker, which should be due to some bottom nitride convert to SiO_xN_y. Moreover, the data retention tail bits can be much improved by O_2 plasma densification on bottom oxide, as shown in Figure 7. The SIMS analysis of bottom nitride film with /without O_2 densification after bottom oxide is shown in Figure 8. The nitrogen peak decreases as the O_2 densification process is applied, which indicates that the bottom nitride film has become SiOxNy-like due to the oxygen incorporation, which with lower trap density [5] and with higher conduction band offset (ΔEc) between bottom nitride and bottom oxide [6], as shown in Figure 9. Therefore, the good retention of MN-ONO can be achieved by O_2 densification on bottom oxide.

CONCLUSION

The Multi-Nitridation ONO has been successfully developed. The retention behaviors of MN-ONO are investigated and the novel model of bottom nitride trap assisted electron thermal emission is demonstrated. The bottom nitridation and bottom oxide re-oxidation process should be precisely controlled to avoid retention fail and keep cell performance.

REFERENCES

[1] K. Kim and J. Choi, IEEE NVSMW 21st. 9 (2006)

[2] B. Govoreanu and D. P. Brunco, *et al, Solid State Electronics*.49 (11) 1841 (2005),

[3] S. Mori. Y. Yamaguchi *et al, IEEE Trans. Electron Devices* 43 (1), 47 (1990)

[4] B. Govoreanu and D. Wellekens *et al, IEDM Technical Digest, pp1-4 (2006).*

[5] G. Van den bosch, A. Furnemont *et al, Proc. NVSMW pp 128-130 (2008).*

[6] G. D. Wilk and R. M. Wallace *et al, Journal of Applied Physics.* 89(10),5243 (2001)

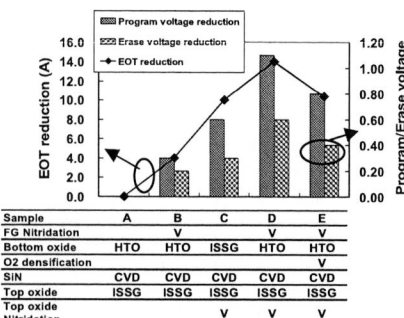

Sample	A	B	C	D	E
FG Nitridation		V		V	V
Bottom oxide	HTO	HTO	ISSG	HTO	HTO
O2 densification					V
SiN	CVD	CVD	CVD	CVD	CVD
Top oxide	ISSG	ISSG	ISSG	ISSG	ISSG
Top oxide Nitridation				V	V

Figure 1: The cell EOT reduction and program/erase improvement comparisons of finger cap in MN-ONO splits and conventional ONO.

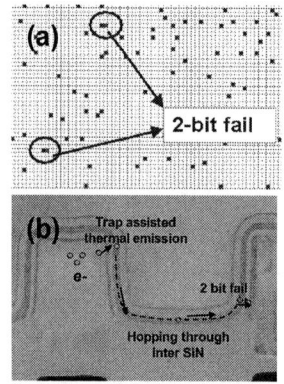

Vth distribution (-3σ ~ 3σ)	
(A) ONO	2.07V
(B) N-ONO (FG Nitridation)	1.68V
(C) ONO-N (Top oxide nitridation)	1.79V
(D) MN-ONO	1.54V
(E) MN-ONO (O2 densification)	1.61V

Figure 2: The 1-pulse-programmed cell Vth distribution width comparisons of MN-ONO splits and ONO of 70nm cell. MN-ONO with tighter Vth distribution than conventional ONO.

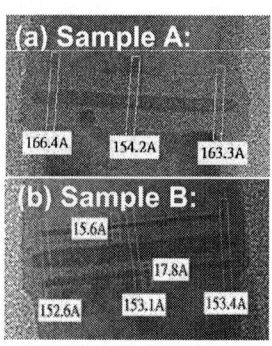

Figure 3: The TEM cross-sectional photograph along BL direction for (a) ONO and (b) MN-ONO. In MN-ONO case, the IPD thickness offset is almost 0A.

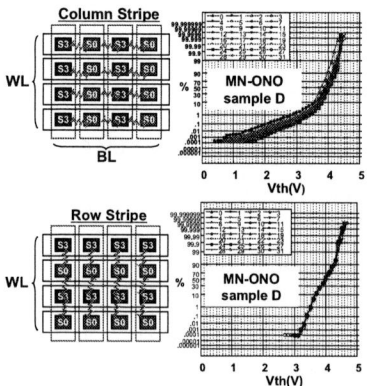

Figure 4: MN-ONO without O$_2$ densification cell string WL0~31 retention failure leakage path check by column stripe and row stripe data pattern at 255°C, 5hr. The S0/S3 indicate different Vt state. Vt of S0 = -3V, Vt of S3 = 4V.

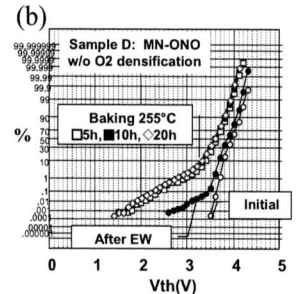

Figure 5. (a) MN-ONO (Sample D) retention failure bit map of checkerboard data pattern after 255°C, 5hr baking, including one bit fail and two bits fail (1-bit-high-Vt / 1-bit –low-Vt) (b) Model of leakage current path in 2 bits failure.

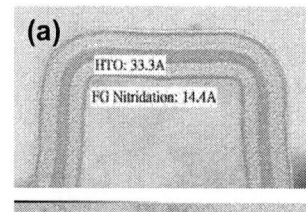

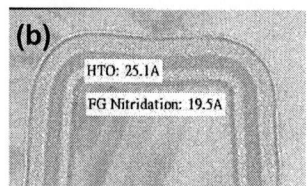

Figure 6: The TEM cross-sectional photograph of the MN-ONO (a) with / (b) without O$_2$ densification for bottom oxide. O$_2$ densification make thicker bottom oxide HTO~8A and thinner bottom nitride~5A due to re-oxidation during O$_2$ densification.

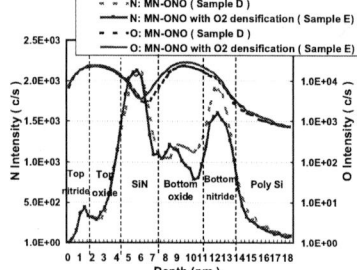

Figure 8: SIMS analysis of MN-ONO film with/without O$_2$ densification.

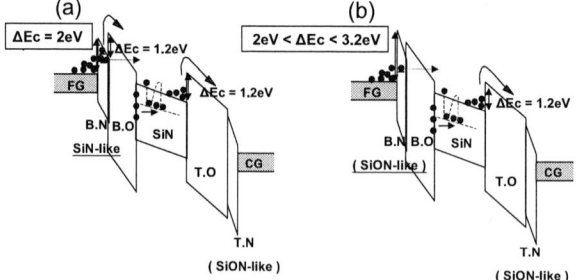

Figure 9: Charge transport mechanism of (a) MN-ONO (Sample D) and (b) MN-ONO with O$_2$ densification (Sample E). The electron trapping can be eliminated by O$_2$ densification due to reduce trap density and with higher conduction band offset.

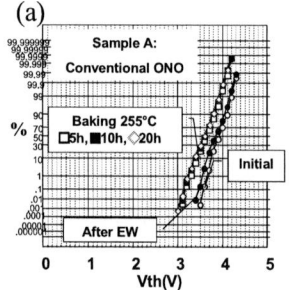

Figure 7: The data retention characteristics of ONO and MN-ONO by checkerboard data pattern. (1) Initial at 70° C (2) Program/Erase 500 Cycles (3) Baking at 255°C for 5,10,20 hrs. MN-ONO with O$_2$ densification can significantly improve data retention.

978-1-4244-2784-0/09 $25.00 © 2009 IEEE

Forming-free HfO₂ Bipolar RRAM Device with Improved Endurance and High Speed Operation

Yu-Sheng Chen[1], Tai-Yuan Wu[1,*], Pei-Jer Tzeng[1], Pang-Shiu Chen[2], Heng-Yuan Lee[1,3],
Cha-Hsin Lin[1], Frederick Chen[1], and Ming-Jinn Tsai[1]

[1]Electronics and Optoelectronics Research Laboratory, Industrial Technology Research Institute Chutung, Hsinchu 310, Taiwan,
[2]Department of Materials Science and Engineering, MingShin University of Science & Technology,
[3]Institute of Electronics Engineering, National Tsing Hua University, Hsinchu 300, Taiwan

* Phone: 886-3-5913750, Fax: 886-3-5917690, E-mail: DoveWu@itri.org.tw

Introduction

Binary oxide based resistive random access memory (RRAM) is an attractive candidate for replacing Flash memory [1], owing to its simple structure and CMOS technology compatibility. However, many different issues should be overcome to fit the requirements of mass production. One of those is the necessity of a forming process to initialize the resistive memory.

It has been reported that for several binary oxides such as CuO and NiO, appropriate oxidation process can eliminate forming [2-4]. But all of these devices suffered unstable resistance switching or insufficient endurance. Forming is a burden for circuit design and testing. In this work, we report the bipolar resistive switching memory using HfO₂ film with a TiN/Ti bi-layer electrode. A robust endurance (> 10^6 cycles) and long data retention (>10 years at 200°C) of this memory device was achieved. The memory also shows fast operation speed (~10 ns), low operation power and capability of multi-level operation.

Device Fabrication

The memory device, consisting of a TiN/Ti/HfO₂/TiN stacked multilayer structure, was fabricated on an 8 inch wafer. The HfO₂ film was deposited by atomic layer deposition (ALD) with a thickness of 3 or 10 nm. Other films were deposited by sputtering systems. Standard lithography and dry etching processes were taken to pattern the memory cells with sizes ranging from 0.36 to 0.96 μm in diameter. Finally, an AlCu layer was deposited and patterned on the top of the device to form the metal pad. The microstructure of the memory device was investigated by cross-sectional transmission electron microscopy (XTEM) and the electrical performances were characterized by an HP4156A semiconductor parameter analyzer and HP81110 pulse generator.

Results and Discussion

Figure 1 shows the XTEM image of the TiN/Ti/HfO₂/TiN stacked layer. The structure of HfO₂ layer with a thickness of 3 nm seemed to be amorphous. The interface between the HfO₂ layer and the Ti layer became obscure; presumably due to inter-diffusion between Ti and HfO₂ layer [6].

In our previous study [5], a specific forming voltage was applied to as-fabricated devices. With the thickness of HfO₂ film more than 5 nm, the memory cell needs this step to initiate the bipolar resistance switch. In this work, however, the initial resistance of the device with 3-nm-thick HfO₂ was much lower (~ 10^3 Ω) than that with 10 nm (Fig. 2). Therefore no forming step is necessary to initiate the resistive switching. Figure 3 shows the bipolar resistance switching of the memory device with cell size of 0.36 μm. The SET and RESET voltage of the initial state are the same as the operation state. The V_{SET} and V_{RESET} for several devices were measured and the accumulated data is shown in Fig. 4. The distribution shows good uniformity and the memory device can operate with low power (within 1.2V). The resistance distribution of the high resistance state (HRS) and the low resistance state (LRS) are presented in Fig. 5. The fluctuation of LRS was tighter than that of HRS, and the resistance ratio of HRS/LRS ranged from 20 to 50. To study the resistance switching mechanism, the typical SET and RESET I-V curves are plotted in double-logarithmic axis (Fig. 6). The slope of the curve increased suddenly during the SET process, which is related to a space charge limited current (SCLC) conduction. This result is consistent with our previous study [5-6]. The resistance switching is attributed to trapping and de-trapping of oxygen vacancies in the HfO₂ film, which are thought to be created by the oxidation of Ti gettering layer. In Fig. 7, the electrical properties of the memory devices with different cell sizes are shown. Only the HRS in device increases as the cell size decreases. Other properties are insensitive to the device size, displaying good scalability.

The switching endurance of the forming-free RRAM device is studied with pulse width of 500 μs, and 10^6 switching cycles was obtained (Fig. 8). The data retention of the memory cells are tested at 200 °C and the results are shown in Fig. 9. Both resistance states in the device remain stable at high temperature and 10-year lifetime can be expected. The device also shows the reversible multi-level operation by controlling the SET current compliance under DC voltage sweep (Fig. 10). The RESET current can also be controlled by changing the SET current compliance. The lowest current for bipolar resistive switching is ~ 20 μA. Finally, the operation speed of forming-free device is also tested with pulse mode as shown in Fig. 11. Resistance switching is possible with 10-ns pulse width and 3-V amplitude. A high speed operation (10 ns) of RRAM device without forming process is demonstrated. The V_{SET} and V_{RESET} with pulse mode are larger than those by DC sweep (~1.5 V). This result may be due to the RC-delay during the measurement.

Conclusions

A forming-free resistive memory of TiN/Ti/HfO₂/TiN with a thin HfO₂ film is demonstrated. The as-fabricated device can be operated without additional forming step to initiate the operation. This device with bipolar operation mode shows high speed (~ 10 ns), robust endurance (> 10^6 times), good data retention (10-year lifetime), enough resistance ratio, and low power consumption. The simple structure and capability of multi-level operation demonstrate RRAM as a high-density memory in the near future.

References

[1] I.G. Baek et al., IEDM Tech. Dig., p587, 2004.
[2] H. B. Lv et al., Elec. Dev. Lett., vol. 29, 1, p47, 2008.
[3] H. B. Lv et al., NVMTS, p52, 2008.
[4] L. Courtade et al., NVMTS, p1, 2007
[5] H. Y. Lee et al., IEDM Tech. Dig., 2008 to be published
[6] H. Y. Lee et al., VLSI-TSA, p146, 2007

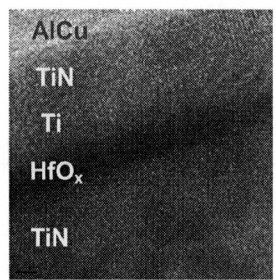

Fig. 1 XTEM image of TiN/Ti/HfO₂/TiN memory device.

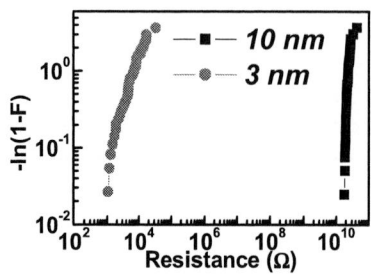

Fig. 2 Initial resistance comparison between HfO₂ thickness of 3 nm and 10 nm of the TiN/Ti/HfO₂/TiN device.

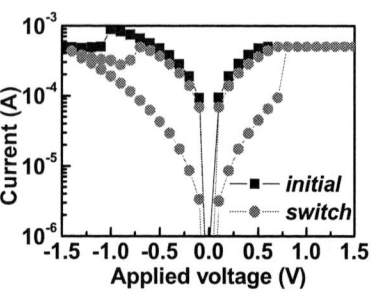

Fig. 3 Bipolar resistance switching characteristic of TiN/Ti/HfO₂/TiN device.

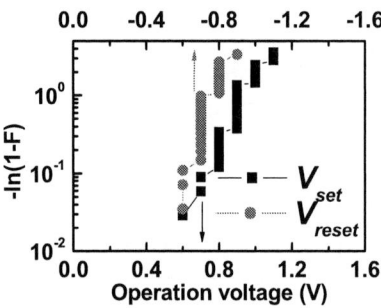

Fig. 4 SET and RESET voltage distribution of the TiN/Ti/HfO₂/TiN memory.

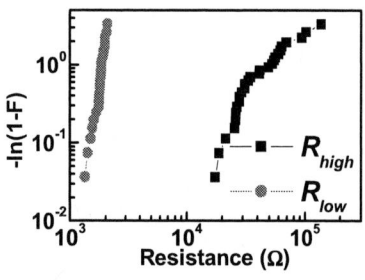

Fig. 5 HRS and LRS distribution of TiN/Ti/HfO₂/TiN forming-free memory device.

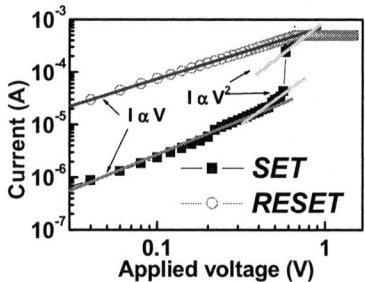

Fig. 6 I-V fitting of HRS and LRS in TiN/Ti/HfO₂/TiN device to SCLC current transport.

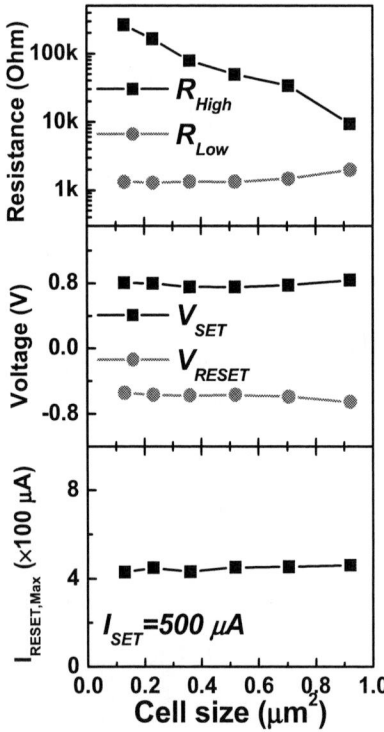

Fig. 7 Cell size dependence of various resistance switching parameters in memory device.

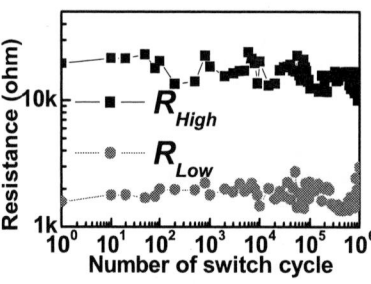

Fig. 8 Switch endurance of 10⁶ cycles in TiN/Ti/HfO₂/TiN forming-free memory device by 500μs pulse.

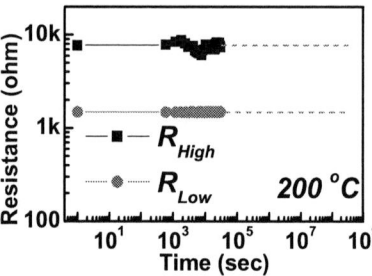

Fig. 9 Data retention of HRS and LRS at 200°C. This result predict 10-year lifetime of stored bit.

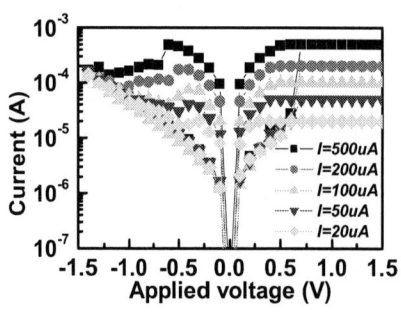

Fig. 10 The multi-level characteristic by controlling the SET current compliance.

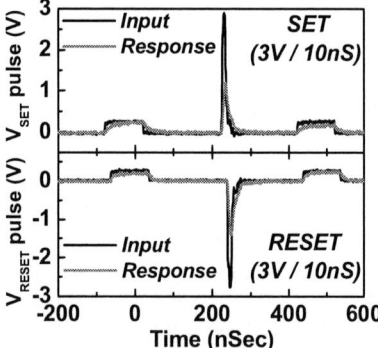

Fig. 11 Test of operation speed with 10-ns pulse width of TiN/Ti/HfO₂/TiN device.

978-1-4244-2784-0/09 $25.00 © 2009 IEEE

34

Multi-Level Phase Change Memory Using Slow-Quench Operation: GST vs. GSST

Der-Sheng Chao, Frederick T. Chen, Yen-Ya Hsu, Wen-Hsing Liu, Chain-Ming Lee, Chih-Wei Chen,
Wei-Su Chen, Ming-Jer Kao, Ming-Jinn Tsai
EOL/Industrial Technology Research Institute
195-4, Sec. 4, Chung Hsing Road, Chutung, Hsinchu, Taiwan, R.O.C.

ABSTRACT

In this paper, we demonstrate the use of a slow-quench waveform for multi-level phase change memory operation and compare the use of $Ge_{21}Sn_{10}Sb_{15}Te_{54}$ (GSST) and $Ge_2Sb_2Te_5$ (GST). A faster multilevel operation is possible with the use of GSST, owing to its faster crystallization speed

INTRODUCTION

Phase-change memory (PCM) is a promising candidate for the next generation nonvolatile memory. Its ability to support increasing bit density is augmented by its ability to support multilevel operation [1]. Typically, multilevel operation is achieved by varying the degree of crystallization. Therefore, the issues concerning multilevel operation are speed, waveform optimization, and retention. We have previously considered GSST as a high-speed phase change material [2]. In addition, we can apply the slow-quench used for SET completion through phase homogenization, as a method of achieving clean, multilevel operation. After assessing the use of slow-quench [3] for sub-microsecond multilevel operation for both GST and GSST, their retention characteristics will be compared.

SLOW-QUENCH DYNAMICS

Figures 1 and 2 give a typical picture of what happens during the slow-quench operation. The temperature profiles were calculated using an electro-thermal finite-element simulation (ANSYS 5.4). First a rectangular voltage pulse (in this case 50 ns) is applied which essentially carries out a RESET operation. This is followed by a trailing time, during which the voltage is ramped to zero. Up to this point, the phase change material temperature is increasing, resulting in the temperature profile of Figure 1. After the heating stops, the cooling begins and after 100 ns, the temperature has dropped significantly (Figure 2). For normal SET operation, this allows crystallization to proceed from the outside in.

EXPERIMENT

Measurements to characterize the multilevel slow-quench operation were performed using the same memory cell and measurement setup as used in Reference 2. The device comprises a tungsten bottom electrode of 0.24 um diameter and a phase-change (PC) layer covered by a TiW top electrode. Figures 3 through 5 show the effect of the initial pulse width on the multilevel operation, using GST as the phase change material. While 50 ns (Figure 3) allows several levels to be distinctly realized, 500 ns (Figure 4) leads to significant recrystallization following the initial melting. This makes it difficult to preserve high resistance levels over a wide programming window. Below 500 ns, there is good modulation of the final resistance by the use of the trailing time (Figure 5).

The slow-quench method is compared with the traditional square pulse method in Figures 8 through 11, for GST vs. GSST. Using a square pulse of varying voltage, it is possible to achieve 4 distinguishable states, utilizing the rising portion of the R-V curve, i.e., varying degrees of RESET. However, the four different voltages can be quite close to one another. In particular, the voltages for the two highest resistance states are within 10% of each other for both GST (Figure 8) and GSST (Figure 9). Referring to the temperature profile of Figure 1, during the RESET voltage application, the largest change

in resistance occurs when the melting region (T>Tm) just touches the dielectric next to the bottom plug. Before or after this point, the change in resistance is relatively negligible. Either most of the current bypasses the amorphous region or else the current does not sample the furthest exterior of the amorphous region.

On the other hand, by using the slow-quench method (Figures 10 and 11), the margin between different trailing times is much better. For GSST, the speed can also be much faster, with a combined (initial pulse + slow quench trailing time) of 100 ns. As discussed earlier [2], the higher crystallization speed of GSST gives it the advantage for SET operations where the phase change resistance is lowered. Since the slow-quench operation's effectiveness also hinges on the re-crystallization process, it also benefits from the use of GSST.

RETENTION

A fast crystallization speed warrants closely checking the retention. The concern is that at temperatures which are elevated above room temperature but possibly can be encountered by the device environment, the crystallization speed is sufficiently high that an initially high resistance state is reduced to a lower resistance state. The crystallization times for GST and GSST were measured at temperatures ranging from 150 °C to 200 °C. Figure 12 shows a plot for deriving the Arrhenius activation energies for GST and GSST using the measured crystallization times at different temperatures. From the slopes we deduce activation energies of 2.53 eV and 1.879 eV for GST and GSST, respectively. The lower activation energy for GSST indicates a lower barrier for crystallization. At temperatures of 80 °C and below, we project that GSST can achieve a 10 year retention. However, GST still has a better retention, for temperatures below 135 °C.

CONCLUSIONS

In this study, we compared the multilevel operation performance for two different phase change materials $Ge_2Sb_2Te_5$ (GST) and $Ge_{21}Sn_{10}Sb_{15}Te_{54}$ (GSST). We also demonstrated the advantage of using a slow-quench operation for multilevel programming. GSST demonstrated a potential for higher speed operation than GST, corresponding to its shorter crystallization time. However, the shorter crystallization time also corresponded to a reduction of retention compared to GST.

REFERENCES

[1] T. D. Happ et al., *2006 Symposium on VLSI Technology*, p. 120-121 (2006).

[2] C-M. Lee et al., *Symposium on VLSI Technology, Systems and Applications, 2007*, p.1-2 (2007).

[3] D-S. Chao et al., Appl. Phys. Lett. 92, 062108 (2008).

ACKNOWLEDGEMENT

The authors would like to gratefully thank the PCM consortium in Taiwan, R.O.C. for its kind support.

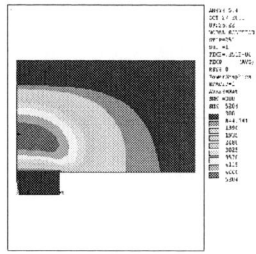

FIGURE 1. Simulated temperature profile at end of 200ns slow quench following a 50 ns pulse.

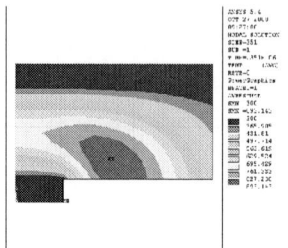

FIGURE 2. Simulated temperature profile after 100 ns cooling after quench.

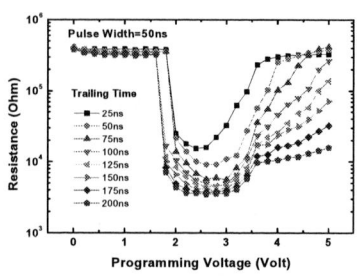

FIGURE 3. R-V with different trailing time for GST T-shape. (P.W.=50ns)

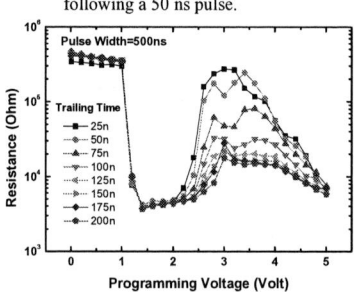

FIGURE 4. R-V with different trailing time for GST T-shape. (P.W.=500ns)

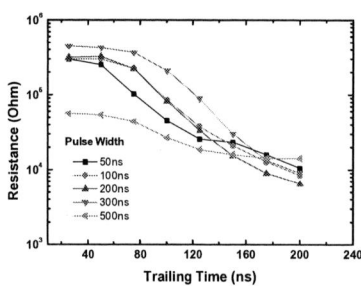

FIGURE 5. Resistance vs trailing time with different pulse width. (GST)

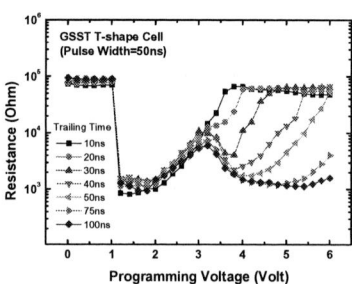

FIGURE 6. R-V with different trailing time for GSST T-shape

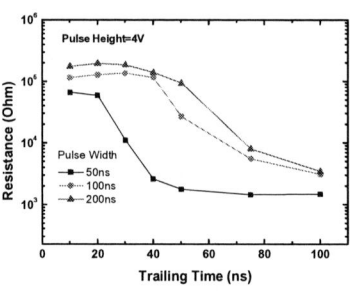

FIGURE 7. Resistance vs trailing time with different pulse width. (GSST)

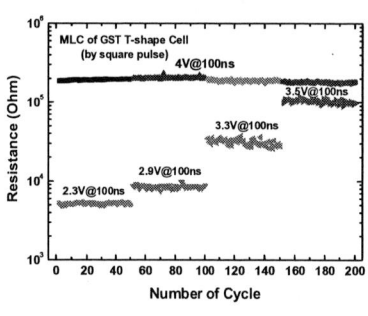

FIGURE 8. MLC of GST T-shape by square pulse

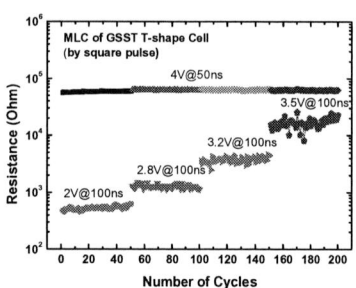

FIGURE 9. MLC of GSST T-shape by square pulse

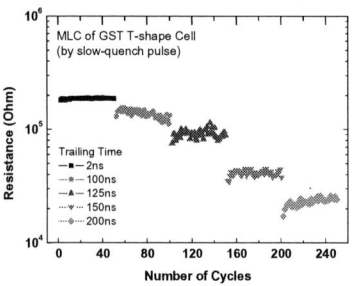

FIGURE 10. MLC of GST T-shape by slow quench pulse.

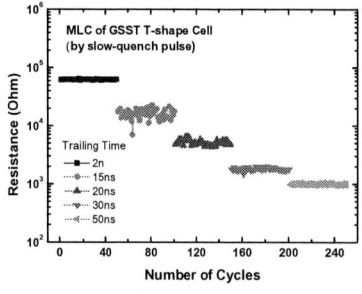

FIGURE 11. MLC of GSST T-shape by slow quench

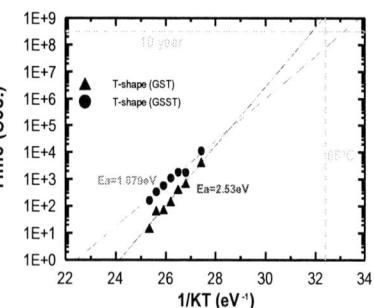

FIGURE 12. The retention of GST and GSST T-shape

978-1-4244-2784-0/09 $25.00 © 2009 IEEE

The Standby Power Challenge: Wake-up Receivers to the Rescue

Jan Rabaey

UC Berkeley, USA

A large fraction of the average power dissipation of many modern multimedia components and mobile devices is spent in standby mode, scanning for potential input activity. Reducing the dissipation of the "always-on" components is essential to the realization of green devices. The common strategy is to duty-cycle the always-on components. While simple to implement, it comes at the expense of latency. A more effective approach that delivers both low standby-power and almost-zero latency is to exploit ultra low-power wake-up receivers. Combining innovative architectures with state-of-the-art CMOS and MEMS technologies, wake-up receivers have been built that consume less than 50 uW in on-mode. Their availability opens a whole new perspective on standby power management. On one end of the spectrum, they enable green devices to operate in a purely reactive mode, that is they are only turned on when input activity happens. On the other side, they allow for substantial improvements in existing communication protocols such as WiFi and Bluetooth. A number of examples will be presented in the talk. One important message that will emerge from the presentation however is that effective standby power management requires a system vision, and that the ad-hoc component-oriented approach of today will rarely be effective.

Realizing Steep Subthreshold Swing with Impact Ionization Transistors

Yee-Chia Yeo

Department of Electrical and Computer Engineering, National University of Singapore (NUS), 117576.
Phone: +65-6516-2298, Fax: +65 6779 1103, E-mail: yeo@ieee.org

Abstract

Recent developments in Impact Ionization Transistors (I-MOS) will be discussed here, including strained impact ionization transistors realized on the nanowire or multiple-gate device architecture. I-MOS devices achieve excellent subthreshold swings well below 5 mV/decade at room temperature. Techniques for enhancing impact ionization rate and reducing the breakdown voltage V_{BD} for device performance improvement will be discussed. Challenges faced by I-MOS will be highlighted. Some challenges may be addressed through the strain and materials engineering. Limitations of the I-MOS will also be discussed.

1. Introduction

Non-scalability of the subthreshold swing in conventional CMOS transistors places a limit on the extent to which the supply voltage V_{DD} can be reduced. This motivates research on alternative electronic devices with a subthreshold swing lower than 60 mV/decade at room temperature. The impact ionization MOS field-effect (I-MOS) transistor has attracted immense interest [1]-[21]. I-MOS devices with sub-5 mV/decade have been demonstrated using various device designs (Fig. 1). In this paper, we review major developments in I-MOS research and development, discuss the key challenges faced, and examine how some of these challenges can be surmounted.

2. Device Concept, Fabrication, and Integration

The basic I-MOS device structure is a gated p-i-n diode with an impact ionization region (I-region) for gate-controlled avalanche of carriers (Fig. 2). For an n-channel I-MOS, the drain is grounded and the source bias V_S is set at a sufficiently large negative value (Fig. 3) before breakdown. When the gate-controlled field in the I-region exceeds a threshold, impact ionization occurs, leading to a large on-state current I_{on} which flows near the channel surface. The I_D-V_G characteristics of an I-MOS device (Fig. 4) shows an abrupt transition between the off- and on-states at the threshold voltage V_{TH}. V_{TH} is lower for a more negative V_S, and decreases with increasing channel doping concentration.

Early I-MOS design [1] employed a lithographically defined I-region adjacent to the channel [Fig. 1(a)]. Scalability and accurate control of the dimension of the I-region is limited by lithographic capabilities. To improve the compactness of the I-MOS, a device structure with spacer-defined I-region and self-aligned source [Fig. 1(b)] was reported [7],[14]. To allow for integration of new materials in the I-region, a further improvement in device/process design was made, where selective epitaxy process was used to form the I-region and the source region [Fig. 1(c)] [8]-[10]. The integration of SiGe or Si:C I-regions was demonstrated, as shown in Fig. 5.

Most I-MOS device structures can be fabricated using a CMOS-compatible process flow. In fact, the I-MOS has been reported to be co-integrated with SOI CMOS [4] and on bulk CMOS [14]. Co-integration of I-MOS and Tunnel FETs has also been reported [17]. The use of a large negative V_S leads to a significant source-to-body leakage in bulk I-MOS devices, which compromises power consumption. However, this can be mitigated with the use of SOI substrates [3],[11],[17]. SiGe-on-insulator substrates have also been used for I-MOS device demonstration [15].

3. Device Characteristics, Enhancement, and Challenges

A. Bandgap/Materials Engineering for Device Enhancement

A large $|V_S|$ is needed to sustain impact ionization for the Si I-MOS, as V_{BD} is still high even for a L_G of ~50 nm (Fig. 3). This is undesirable. Materials with smaller bandgap E_G or effective masses could be used to reduce V_{BD}. Strain effects have been exploited to reduce E_G or effective masses further. The use of SiGe I-region reduces the threshold impact ionization energy and $|V_S|$ and realizes larger electron-hole pair multiplication rate [11] than a Si I-region. For a given $|V_S|$, I_{off} is higher for SiGe I-region. However, to achieve a matched V_{TH}, a higher $|V_S|$ is used for Si I-region. A higher $|V_S|$ gives a higher field in the I-region for reduced $|V_T|$ but also increases I_{off} (Fig. 6).

Strain-induced E_G reduction in the I-region for I-MOS performance enhancement has been reported using Si:C source [12]. *In situ* doped source region was also incorporated. Besides use of novel materials and strain engineering in the I-region, a multiple-gate architecture [Fig. 1(d)] can be used to enhance the gate controlability and to increase the lateral electric field in the I-region for higher impact-ionization rate [11],[12]. In this way, V_{BD} could be reduced, leading to a lower $|V_S|$. The increased gate-to-channel coupling effect enhances carrier multiplication, and improves the drive current.

B. Reliability Issues

A major problem faced by the I-MOS is device reliability. Impact ionization generates hot carriers in the vicinity of the gate dielectric, leading to rapid device degradation. Repeated sweeps to obtain the I_D-V_G plot lead to a significant increase in the subthreshold swing. Structural modifications to the I-MOS to move the impact ionization region away from the gate dielectric did not improve the reliability significantly [8]. A depletion mode operation, instead of inversion mode, has recently been shown to improve I-MOS device reliability through reduction of hot carrier injection [13]. Lower V_{BD} and higher I_{on} were also reported.

C. Rate of Carrier Generation

A trade-off between I_{off} and switching delay exists for I-MOS devices (Fig. 7) [19]. This is a direct consequence of the self-amplifying carrier multiplication phenomenon. It takes time in the order of ps (longer than a conventional MOSFET) for carrier multiplication to increase the drain current by ~5 orders of magnitude from I_{off} to I_{on}. There is also a likelihood of carrier extinction, where the number of carriers goes to zero (Fig. 8). This is more likely at lower fields. Random distribution of the switching time due to stochastic carrier multiplication is also a fundamental problem.

3. Summary

I-MOS devices have been intensively investigated in recent years. Steep subthreshold swings have been widely reported. The large $|V_S|$ required can be reduced substantially with the integration of materials with narrower E_G and with strain engineering. However, I-MOS still suffers from fundamental issues such as device reliability and problems related to the stochastic carrier multiplication process.

Acknowledgment. Grant NRF-RF2008-09 from National Research Foundation, and discussions with E.-H. Toh, C. Shen, and G. Samudra.

References

[1] K. Gopalakrishnan *et al. IEDM 2002*, pp. 289.
[2] K. Gopalakrishnan *et al. IEEE TED 52*, pp. 69, 2005.
[3] F. Mayer *et al.*, *SSE 51*, pp. 579, 2007.
[4] F. Mayer *et al.*, *ESSDERC 2006*, pp. 303.
[5] W.-Y. Choi *et al.*, *IEDM 2004*, pp. 203.
[6] W.-Y. Choi *et al.*, *JJAP 46*, pp. 122, 2007.
[7] E.-H. Toh *et al.*, *IEEE EDL 29*, pp. 189, 2008.
[8] E.-H. Toh *et al.*, *IEDM 2005*, pp. 971.
[9] E.-H. Toh *et al.*, *IEEE TED 54*, pp. 2778, 2007.
[10] E.-H. Toh *et al.*, *IEEE EDL 27*, pp. 975, 2006.
[11] E.-H. Toh *et al.*, *IEDM 2007*, pp. 195.
[12] E.-H. Toh *et al.*, *IEEE EDL 29*, pp. 731, 2008.
[13] C. Onal *et al.*, *IEEE EDL 30*, pp. 64, 2009.
[14] C. Charbuillet *et al.*, *ESSDERC 2006*, pp. 299.
[15] E.-H. Toh *et al.*, *APL 91*, 153501, 2007.
[16] A. M. Ionescu *et al.*, *VLSI-TSA 2008*, pp. 72.
[17] W.-Y. Choi *et al.*, *IEDM 2005*, pp. 975.
[18] U. Abelein *et al.*, *IEEE EDL 27*, pp. 65, 2007.
[19] C. Shen *et al.*, *IEDM 2007*, pp. 117.
[20] C. Shen. *et al.*, *SSDM 2007*, pp. 608.
[21] F. Mayer *et al.*, *IEEE TED 55*, pp. 1373, 2008.

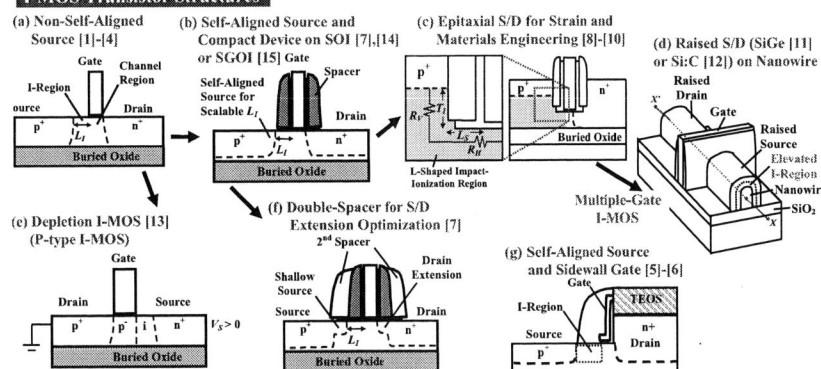

Fig. 1. Schematic of (a) non-self-aligned planar I-MOS [1]-[4], (b) a planar I-MOS with self-aligned source [7], [14], [15] (c) an IMOS with L-shaped I-region comprising epitaxial S/D for strain and materials engineering [8]-[10], (d) a multiple-gate nanowire I-MOS [11],[12] (e) a depletion I-MOS [13], (f) a double-spacer IMOS which enables further device optimization [7], and (g) I-MOS with a sidewall gate [5]-[6].

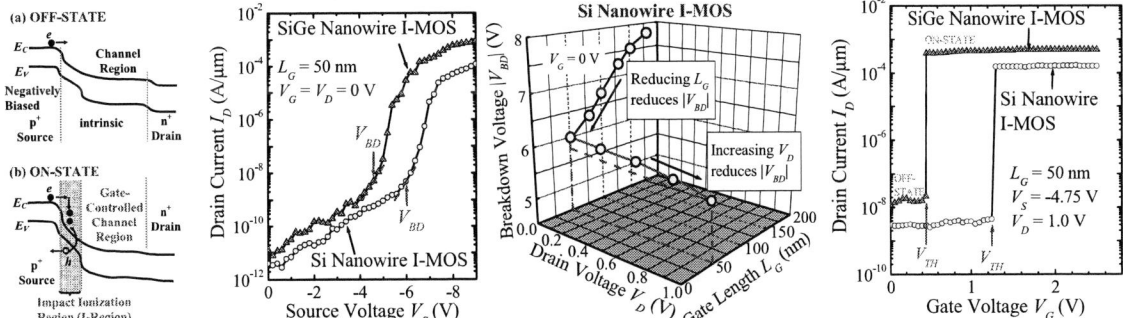

Fig. 2. Energy band diagram for an n-channel I-MOS in (a) off-state and (b) on-state where impact ionization occurs in the I-region.

Fig. 3. I_D-V_S plot (left) for nanowire I-MOS FETs with $V_G = V_D = 0$, showing the breakdown characteristics of Si and $Si_{0.75}Ge_{0.25}$ p-i-n structures. V_{BD} is defined at $dI_D/dV_S = 1$ decade/V. When L_G is reduced, a larger field exists in the I-region, leading to reduced V_{BD} (right). Increasing the V_D also decreases V_{BD} moderately, and this is termed Drain-Induced Breakdown Voltage Lowering (DIBVL).

Fig. 4. Abrupt transition between the off- and on-states is seen from the I_D-V_G plot for nano-wire I-MOS FETs. The $Si_{0.75}Ge_{0.25}$ nanowire I-MOS shows a higher drain current in the on-state.

(a) I-MOS Process Flow

- Active Formation
- Gate Stack Formation
- Masked Drain Extension Implant
- Spacer Formation
- Masked or Unmasked Selective Epitaxy (e.g. Si, SiGe, Si:C) in I-Region
- Doped Source may be epitaxially grown or implanted
- Masked Heavy Drain
- Dopant Activation and Metallization

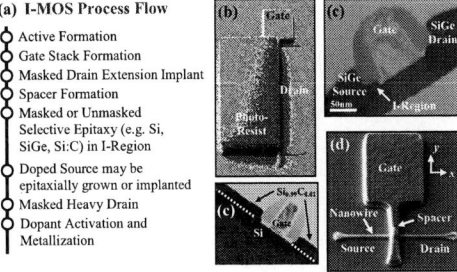

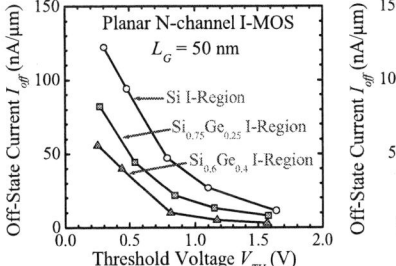

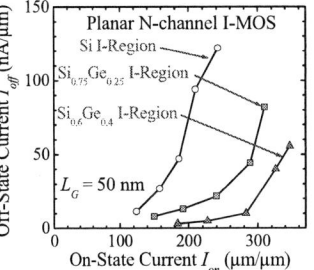

Fig. 5. (a) An example of an I-MOS process flow, in which (b) a masked drain extension implant is performed. This process allows for the formation of (c) SiGe or (d) Si:C I-region and source. (d) SEM image of a nanowire I-MOS device with SiGe S/D.

Fig. 6. At matched V_{TH}, the Si I-region gives a higher I_{off} than a SiGe I-region (left). The I_{off}-I_{on} plot measured at various $|V_S|$ for the n-channel devices with L_G of 50 nm are depicted on the right. At matched V_T or the same I_{off}, I-MOS devices with SiGe I-region shows enhanced I_{on}.

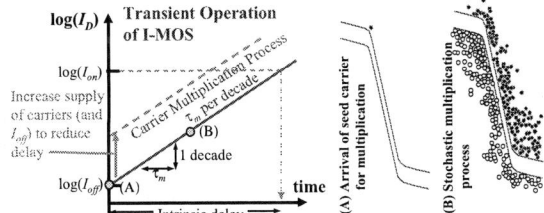

Fig. 7. Transient operation in I-MOS, where I_D grows exponentially with time at a rate of τ_m per decade. To reduce the delay to reach I_{on}, the supply of carriers or I_{off} has to be increased or a higher field has to be used.

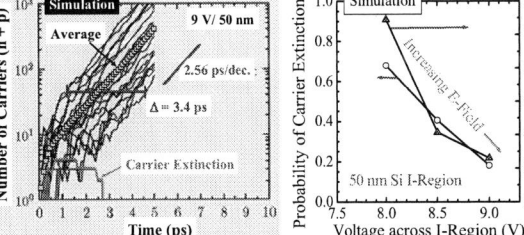

Fig. 8. An electron in a high-field I-region undergoes stochastic carrier multiplication, where the carrier number grows with time (20 Monte Carlo runs shown) (left) [19]. The growth rate increases with electric field (right) [19].

978-1-4244-2784-0/09 $25.00 © 2009 IEEE

Optimizing Tunnel FET Performance – Impact of Device Structure, Transistor Dimensions and Choice of Material

Joachim Knoch

TU Dortmund University, Emil-Figge Strasse 68, 44227 Dortmund, Germany

joachim.knoch@tu-dortmund.de, Tel.: +49-(0)231-755-3966, Fax: +49-(0)231-755-4450

INTRODUCTION

In recent years tunnel FETs (TFETs) have attracted a great deal of attention [1-9]. The reason for this is that TFETs potentially allow beating the 60mV/dec limit and thus eventually enable lowering the power consumption of ICs. However, TFETs usually exhibit an on-state performance inferior to a conventional MOSFET. Moreover, in order to obtain a superior off-state TFETs must exhibit subthreshold swings substantially smaller than 60mV/dec over several orders of magnitude in current. In the present paper the impact of device structure, dimensions and the choice of material on the performance of TFETs will be discussed. In particular, the use of heterostructures and one-dimensional nanowires will be analyzed in detail.

WORKING PRINCIPLE AND ISSUES OF TUNNEL FETS

Fig. 1 shows the conduction and valence bands along current transport direction in a TFET for two different V_{gs} in the device's off-state (gray) and on-state (black). In the off-state very low leakage currents can be obtained since the TFET is a reversed biased p-i-n diode. Once the valence band in the channel is moved above the conduction band in source a current will flow due to band-to-band tunneling (BTBT) and potentially a subthreshold swing S<60mV/dec can be achieved. However, due to the presence of the BTBT barrier, on-currents of TFETs are always lower compared to conventional FETs. In addition, TFETs exhibit a non-linear current increase for small bias as shown in Fig. 2 (b) [7]. This non-linearity is caused by a charge injection into the channel from the drain side giving rise to a strongly bias dependent charge in the channel which in turn results in drain-induced barrier thinning (DIBT) as is displayed in Fig. 2 (a). DIBT also leads to a shift of V_{th} (see Fig. 2 (c)).

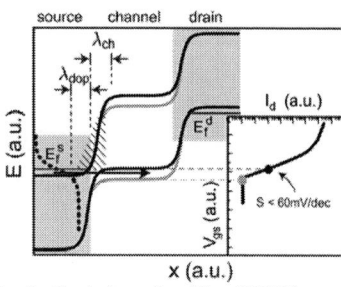

Fig. 1. Conduction/valence band in a TFET for two different V_{gs}. The inset shows a log-scale plot of the transfer characteristics.

In order to optimize the performance of TFETs it is instructive to calculate an analytical expression for the BTBT probability. Approximating the BTBT barrier with a triangularly shaped potential (see dashed area in Fig. 1) the WKB approximation can be used resulting in $T_{WKB} \propto \exp\left[-4(\lambda_{dop} + \lambda_{ch})\sqrt{2m^*E_g^{1.5}}/3\hbar(\Delta\Phi + E_g)\right]$ which in turn allows computing I_d and thus also the subthreshold swing $S = \ln(10)\left[\partial I_d/\partial V_{gs} \cdot 1/I_d\right]$. Here λ_{dop} and λ_{ch} are the screening lengths

in the contact and the channel due to the doping concentration and the channel layout, respectively. $\Delta\Phi$ is the energetic window for BTBT and E_g the band gap. Then calculating

$$\frac{\partial I_d}{\partial V_{gs}} \propto \left(\frac{\partial T_{WKB}}{\partial E_v^{ch}}F(E_v^{ch}) + T_{WKB}\frac{\partial F(E_v^{ch})}{\partial E_v^{ch}}\right) \quad (1)$$

(where $F(E_v^{ch})$ is an incomplete Fermi integral and E_v^{ch} is the valence band edge in the channel) it is obvious that there are two contributions allowing for S<60mV/dec [5]. Changing T_{WKB} by increasing V_{gs} results in an exponential increase of I_d independent of the injection from the source Fermi function that gives rise to the S=60mV/dec limit in conventional FETs [7]. The second term in (1) represents a band-pass filter yielding an effectively cooled source Fermi function. It this filtering that allows to obtain subthreshold swings smaller than 60mV/dec over several decades of current change. To make the second contribution the dominating one, the BTBT probability should hence be as close to unity as possible.

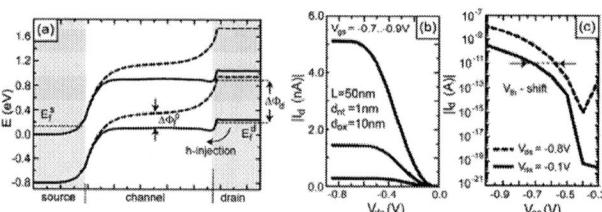

Fig. 2. Conduction/valence band in a TFET for two V_{ds}. (a). Output (b) and transfer (c) characteristics for a wrap-gate nanowire TFET.

OPTIMIZATION I – OXIDE AND CHANNEL THICKNESS

Since the effective screening length λ_{ch} is a function of the gate oxide d_{ox}, channel thicknesses d_{ch} and the device structure, λ_{ch} can be made smaller and therefore T_{WKB} larger if $d_{ox,ch}$ are being decreased [6-9]. The effect of scaling $d_{ox,ch}$ down has recently been observed in SOI TFETs as well as in Si nanowire TFETs yielding higher on-currents and steeper subthreshold swings [8,9]. In particular, in TFETs based on 1D nanowire/tubes the so-called quantum capacitance limit (QCL) can be reached in the case of a wrap-gate device structure and very thin d_{ox}. In the QCL, the gate gains complete control over the bands in the channel which effectively eliminates DIBT and regular output characteristics are obtained (see left inset in Fig. 3) [7]. We have calculated the characteristics of a TFET inverter based on a 1nm diameter nanotube channel in a wrap-gate configuration for three different gate capacitances (shown in the inset of Fig. 3). From this the voltage-dependent gate capacitance was extracted and the gate delay τ was calculated according to $\tau = \int_0^{V_{dd}} dV\, C_g(V)/I_d(V)$. Fig.3 shows that τ strongly decreases for larger geometrical oxide capacitances C_{ox}, i.e. when approaching the QCL. The low τ-values obtained for the largest C_{ox} originate from the increased I_d^{on} as well as the absence of a non-linearity in the output characteristics (see inset) and are therefore a result of being in the QCL, making this the desired operating mode of TFETs [6,7].

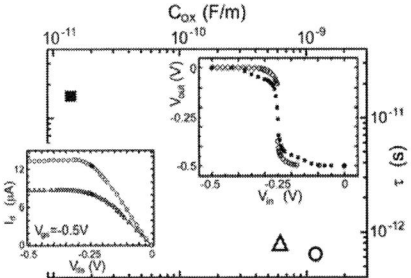

Fig.3. Gate delay versus C_{ox}. The insets show the output characteristics for the two largest C_{ox} and the inverter characteristics.

OPTIMIZATION II – DOPING THE SOURCE CONTACT

A large dopant concentration in the source contact is required in order to screen the effect of the gate rendering λ_{dop} as small as possible. Recently it has been shown that increasing the source dopant concentration in SOI TFETs increases the on-current of the devices (see Fig. 4 (a)) [8]. However, since in single-gate SOI devices $\lambda_{ch} > \lambda_{dop}$ the observed increase of I_d was attributed to band gap narrowing caused by the heavy doping. Looking at the expression for T_{WKB} it is apparent that lowering the band gap E_g increases the BTBT probability making this an effective approach. However, care has to be taken when doping the source contact: if the Fermi energy gets too large the band pass filter consisting of the band gap in source and channel cuts out the wrong portion of the source Fermi function. In other words the filter resides in the exponential tail of the Fermi function and hence, if the BTBT probability is large the conduction and valence band look like in a conventional FET (see Fig. 4) and one obtains S=60mV/dec [5]. The optimum energetic position of the source Fermi energy is very close to the (conduction) band which, however, is inconsistent with the requirement of a large carrier density within the source contact in order to obtain efficient screening. Finding an appropriate trade-off is essential particularly for TFETS in low effective mass materials.

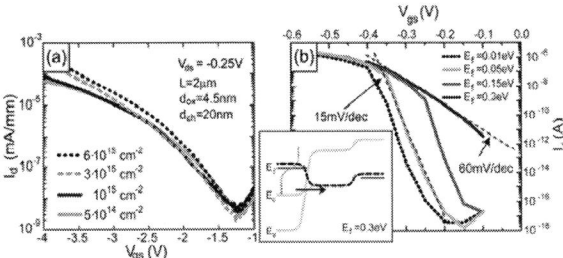

Fig. 4. (a) Transfer characteristics of an SOI TFETs for various doping concentrations. (b) shows the effect of a large Fermi energy

OPTIMIZATION III – HETEROJUNCTION TFETs

To realize large I_d^{on} and steep subthreshold swings the spatial extend, i.e. $\lambda_{dop} + \lambda_{ch}$, of the source-channel junction has to be made as small as possible. III-V compound semiconductors allow the growth of so-called type II hetero-interfaces with a staggered or even broken band line-up (see e.g. Fig. 5 (b)). A broken line-up seems to be particularly appealing since in this case the BTBT barrier tends to zero and

optimum device performance could be expected. We have therefore simulated II-type hetero-interface TFETs in the InAs/Al$_x$Ga$_{1-x}$Sb system. When varying the composition x the band line-up can be tuned from staggered to broken [10]. A vertical AlGaSb nanowire on top of an InAs substrate with a wrap-gate is considered (as depicted in Fig. 5 (a)). The charge in and current through the device is calculated with the non-equilibrium Green's function approach on a finite difference grid [6,7]. Fig. 5 (c) shows exemplarily three transfer characteristics for the cases x=1, an optimum composition of x=0.47 and x=0. The extracted S-values for several x are plotted in Fig. 5 (d). Obviously, the optimum device performance in terms of the subthreshold swing is *not* obtained for a broken band line-up. A staggered line-up close to the broken case on the other hand yields the best performance with S≈30mV/dec over 3-4 decades of current and a large I_d^{on} (see Fig. 5 (c)). (Note that the x=2 data point is artificial. It illustrates the evolution of S going from a type II to a type I hetero interface). Due to their low effective mass which enables to reach the QCL and the possibility of type-II hetero-interfaces, III-V semiconductor appear as very promising material for the realization of optimized TFET devices.

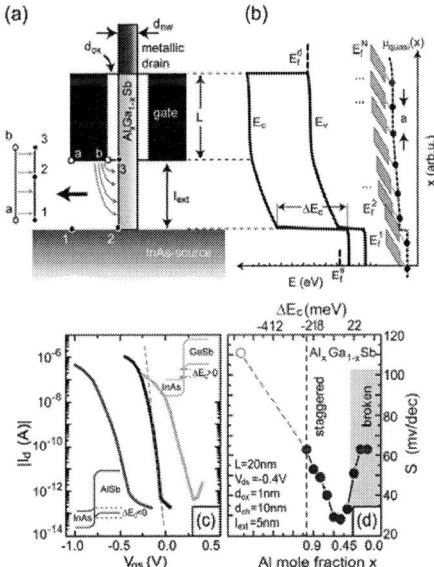

Fig. 5. (a) Schematics of the vertical nanowire TFET. Conformal mapping is used to take the source underlap region into account. (b) shows the conduction/valence band profile. Note the hetero-interface at the InAs substrate. (c) Transfer characteristics for three different x. (d) extracted S values as a function of composition x.

REFERENCES

[1] J. Appenzeller et al., Phys. Rev. Lett, vol. 93, pp. 196805, 2004.
[2] K.K. Bhuwalka et al., Jap. J. Appl. Phys., vol. 45, pp. 3106, 2006.
[3] K. Boucart and A.M. Ionescu, IEEE Trans. Electron Dev., 54, pp. 1725, 2007.
[4] W.Y. Choi et al., IEEE Electron Dev. Lett., vol. 28, pp. 743, 2007.
[5] J. Knoch et al., Solid-State Electron.., vol. 51, pp. 572, 2007.
[6] J. Appenzeller et al., IEEE Trans. Electron Dev., vol. 55, pp. 2827, 2008.
[7] J. Knoch and J. Appenzeller, phys. stat. sol. a, vol. 205, pp. 679, 2008.
[8] C. Sandow et al., Dev. Res. Conf. Conf. Digest, pp. 79, 2008.
[9] M.T. Bjoerk et al., Appl. Phys. Lett., vol. 92, pp. 193504, 2008.
[10] R. Magri et al, J. Appl. Phys., vol. 98, pp. 043701, 2005.

Nanoelectromechanical Systems for Ultra-Low-Power Computing and VLSI

Philip Feng

Caltech, USA

Nanoscale devices with mechanical degrees of freedom offer compelling characteristics that make them very attractive for mechanical and quantum logic devices. As we are able to create nanoelectromechanical systems (NEMS) with unprecedented feature sizes, advanced complexity and functionality, and high yield and control (at wafer-scale), they become increasingly interesting for low-power logic and memory, as well as become more meaningful for VLSI. Partly this is driven by NEMS devices' unique merits such as exceptionally large on/off ratio, non-leakage, ultralow switching power, fast speed, and temperature insensitivity. In parallel, this is also an intriguing effort in the quest for the ultimately energy-efficient implementation of logic and computing. In this talk, I shall introduce the Caltech research effort towards these goals, including the recent demonstrations of several generic prototypes of nanoscale electromechanical switching devices, their characteristics and performance, progress on engineering such building blocks for NEMS-based logic and memory, all-mechanical and hybrid NEMS-CMOS, along with discussions and perspectives of technological promises and challenges.

978-1-4244-2784-0/09 $25.00 © 2009 IEEE

Single-Metal Dual-Dielectric (SMDD) Gate-First CMOS Integration Towards Low V_T and High Performance

L-Å Ragnarsson, T. Schram, E. Röhr, F. Sebaai, P. Kelkar, M. Wada[†], T. Kauerauf, M. Aoulaiche, M. J. Cho, S. Kubicek,
A. Lauwers, T. Y. Hoffmann, P. P. Absil and S. Biesemans

IMEC, Kapeldreef 75, BE-3001, Leuven, Belgium (Email: Lars-Ake.Ragnarsson@imec.be)
[†]Dainippon Screen Mfg. Co., Ltd. 480-1, Takamiya, Hikone, Shiga 522-0292, Japan

ABSTRACT

This paper overviews integration challenges of low-V_T gate-first CMOS featuring one metal gate electrode and one host dielectric with Al_2O_3 and La_2O_3 cap-dielectrics for pMOS and nMOS respectively. The advantages and disadvantages of employed low EOT low V_T enabling technologies are compared with respect to processing simplicity as well as device performance and reliability. The latest state-of-the art SMDD device results are reported.

INTRODUCTION

Metal-gated (MG) CMOS devices with high-κ dielectrics are currently being introduced into manufacturing. High performance, low V_T devices have already been demonstrated using gate-first [1]-[5] and gate-last technologies [6], [7]. However, it remains an urgent goal to reduce the complexity of the process flows further. Table 1 compares the approximate number of additional steps needed to implement metal-inserted-poly-Si (MIPS) CMOS using different approaches. In the least complex solution, one MG and one high-κ dielectric (SMSD) is used, whereas in the most complex solution (DMDD), separate metals and high-κ's are used for the n and pMOS respectively. DMDD offers high flexibility and low V_T is relatively easy to achieve [3] while with true SMSD, low V_T remains to be demonstrated. In this paper, the focus is on SMDD, which offers a good trade-off between additional process steps and low V_T challenge (Table 1). Furthermore, the use of one MG simplifies the gate-stack etch significantly over e.g. DMDD. Note that unlike other approaches [2], SiGe is not used to improve on the pMOS V_T. Comparisons are made between several possible SMDD flows with respect to process complexity and device performance. By using a SMDD-flow with Al_2O_3 and La_2O_3 cap-dielectrics below or on top of HfO_2 excellent unstrained device performance comparable to reported DMDD results [3] are demonstrated.

EXPERIMENTAL

The CMOS devices for this study were fabricated on 300 mm (100) rotated-notch Si-wafers using six MIPS integration flows as shown in Figure 1. The differences between the flows lie in the location of the cap-dielectrics and the order in which they are deposited. Figure 1 also shows a simplified schematic of process flow 3. The process to selectively remove the cap-dielectrics is soft-mask based (resist-only).

Flows 1, 2, 5 and 6 have both cap-dielectrics located either below or above the host. Consequently, one of the cap-dielectrics has to be selectively removed from the other. In flow 1 for example, the Al_2O_3 must first be removed from the interface layer (IL) on the nMOS side

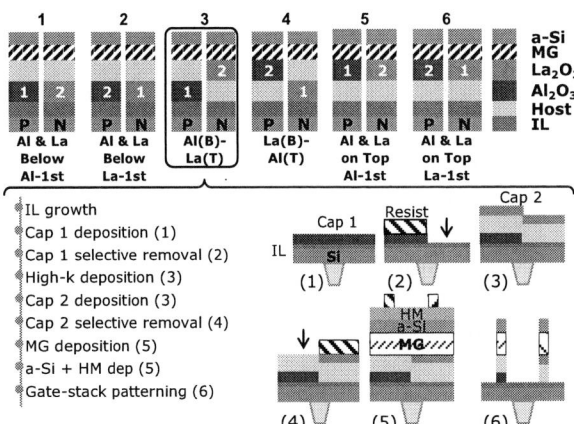

FIGURE 1. MIPS SMDD PROCESS FLOW OPTIONS AND FLOW 3 SCHEMATIC WITH Al_2O_3 BELOW HOST AND La_2O_3 ABOVE. SOFT-MASK (RESIST-ONLY) PATTERNING IS USED TO SELECTIVELY REMOVE THE CAP-DIELECTRICS.

and then the La_2O_3 must be selectively removed from the already-deposited Al_2O_3 on the pMOS side. In contrast, in flows 3 and 4 (Figure 1), the cap-dielectrics are separated by the host.

The host-dielectric was 1.2-1.5 nm HfO_2 grown by ALD or 1.8 nm HfSiO grown by MOCVD on a 0.8 nm interfacial SiO_2. The V_T-values of the p and nMOS devices were adjusted by 0.9 nm ALD Al_2O_3 and 0.7 nm La_2O_3 cap-dielectrics [3] respectively. An optional post deposition anneal (PDA) in N_2 was done on some wafers to promote intermixing between the cap and host dielectrics. The metal gate was 7 nm TaN + 3 nm TiN grown by PVD. For the soft mask approach, a resist with a BARC for optimum adhesion was used [5]. All results are for devices without strain. Comparisons are done to SMSD wafers (no selective removal) with the same cap dielectrics.

The impact from the different strip, etch and clean chemistries as well as stack options are discussed in the following sections with respect to selective removal, gate-stack integrity and performance.

RESULTS AND DISCUSSION

Cap-dielectric Removal Selectivity and Efficiency

The cap-dielectric removal efficiency was tested by depositing 1 nm of Al_2O_3 and then removing it by using developer with or without additional etch in diluted HCl (dHCl). Figure 2 (a) shows the remaining Al concentration compared to 1 ALD cycle of Al_2O_3. It is clear that the removal is more efficient on SiO_2 than on HfSiO. The remaining Al contamination with Al_2O_3 on HfSiO is comparable to 1 cy Al_2O_3 even with the extra etch in diluted HCl. Similar results are found with deposition on HfO_2 and with La_2O_3. Figure 2 (b) shows the impact from the exposure to dHCl, developer and resist deposition and strip (RS) on the interface trap density (N_{IT}). The quality of the interface is largely unaffected by the various chemistries. The exposure of the HfSiO to a dry RS (O_2 plasma) results in reduced N_{IT}, but this is related to an interface oxidation resulting in EOT increase. The impact of the various chemistries on TDDB and BTI [8], [9] were found to be small.

TABLE 1) PROCESS FLOW COMPLEXITY FOR MIPS CMOS. ADDITIONAL PROCESS STEPS ARE IN COMPARISON TO POLY-SION. GATE-STACK MATERIALS INCLUDE INTERFACE, DIELECTRICS, METALS AND POLY-SI.

	SMSD	SMDD	DMSD	DMDD
	Single-Metal Single-Dielectric	Single-Metal Dual-Dielectric	Dual-Metal Single-Dielectric	Dual-Metal Dual-Dielectric
Additional Process Steps	2	10	12	18
Gate-Stack Materials	4	5	5	10
Low V_T Challenge	Very Difficult	Moderate	Difficult	Easy
Gate-stack Etch Challenge	Easy	Moderate	Difficult	Difficult

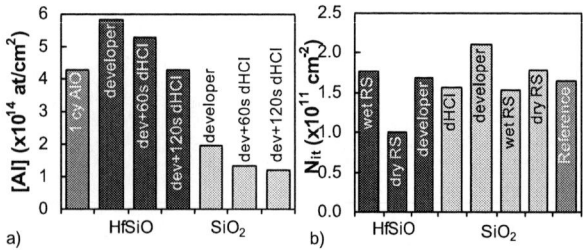

FIGURE 2 (A) RESIDUAL AL CONCENTRATION ON HFSIO AND SIO$_2$ AFTER DEPOSITION OF 1 NM OF AL$_2$O$_3$ AND SELECTIVE ETCH COMPARED TO 1 CY AL$_2$O$_3$ (dHCl = 1:10 HCl). (B) INTERFACE STATE DENSITY (N$_{IT}$) FROM CHARGE-PUMPING ON UNCAPPED HFSIO/SIO$_2$ AFTER RESIST AND STRIP (WET OR DRY RS) AND EXPOSURE TO dHCl OR DEVELOPER.

The resulting long channel V_T-values are shown in Figure 3 (a). It illustrates first, that using a cap-dielectric below the host gives a larger shift. Furthermore, the recovery of the V_T is complete only when the cap is below the host. Figure 3 (a) also shows that the Al$_2$O$_3$ is insensitive to RS while some of the shift is lost in the La$_2$O$_3$ case. To compensate, the La$_2$O$_3$ thickness needs to be increased. Note also that the V_T-recovery is slightly worse with full processing, i.e. resist deposition and development, selective etch and resist strip (RES) than blanket wafer etch.

From a processing and V_T-shift point-of-view, it appears clear that the preferred cap location is below the host dielectric. However, TDDB results show reduced t_{BD} with the cap-dielectric below [8]. Furthermore, the mobility (not shown) is somewhat lower when the cap is located below compared to above the host.

For flows with both caps below (1, 2) or above (5, 6) the host, the selective removal is done in a similar way. However, Al$_2$O$_3$ and La$_2$O$_3$ both etch in HCl which makes selective removal difficult. Fortunately, at low pH the La$_2$O$_3$ etch rate is considerably higher than for Al$_2$O$_3$. Thus, by diluting the HCl to 1:1000, a 400:1 etch rate ratio between La$_2$O$_3$ and Al$_2$O$_3$ was achieved. With this approach, CMOS devices were realized using all of the various SMDD flows.

Overall Performance Comparisons

The achieved V_T-values as shown in Figure 3 (b) are between 0.2 and 0.4 V in the nMOS case and between -0.4 and -0.5 V for pMOS devices. The La$_2$O$_3$ thickness-loss was overestimated in flow 3, which resulted in too low V_T. This can easily be solved by reducing the La$_2$O$_3$ thickness.

Figure 4 (a) compares the EOT-scaling of flows 3, 5 and 6. While the differences between flows 5 and 3 are relatively small, flow 3 still offers the lowest EOT (12/13Å for N/PMOS at J_G<1 mA/cm^2). Furthermore, as shown in Figure 4 (b) the mobility is compromised in flows with both caps below or on top of the host: it is always lower for the device with the second cap (for example, in flow 1 in which La$_2$O$_3$ was deposited last is the nMOS mobility low).

The final device performance is summarized in Figure 5, which shows (a) N and PMOS I_{on}-I_{off} characteristics for flows 3, 5 and 6.

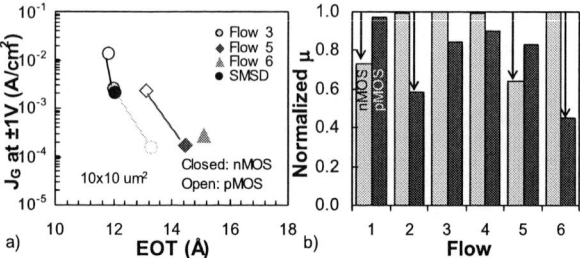

FIGURE 4 (A) LEAKAGE CURRENTS VS EOT FOR FLOWS 3, 5 AND 6 COMPARED TO SMSD WAFERS. BEST EOT SCALING IS OBSERVED WITH FLOW 3. (B) THE MOBILITY FOR N AND PMOS DEVICES FROM CMOS FLOWS 1-6 NORMALIZED TO SMSD DEVICES. THE OVERALL IMPACT ON THE MOBILITY IS SMALLER FOR FLOWS 3 AND 4.

Due to the very low V_T for flow 3, the I_{off} is higher than for flow 6. The best combined n and pMOS performance is achieved with flow 3. The drive currents at I_{off}=100 nA/µm is I_{on}=870 and 440 µA/µm for the n and pMOS respectively. Figure 5 (b) shows the very good ring-oscillator delays for the compared flows.

CONCLUSIONS

High performing Single-Metal Dual-Dielectric (SMDD) CMOS is demonstrated using six different approaches. Common to all approaches is the use of Al$_2$O$_3$ and La$_2$O$_3$ cap-layers to adjust the V_T, and soft masking to selectively remove the cap-dielectrics from the complementary areas. While all flows yield working CMOS, better overall performance is found when the cap-dielectrics are separated by the host. With this approach, low EOT (~12 Å), low n and pMOS V_T (0.2V/-0.4V) and high unstrained device performance ($I_{on,n}$ ~ 870 µA/µm, $I_{on,p}$ ~ 440 µA/µm at I_{off}=100 nA/µm) are demonstrated.

ACKNOWLEDGEMENTS

We acknowledge the sub-32 nm IMEC Industrial Affiliate Program members and the PULL-NANO project for their technical and financial support; the IMEC pilot line (p-line) for the device fabrication; and amsimec for measurement support.

REFERENCES

[1] M. Chudzik et al, *Symp, VLSI Technology*, p 194, 2007
[2] H. R. Harris et al, *Symp, VLSI Technology*, p 154, 2007
[3] S. Kubicek et al, *IEDM Tech. Dig.*, p. 49, 2007
[4] S. Kubicek et al, *Symp. VLSI Technology*, p.130, 2008
[5] T. Schram et al, *Symp. VLSI Technology*, p. 44, 2008
[6] C. Auth et al, *Symp. VLSI Technology*, p. 128, 2008
[7] S. Natarajan et al, *IEDM Tech. Dig.*, p.941, 2008
[8] T. Kauerauf et al, *Proc. Int. Reliab. Phys. Symp*, 2009
[9] M. Aoulaiche et al, Proc. *Int. Reliab. Phys. Symp*, 2009

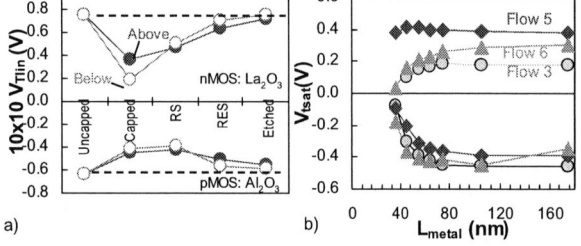

FIGURE 3 (A) LONG CHANNEL V_T FOR DEVICES WITH LA$_2$O$_3$ / AL$_2$O$_3$ CAPS POSITIONED ABOVE OR BELOW THE HFSIO. THE V_T IS LOWER AND RECOVERY IS BETTER WITH THE CAPS BELOW. (B) $V_{T,SAT}$ ROLL-OFF CHARACTERISTICS FOR CMOS FLOWS 3, 5 AND 6.

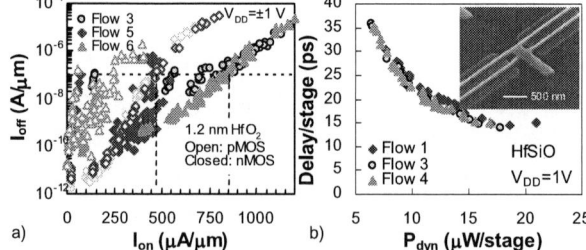

FIGURE 5 (A) N AND PMOS ION-IOFF FOR FLOWS 3, 5 AND 6. (B) RING-OSCILLATOR DELAYS FOR FLOWS 1, 3 AND 4. INSET: ONE OF THE RING-OSCILLATOR INVERTERS.

978-1-4244-2784-0/09 $25.00 © 2009 IEEE

Low Capacitance Approaches for 22nm Generation Cu Interconnect

T.I. Bao, H.C. Chen, C.J. Lee, H.H. Lu, S.L. Shue and C.H. Yu
Taiwan Semiconductor Manufacturing Company
8,Li-Hsin Rd. 6, HsinChu Science Park, HsinChu, Taiwan 300-77, R.O.C
(phone) +886-3-5636688 Ext. 7125387, (E-mail) tibao@tsmc.com

ABSTRACT

Various integration approaches, including homogeneous porous Low-k and air gaps, for low-capacitance solution were investigated for 22nm Cu interconnect technology and beyond. For homogeneous Low-k approach, K=2.0 Low-k material is successfully integrated with Cu. Up to 15% line to line capacitance reduction compared with LK-1 (K= 2.5) was demonstrated by a damage-less etching and CMP process. For air gap approach, a cost-effective and Selective air gaps formation process was developed. Air gaps are selectively formed only at narrow spacing between conduction lines without additional processes.

INTRODUCTION

As feature size continues to shrink, the need for reduction in line-to-line capacitance to minimize interconnect RC delay and power consumption becomes more critical. In this paper we revealed the low capacitance (low-C) solutions to BEOL of 22nm and beyond. The conventional approach is the introduction of homogeneous porous low-k material. Reduction of k damage due to Etch and CMP processes becomes more important with more porous materials. Another approach is air gap. This attracted attentions since keeping same low-k materials with more volume of air gaps gives lower effective k. Among various techniques to incorporate air gaps, thermal or UV degradable sacrificial layer method is one of the low-cost approaches. Collapse of air gap in wide metal space is the showstopper for this approach. In this paper, a novel selective formation of air gap in narrow space only is proposed to solve the problem.

HOMOGENEOUS LOW-K MATERIAL

Film properties and k damage by integration processes

Both CVD and spin-on (SOD) porous organo-silica (SiOCH) low-k dielectrics are compared. By adding porogen into sol-gel solution, porous SOD SiOCH film with similar bonding structures as CVD SiOCH film can be achieved. The dielectric constant and mechanical strength (hardness, H in GPa) can be optimized by various porogen concentrations and function groups R as shown in Fig. 1.

Key properties of low-k film with k<2.2 (LK-2) were summarized in Table 1 with comparison to k~2.5 (CVD LK-1). Fig.2 illustrates the typical FTIR of LK-1 and LK-2. Via first approach is adapted for Cu damascene integration with traditional PVD/ECP, CMP and profile optimized ETCH. Fig.3 and Fig.4 show the electrical performance is only 15% RXC improvement with one order of magnitude larger in line-line leakage current. Raphael simulation suggests the film is damaged with effective k=2.3. The reductions of k damage becomes the main task of this homogeneous low-k materials. Key modules that induce damage will be discussed separately.

K damage by ETCH process

The ETCH damage mechanism is shown in Fig.5. Improper wet chemicals may result in more severe line-line leakage degradation, as shown in Fig.6. In addition to choose proper wet chemicals, optimized ASH and recovery, and pre-metal cleaning steps were also used to reduce the damage to LK-2. Roughness is effectively reduced from SEM inspection (Fig.7). The line-line leakage was recovered by one order of magnitude and around 14% capacitance reduction, as shown in Fig.8 and Fig.9 respectively. After thermal budget of following processes, the improvement of capacitance reduced to around 3%. Thus the integrated capacitance of LK-2 is at least 15% lower than that of LK-1. Fig.10 (Kelvin VIA) and Fig.11 (VIA chain) illustrate 100% yield, which demonstrate no degradation on Cu surface of VIA bottom. In addition, 10% VIA resistance reduction was found.

K damage by CMP process

Direct CMP is hindered by the easy hydrophilisation tendency of porous low-k surface. As illustrated in Fig.12, the hydrophilisation results in the hydrophilic surface. Thus the dielectric constant increased.

We proposed a new designed slurry chemistry to reduce the reaction rate of hydrophilisation. And, at the same time, an additive is mixed into slurry to convert hydrophilic Si-OH to hydrophobic Si-X (X is hydrophobic group).

Fig.13 shows the absorption peaks in FTIR of prime LK-2 and after CMP by both old and new slurries. Fig.14 shows the improvement in k-shift of two blanket LK-2s (k=2.2 and k=2.0) after new slurry CMP. Accordingly line-to-line capacitance was reduced around 20% as shown in Fig.15.

AIR GAP

Techniques to incorporate air gaps

In this paper, an UV degradable sacrificial layer technique is used to generate air gap. A novel selective formation of air gaps is compared with an non-selective one. Fig.16 and Fig.17 illustrate the process flows of "Selective Air-Gap" and "non-Selective Air-Gap", respectively. For both approaches, an AGF (Air Gap Film) and a capping low-k film are first deposited in a CVD chamber, and then followed by the usual Cu damascene processes. AGF is a kind of organic compound, and decomposable by an UV curing process.

Distinct from non-selective approach, a novel Curing with Globally Encapsulated Layer (CGEL) is then conducted after metal CMP to form the Selective Air-Gap. The AGF in narrow line-line spacing is decomposed and evacuated through the capping low-k and the GEL. By choosing the proper GEL, air gap formation occurs only at narrow line-line spacing, such as ground rule, but not occurs at wide spacing. Therefore, the film collapse at wide spacing can be minimized.

Results and Discussion:

978-1-4244-2784-0/09 $25.00 © 2009 IEEE

The AGF removal efficiency is examined with two types of CVD dielectric GEL on blanket wafers. One is the film with higher molecular diffusivity; another is with lower diffusivity. As shown in Fig.18 (FTIR), and Fig.19 (shrinkage), the GEL with lower diffusivity hinders the removal of AGF with removal percentage less than 40%. In a few extreme cases, severe peeling is observed (Fig.20).

For the patterned structures, AGF removal process is more complicated than the blanket ones. On the one hand, without GEL, AGF removal percentage is higher than that with GEL. The low-k capping collapses at wide line-line spacing (Fig.21a). With GEL, no Air-Gap is formed at wide line-line spacing (Fig.21b), and therefore, no collapse found. On the other hand, the ground-rule space for both cases show Air-Gap formation (Fig.21c). Quantitative results are summarized in Fig.22 and Fig.23. TEM pictures of two metal levels with dense lines on top of large spacing and Via chain are shown in Fig.24 and Fig.25.

The electrical data of Selective Air-Gap with k=2.55 low-k capping are shown in Fig.26, Fig.27. Line-line leakage is comparable to that of "non Air-Gap" scheme, and about 20% reduction in capacitance is obtained. Effective k value 2.07 of Air-Gap+low-k is extracted by Raphael simulation with real line-line structure (Fig.28). Comparison of high temperature breakdown voltage at wafer level is also depicted in Fig29.

DISCUSSION AND SUMMARY

Cost-effective selective air gap process and homogeneous porous low-k process are proposed for low-capacitance solutions of 22nm generation and beyond. For homogeneous porous low-k approach, less damaged ETCH and CMP processes with recovery of carbon content and the reduction of hydrophilisation reaction are the keys to maintain the low capacitance performance.

For air gap approaches, the benefit is using the same low-k material as previous generation. New integrations scheme, new material, such as AGF, and additional processes steps becomes the new source of defect and failure. In this paper, a cost-effective and Selective Air-Gap formation process is proposed. This solves part of integration failures for this type of air gap approaches.

REFERENCES

[1] T.I. Bao et al., pp583, IEDM 2002
[2] C.M. Yang et al., pp1099, Adv. Mater. 2001.
[3] Y. Oku et al., S6.1, IEDM 2003
[4] S.M. Rossnagel et al, S4.4, IEDM 2005.
[5] S. Kondo et al., pp164, Proc. of IITC 2006.
[6] Winston S. Shue, pp175, Proc of IITC 2006.
[7] K. Mori et al., pp99, Proc. of IITC 2008.
[8] S. Ogawa et al., pp102, Proc. of IITC 2008.
[9] L.G. Gosset et al., IITC, pp65 (2003)
[10] R. Daaman et al., IITC, pp 240 (2005).
[11] T. Harada et al., IITC, pp 15 (2006).
[12] Y.N. Su et al., IEDM, pp91 (2005).
[13] S. Sankaran, IEDM, S13.2 (2006).
[14] R.J.O.M. Hoofman, AMC, VA1 (2007).

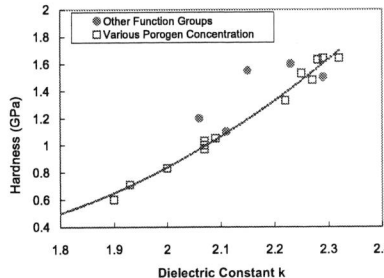

Fig.1 Hardness vs k with various porogen concentrations and function groups

Item/Film	LK-1	LK-2			
	CVD	CVD1	CVD2	SOD1	SOD2
K	2.55	2.2	2.1	2.0	2.1
H (GPa)	1.4	0.8	0.9	1	0.7
Pore Size (A)	<12	<20	<15	<20	<25
Leakage Current (A/cm2) @ 1MV/cm	<1E-9	<1E-9			
Breakdown Voltage (MV/cm)	>5	>4			

Table 1 Films property comparison (measured by Hg-probe, nano-indentor, ellipsometry and XRR/SAXS respectively).

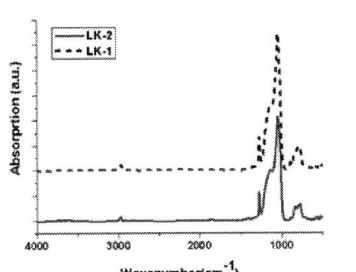

Fig.2 FTIR shows similarity between LK-1 and LK-2.

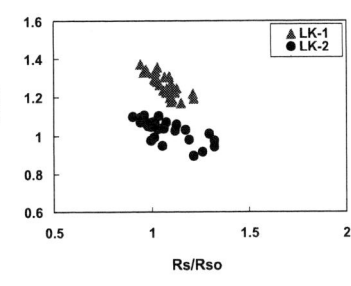

Fig.3 Line-Line (spacing 70nm) coupling capacitance vs. Rs (width 70nm) demonstrates around 15% reduction in capacitance when Rs/Rso. =1.

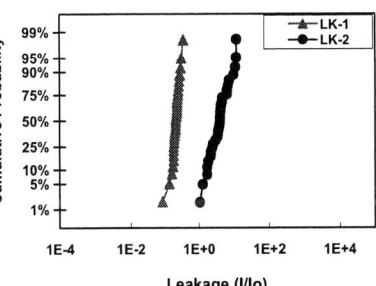

Fig.4 70nm Spacing Line-Line leakage

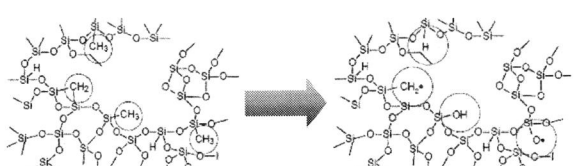

Fig.5 Etch induced damage

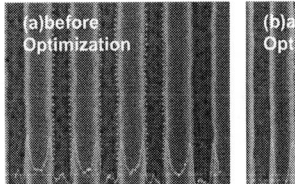

Fig.7 SEM image on trench bottom surface (a) before and (b) after optimization.

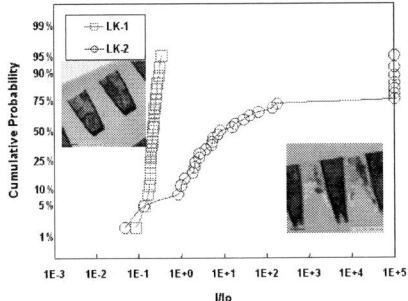

Fig.6 Improper Wet clean chemicals worsen line-line leakage

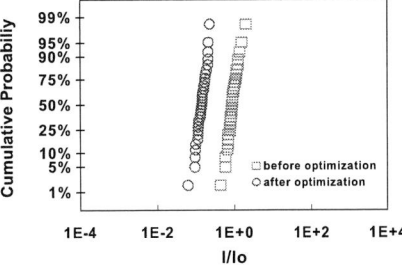

Fig.8 70nm Spacing Line-Line leakage was improved after optimization.

978-1-4244-2784-0/09 $25.00 © 2009 IEEE

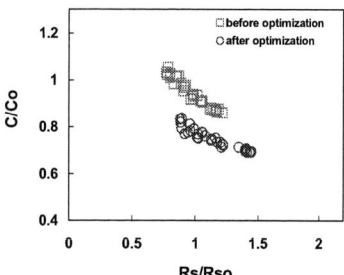

Fig.9 Around 14% reduction in capacitance was achieved after optimization.

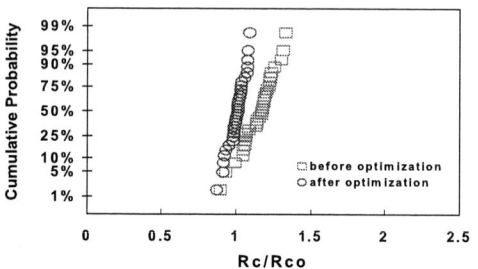

Fig.10 Kelvin Via resistance is reduced after optimization.

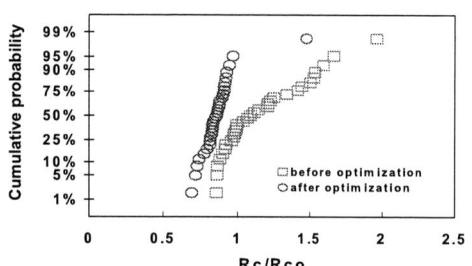

Fig.11 Rc of VIA chain is reduced after optimization.

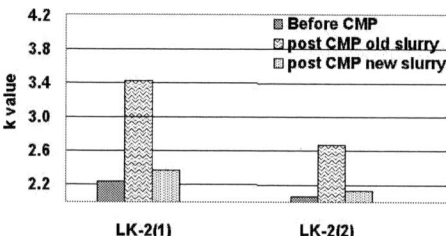

Fig.14 New slurry improved in k-shift of blanket LK-2s.

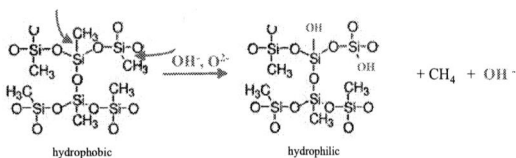

Fig.12 Mechanism of CMP Damage on LK-2

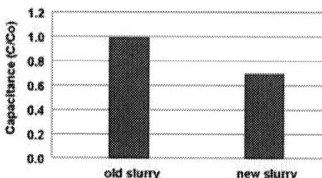

Fig.15 around 20% reduction of line-to-line (spacing 70nm) capacitance was achieved

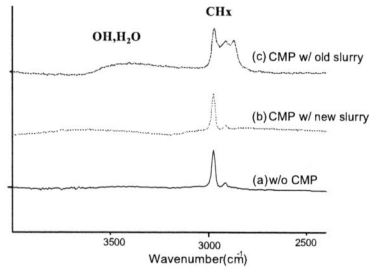

Fig.13 LK-2 FTIR of pre CMP, after CMP with old and new slurries.

978-1-4244-2784-0/09 $25.00 © 2009 IEEE

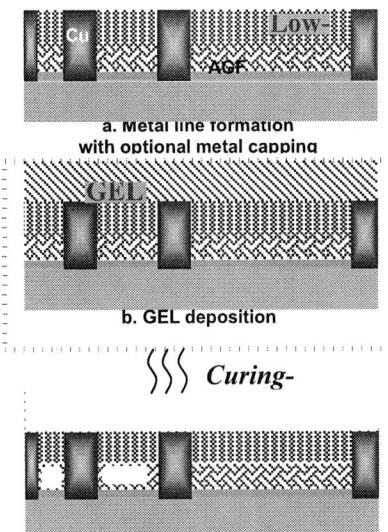

Fig.16 Process flow with optimized GEL

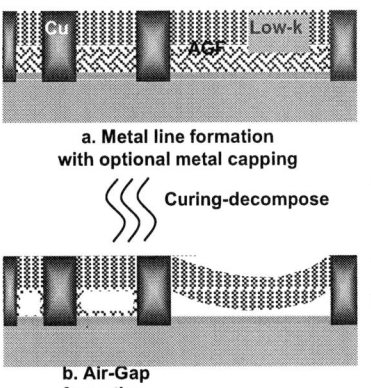

Fig.17 Air-Gap Process flow without GEL

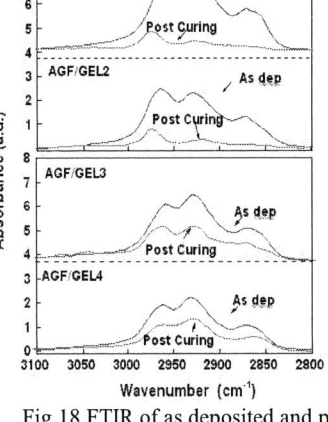

Fig.18 FTIR of as deposited and post cured AGF with various GELs.

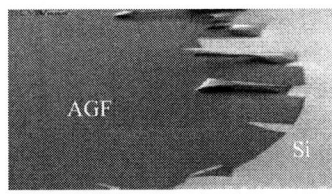

Fig.20 Peeling results from extreme low molecular diffusivity GEL.

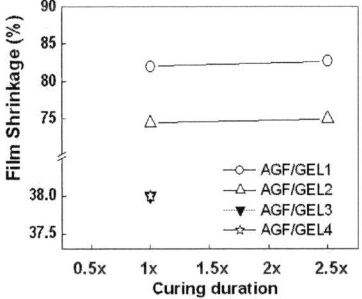

Fig.19 Shrinkage vs. curing duration of AGF with various GELs.

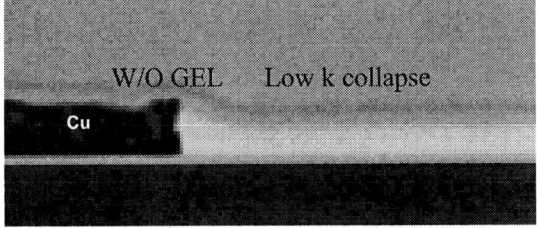

Fig.21a Without GEL wide spacing IMD collapse after AGF is removed. As-deposited thickness of AGF is the same as capping low-k.

Fig.21b Instead of collapse, only minor dishing occurs for wide spacing IMD with GEL.

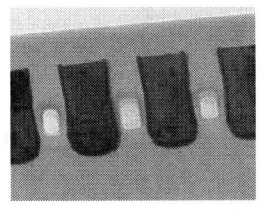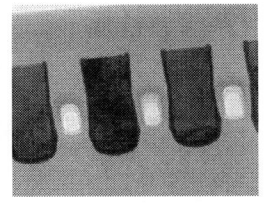

Fig.21c Air-Gap formation at IMD spacing < 70nm without GEL (left) and with GEL (right).

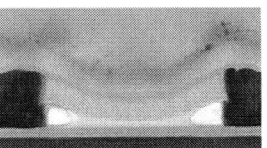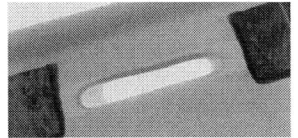

Fig.22a Without GEL 0.4um spacing shows severe dishing (left). With optimized GEL dishing can be well controlled (right).

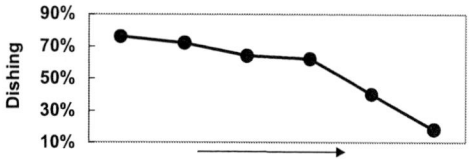

Higher GEL density/thickness

Fig.22c Dishing level of 0.4um spacing is examined for various GELs.

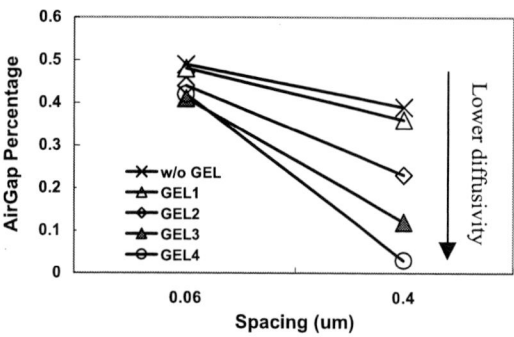

Fig.23 Percentage of Air-Gap formation vs. spacing with various molecular diffusivity of GEL.

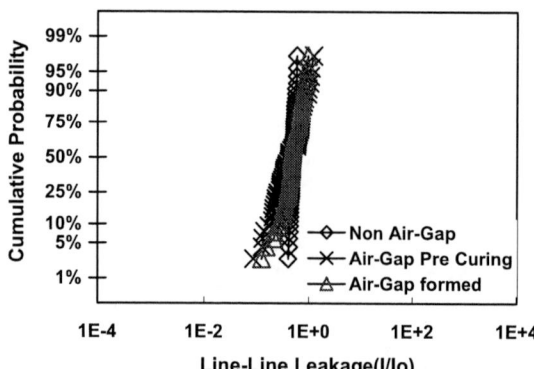

Fig.26 Similar 60nm spacing Line-Line leakage performance of non-Air-Gap and Air-Gap approaches.

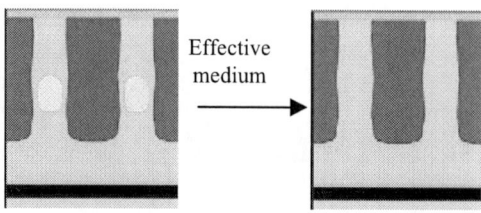

Fig.28 Effective k value of Air-Gap+low-k is extracted by Raphael simulation on duplication of real line-line structure.

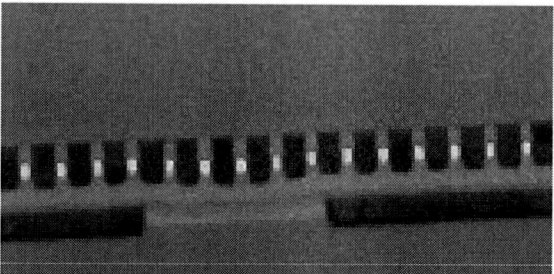

Fig.24 Dual damascene Air-Gap formation on top of single damascene without Air-Gap formed.

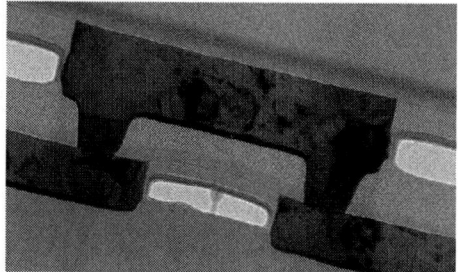

Fig.25 Air-Gap formation at Via Chain

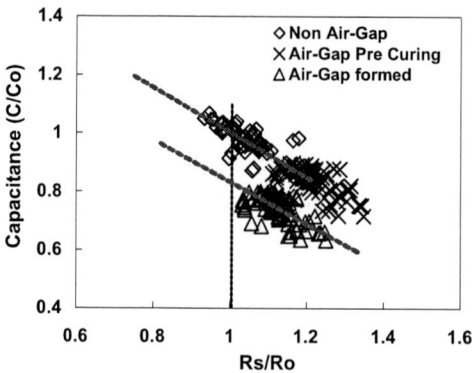

Fig.27 Capacitance vs. Rs demonstrates around 20% capacitance reduction by Air-Gap formation

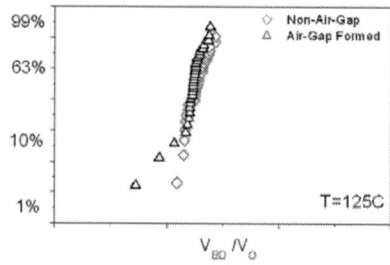

Fig.29 Wafer level breakdown voltage comparison at 125°C.

978-1-4244-2784-0/09 $25.00 © 2009 IEEE

A V_{FB} tunable Single Metal Single Dielectric approach using As I/I into TiN/HfO$_2$ for 32nm node and beyond

J. Petry[1,*], G.Boccardi[1], R. Singanamalla[1], C.S.Liu[2], K. Xiong[1], P.Escanes[1], J.-L. Huguenin[1], J.Tseng[2], L.Van Nimwegen[3], F.Voogt[3], C.W.T.Bulle-Lieuwma[4], M.Müller[1]

[1]NXP-TSMC Research Center, Kapeldreef 75, B-3001 Leuven, Belgium, [2]TSMC, 8, Li-Hsin Rd. 6, Hsinchu Science Park, 300-77 Hsinchu, Taiwan, [3]NXP Semiconductors, Nijmegen, The Netherlands, [4]Philips Research, Eindhoven, The Netherlands

[*]tel.: 32-16-288 336, Fax 32-16-287 588, email : jasmine.petry@nxp.com

ABSTRACT

Easily integrable cost effective gate first Single Metal Single Dielectric (SMSD) solution based on As implantation into TiN/HfO$_2$ with ~ 1 nm EOT is presented. A consistent n-type shift of 250 mV down to 35 nm L_g is obtained by As I/I compared to the reference stack. Symmetrical threshold voltages (~ ±0.5 V) are met for the bulk planar devices using this technique, which would corresponds to low-V_T (±0.2V) target for the FD FETs. The possible counter-doping effects were evaluated electrically and physically with back-side SIMS. It was found to be negligible implying negligible concentration of As in the channel region. As I/I technique opens up possibility of multiple V_T tuning without adding any process complexity.

INTRODUCTION

The conventional Poly-Si/SiON gate stacks are currently being replaced by high-k/metal gate stacks [1, 2]. Metal-gate stacks with Effective Work Functions (EWF) of ±200 mV from the Si midgap can meet the high-V_T targets (±0.5V) for the planar bulk FETs and low-V_T target (±0.2V) for the fully depleted (FD) FETs [3]. Because of its process simplicity, gate first Single Metal Single Dielectric based CMOS integration is most desirable and attractive for its cost effectiveness to meet the above mentioned V_T targets. In this paper we will present an ion implantation (I/I) based solution. I/I techniques have been shown to influence the EWF/V_{FB} of the gate stack [4, 5]. In the proposed CMOS solution, we first optimize the TiN thickness to make it closer to p-type EWF [6]. Secondly, we tune the EWF to n-type by means of As ion I/I for n-FET. As I/I in the gate stack may induce counter-doping in the Si channel region which can lead to undesirable buried channel effect. I/I energies and doses were optimized to alleviate these effects. This will be analyzed by means of V_T-V_{BS} measurements, sub-threshold slope, V_T roll-off for the As implanted stacks compared to the un-implanted ones.

RESULTS AND DISCUSSION

The gate dielectric is formed by a 2 nm ALD HfO$_2$ deposited on a thin chemical oxide. 7 nm PVD TiN formed the metal gate electrode. This is followed by a thin Si cap deposition. P-like TiN gate stacks are implanted by As, while complementary devices are masked using a standard well mask. The As ions were implanted at energies ranging from 6 to 8 keV and doses between 2E15 to 4E15 at/cm^2. Subsequently a-Si deposition completes the gate stack followed by a conventional CMOS process with junctions activation using a standard high temperature spike annealing and S/D regions silicidation using NiPt. The CMOS flow proposal has no added process complexity with respect to conventional poly-Si technology.

RESULTS AND DISCUSSION

Because of its p-type EWF, the optimized 7 nm TiN was chosen as starting metal gate. A cross-section TEM of a 35nm-long transistor is shown in Fig.1. Fig. 2 shows the excellent drive of 360μA/μm at an I_{OFF} of 1E-9 A/μm for p-FETs. As seen in CV curves (Fig.3), the

implantation of As into this TiN shifts it n-type: 4E15 at.cm^{-2} As implanted at 6keV lead to 250mV V_{FB} shift towards n-type, without any evidence of gate stack and Si/SiO$_2$ interface degradation, as seen on the CV curves. I_D-V_G curves in Fig.4 show the corresponding V_T reduction of ~ 250 mV leading to a V_T of ~ 0.5 V for long channel n-FET. A consistent shift of 250mV is also seen in the J_G-V_G curve for the 6 keV As implanted, which indicates the absence of gate leakage degradation with respect to the reference (inset of Fig.4). Both the stacks show excellent sub-V_T slope of 66 mV/dec confirming a good Si/SiO$_2$ interface and ruling out the buried channel formation by possible As I/I penetration into the channel. Furthermore, G_m.CET shows a very limited degradation restrained to the weak inversion regime (Fig.5) which might be linked to an acceptable increase in the interface state density after I/I as seen in charge pumping measurements. In order to evaluate the presence of As in the Si channel, we have performed back-bias measurements. Results are shown in Fig.6. V_T varies similarly with V_{BS} for the reference and the 6keV-implanted TiN. In this case, the shift in V_{FB} is thus believed to be related to a true gate stack effect and not due to undesired channel counterdoping phenomena. As shown in Fig. 7 the V_T reduction by As I/I is maintained down to the 35 nm gate lengths with equivalent roll-off. Looking more closely at the effect of implant conditions, we found that up to 7keV As I/I energy, no degradation of the peak transconductance is observed, as seen in Fig.8. This is in line with backside SIMS analysis, which shows a qualitative and quantitative change in the As profile at the Si level, from 6 to 8 keV implant energy, as seen in the inset of Fig. 8. Concerning the gate leakage a slight increase is seen for the As implanted TiN, which is entirely explained by a EOT reduction of about 1Å (Fig. 9).

CONCLUSIONS

We demonstrated a cost-effective gate first SMSD CMOS process flow based on As I/I into an optimized p-like TiN/HfO$_2$ gate stack with ~ 1 nm EOT leading to symmetrical V_T (~ ±0.5 V) on planar bulk FETs. No evidence of channel counterdoping was observed in sub-V_T-slope, C-V curves, V_T roll-off behavior, back bias sensitivity and backside SIMS analysis. This technique opens up promising options for multi-V_T engineering in sub-32nm FD device technology via tuning of the effective work function without added process complexity.

REFERENCES

[1] Kubicek et al. IEDM 07 p.49
[2] K. Mistry et al, IEDM 07, p.247.
[3] ITRS Roadmap
[4] R. Singanamalla *et al*, JJAP 46, pp. L320, 2007.
[5] R. Singanamalla *et al*, EDL 28(12), pp. 1089, 2007.
[6] Kadoshima et al., p.48 VLSI 2008.
[7] Nicollian & Brews, *MOS physics and Technology*, ed. Wiley, 2003

AKNOWLEDGMENTS

This research is partially supported by the European Commission's Information Society Technologies Program under PULLNANO project contract No. IST-026828.

978-1-4244-2784-0/09 $25.00 © 2009 IEEE

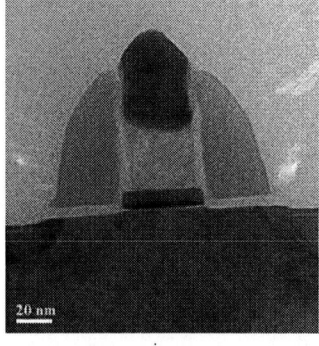

Figure1: Cross-section TEM picture of a ~ 35 nm long device.

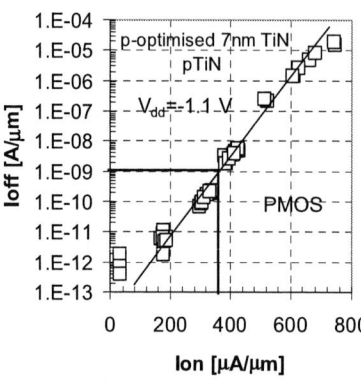

Figure 2: PMOS I_{ON}-I_{OFF} curve for 7nm TiN, measured at V_d= -1.1V. Excellent I_{ON} of 360 µA/µm at an I_{OFF} of 1 nA/µm.

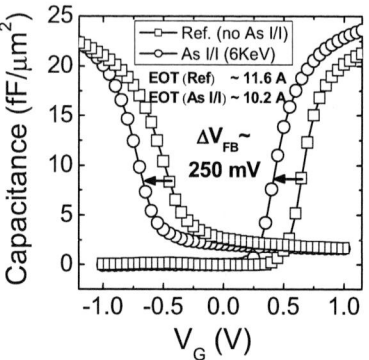

Figure 3: Inversion and accumulation CVs for 7nm TiN with and without As implantation at (6 keV + 4E15 cm^{-2}).

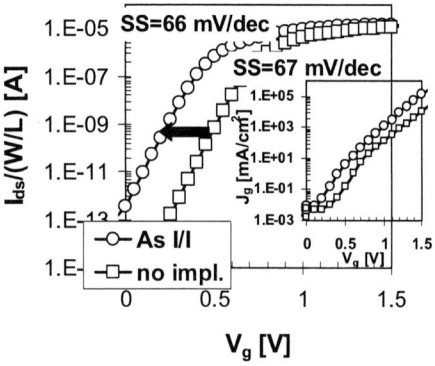

Figure 4: Normalized I_dV_g and J_gV_g curves on WxL = 1x1µm^2 transistors for the reference and As I/I gate stack at (6 keV + 4E15 cm^{-2}).

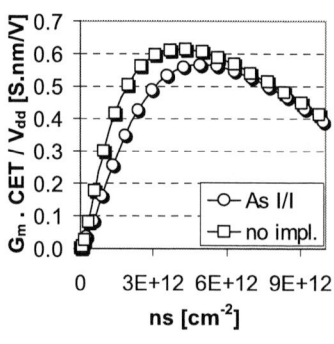

Figure 5: CET normalized transconductance for the optimized TiN/Si stack, un-implanted and implanted with As at 6keV, 4E15 cm^{-2}.

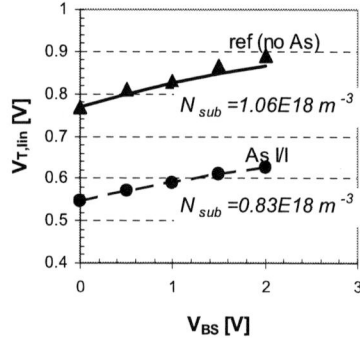

Figure 6: V_{Tlin} as a function of back bias for the reference and the As implanted TiN at 6keV, 4E15 cm^{-2}.

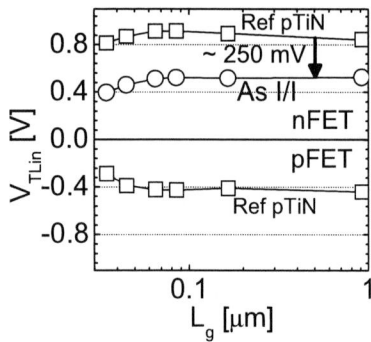

Figure 7 p and n-$V_{T,lin}$ as a function of L_g, for optimized 7nm TiN/Si metal stack, implanted by As at 6 keV, 4E15 cm^{-2} for n-type.

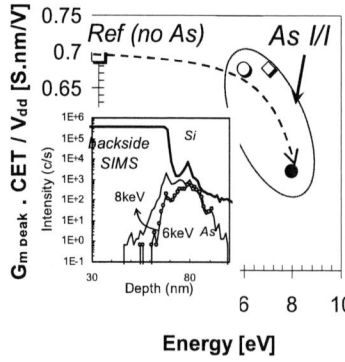

Figure 8: Normalized peak transconductance as a function As I/I energy. Inset: backside SIMS of the stack with As implanted at 6 and 8keV

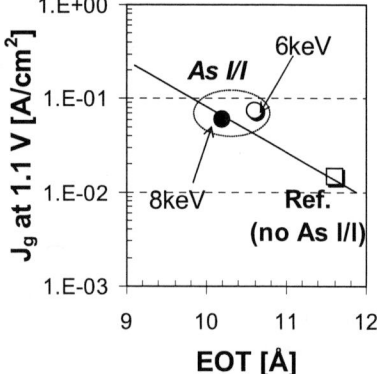

Figure 9: Gate leakage as a function of EOT for reference (no As I/I) and As implanted TiN.

978-1-4244-2784-0/09 $25.00 © 2009 IEEE

La-doped metal/High-K nMOSFET for Sub-32nm HP and LSTP Application

C. S. Park, J. W. Yang, M. M. Hussain, C. Y. Kang, J. Huang, P. Sivasubramani, C. Park, K. Tateiwa[1],
Y. Harada[1], J. Barnett, C. Melvin, G. Bersuker, P. D. Kirsch, B. H. Lee, H. H. Tseng and R. Jammy

SEMATECH, 2706 Montopolis Drive, Austin TX 78741, [1]Panasonic

Abstract: This paper presents results on nMOSFETs with the La-doped high-k/metal gate stack to see its suitability for sub-32nm LSTP and HP applications. The 32nm gate length transistors exhibit an excellent I_{on}-I_{off} characteristic, and the PBTI results meet the 32nm technology node requirement. Furthermore, for the first time, V_t variation in the La-doped high-k/metal gate stack devices is investigated. The results suggest that employing the metal electrode suppresses V_t variability while no additional parameter fluctuations due to La-doping of the high-k dielectric were observed.

Introduction

High-k/metal gate stack has been introduced for the 45 nm node [1] and it has been extensively demonstrated for the 32nm node and beyond [2]. The La-doped high-k/metal gate stack has been considered as one of the nMOSFET gate stack options [3,4] but its suitability for the sub-32nm technology has not been studied. In this paper, we show results on the La-doped high-k/metal gate nMOSFET for the sub-32nm LSTP and HP applications. V_t variation of the La-doped high-k/metal gate stacks is also discussed for the first time.

Experiments

nMOSFETs with the La-doped high-k/metal gate stack were fabricated. La-doping into Hf-silicate was performed by utilizing a thin LaO_x cap layer deposited on the high-k film. A TaN film was used as a gate electrode followed by a poly-Si cap. An optimized dry etch process was used to define the 32nm gate length as seen in Fig. 1(a). A conventional spike anneal over 1000°C was applied for the S/D activation and finally a forming gas anneal was performed. The conventional n+ polysilicon nMOSFET was fabricated using 2.0 nm of SiO_2 as a reference.

Discussion

A TEM image in Fig. 1(b) shows a well defined 32nm gate length gate stack. I_d-V_g and I_d-V_d data demonstrate well behaved 32nm nMOSFET characteristics (Figs 2 and 3). A sub-threshold swing of about 80 mV/dec was obtained. V_t is found to be about 0.45V and 0.33V in long channel and 32nm channel length devices, respectively (Fig. 4). Inset in Fig. 4 also shows DIBL in the 32nm device that meets the node requirement. EOT of about 0.95nm was extracted from the C-V characteristics of nMOSFETs with and without La-doping, Fig. 5. La-doping does not affect the EOT values but rather results in

lower Vt. The J_g-V_g data of the La-doped nMOS gate stacks in Fig. 6 shows the optimization (zone B) can even further improve J_g of La-doped high-k gate stacks: Ig can be up to four orders of magnitude smaller than that of the SiO_2/polySi gate stack. An excellent electron mobility of 90% of the universal value at E_{eff}=1MV/cm is also achieved (not shown here). Fig. 7 shows the nMOSFET I_{on}-I_{off} characteristics measured at V_{dd}=1.0V. La-doped high-k nMOSFETs are demonstrated to be suitable for the HP and LSTP applications as they show excellent I_{on}= >800µA/µm and 1100µA/µm, respectively, at Ioff= 1nA/1µm. These excellent results are achieved without any additional performance boosting techniques. Fig. 8 shows an interface states density of the La-doped high-k nMOSFETs measured by the charge pumping method. PBTI tests (at 125°C) of the La-doped high-k nMOSFETs were performed under the high stress voltage conditions (Fig. 9). The observed degradation is comparable to the historical results [4]. Fig. 10 shows a lifetime projection suggesting that the La-doped high-k gate stack is expected to meet the 10 year device lifetime requirement. Since the 'local' variability due to intrinsic parameter fluctuations presents a major challenge to the scale down CMOS devices, the variability of the La-doped high-k nMOSFET were examined and compared with the SiO_2/polySi gate stack and non-doped high-k/metal gate stack, Fig. 11. The polysilicon gate stack devices show higher V_t variation comparing to that of the metal gate devices. The results are consistent with the previous study [5]. V_t variability comparison of the nMOSFETs with and without the La-doped high-k also suggests that no additional parameter fluctuation due to La-doping occurs (Fig. 12).

Conclusion

32nm gate length La-doped high-k nMOSFETs have been demonstrated to exhibit well-behaved device characteristics. Excellent I_{on}-I_{off} and PBTI results suggest La-dope high-k/metal gate stack is suitable for the sub-32nm LSTP and HP application. It has been shown that the La-doped high-k/metal gate stack suppresses V_t variability compared to the SiO_2/polySi gate, and La-doping does not cause additional parameter fluctuations.

References

[1] K. Mistry, IEDM2007, p.247.
[2] X. Chen et al, VLSI2008, p. 88.
[3] H. Alshareef et al, VLSI2006, p. 21.
[4] P. D. Kirsch, et al, IEDM2006, p. 629.
[5] J. W. Yang et al, SSDM 2008, p. 866.

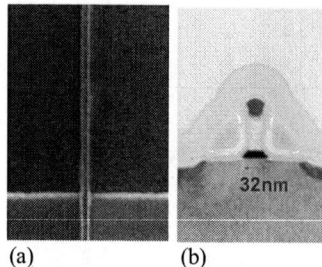

(a)   (b)

Fig. 1 (a) SEM image after gate etching of La-doped high-k/TaN gate stack (b) finally defined 32nm gate length La-doped high-k nMOSFET gate stack.

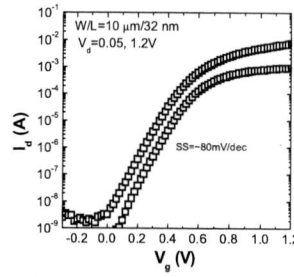

Fig. 2 I_d-V_g characteristics of 32nm gate length La-doped high-k nMOSFET device.

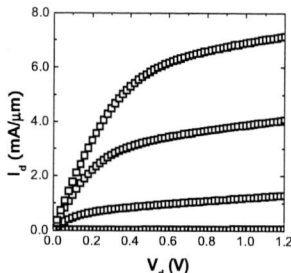

Fig. 3 I_d-V_d characteristics of 32nm gate length La-doped high-k nMOSFET device

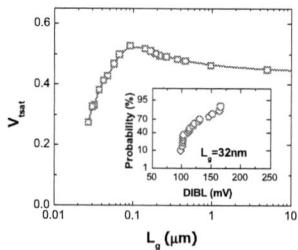

Fig. 4. L_g-V_{tsat} of La-doped high-k nMOSFET. Inset shows DIBL at a short channel device with L_g=32nm.

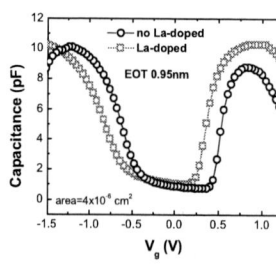

Fig. 5. High frequency C-V of La-doped high-k and undoped high-k nMOSFETs.

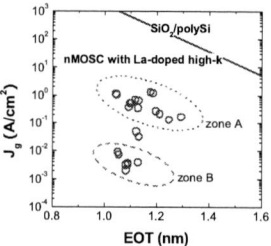

Fig. 6 J_g-EOT plot of La-doped high-k nMOS capacitors.

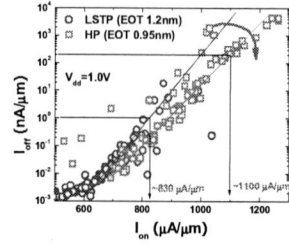

Fig. 7 I_{on}-I_{off} characteristics of La-doped high-k nMOSFETs.

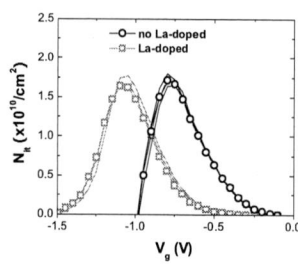

Fig. 8 N_{it} of La-doped high-k and undoped high-k nMOSFETs measured by charge pumping method.

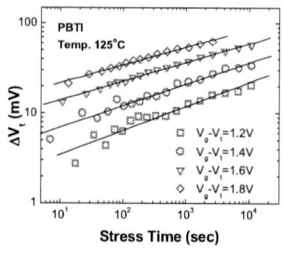

Fig. 9 PBTI characteristics of La-doped high-k nMOSFETs.

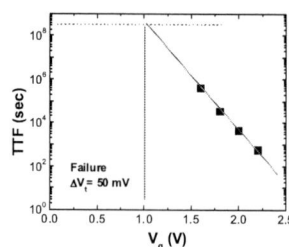

Fig. 10 Lifetime extraction from PBTI test meets 10-year lifetime requirement, $V_{32nm\ technology}$=1V.

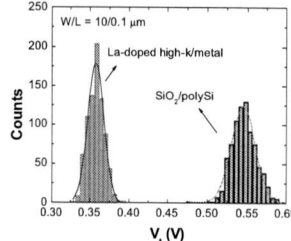

Fig. 11 V_t variability comparison of La-doped high-k/metal gate stack and SiO$_2$/polySi gate stack

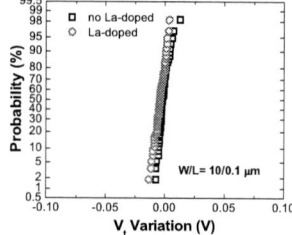

Fig. 12 V_t variability comparison of La-doped high-k/metal gate stack and undoped high-k/metal gate stack.

978-1-4244-2784-0/09 $25.00 © 2009 IEEE   54

Extending spectroscopic ellipsometry for identification of electrically active defects in Si/SiO$_2$/high-*k*/metal gate stacks.

J. Price, G. Bersuker, P. S. Lysaght, and H.-H. Tseng.

SEMATECH, 2706 Montopolis Dr. Austin, TX, 78741

ABSTRACT

This paper presents a new method utilizing spectroscopic ellipsometry (SE) to non-invasively identify the oxygen vacancy defects located in the bottom interfacial SiO$_2$ layer (BIF) of the scaled high-*k*/metal gate stacks. Discrete absorption features within the bandgap of the SiO$_2$ BIF are identified, and their relation to both intrinsic and process-induced defects is proposed. Sensitivity to changes in these defects with different process conditions is demonstrated, along with evidence suggesting that these same defects may contribute to the mechanism associated with the V$_{fb}$ roll-off phenomenon.

INTRODUCTION

Stabilization of the V$_{fb}$ with aggressively scaled EOT must be achieved while accommodating various process integration schemes. A new model was recently proposed suggesting that oxygen vacancy defects within the SiO$_2$ BIF contribute to V$_{fb}$ roll-off [1]. This was experimentally supported by the observation that as the EOT scales with decreasing BIF thickness, V$_{fb}$ roll-off is enhanced along with a corresponding increase in the interface trap density N$_{it}$ (Fig. 1). Therefore, to tune the effective work function accurate physical characterization of these defects is required. Spectroscopic ellipsometry data may provide a unique insight into the nature of the performance-affecting defects although this technique has never been previously employed for this purpose. This study demonstrates effectiveness of SE in addressing critical properties of the dielectric/substrate interfaces.

EXPERIMENTAL

All sub-bandgap (sub-E$_g$) absorption measurements were performed with a J. A. Woollam VuV spectroscopic ellipsometer. For each measurement, the imaginary part of the dielectric function, ε_2, was obtained by a point-by-point extraction method [2].

RESULTS AND DISCUSSION

The recently proposed phenomenological model provides a mechanism that explains the V$_{fb}$ roll-off as an enhancement of positively charged oxygen vacancies within the BIF as it physically scales below a certain critical value [1]. The following results demonstrate a strong correlation between each attribute of this model and the measured sub-E$_g$ absorption features. First, the characterization of sub-E$_g$ absorption features for Si/SiO$_2$/high-*k* gate stacks previously demonstrated that such optically active defects are physically located in the BIF [2]. These defects are summarized in Table 1. Figure 2 confirms the interfacial specificity of these defects by plotting the absorption (ε_2) of a 10 nm thick HfO$_2$ film deposited on both Si and (fused) silica substrates annealed in N$_2$ at 900 °C. The

absence of sub-bandgap absorption features in the silica/HfO$_2$ sample indicates the observed features in Si/SiO$_2$/HfO$_2$ are not bulk HfO$_2$ defects and must, therefore, originate from the Si substrate (3.4 eV, 4.25 eV, 5.3 eV) and the BIF (2.9 eV, 3.9 eV, and 4.75 eV). Additionally, sub-E$_g$ results for an intentionally prepared oxygen-deficient HfO$_2$ sample (Fig. 3) identify a significant enhancement in the amplitude of the 2.9 eV peak suggesting a relation to oxygen vacancies. Electron Spin Resonance (ESR) measurements also find an enhancement in P$_b$ and E' centers for this oxygen-deficient sample [3]. Finally, Figure 4 identifies how the 2.9 eV defect peak increases as the SiO$_2$ BIF thickness is scaled from 4 nm to 1 nm, indicating more defects are present in the thinner SiO$_2$ BIF. ESR measurements (Fig. 5) on identically prepared samples also identified an increase in both P$_b$ and E' centers for the thinner BIF sample [3]. Indeed, these results confirm that the optically active defects originate within the SiO$_2$ BIF, are associated with oxygen vacancies, and are enhanced as the SiO$_2$ becomes thinner; consistent with each attribute of the proposed model.

To confirm the model predictions that BIF oxygen vacancy defects contribute to the V$_{fb}$ roll-off, sub-E$_g$ results are compared to different process-specific changes associated with the observed roll-off dependencies. A direct comparison between the results shown in Figure 4 and in Figure 1 verifies how the 2.9 eV amplitude correlates with the V$_{fb}$ roll-off and scaled BIF thickness. In addition, different anneal conditions have also been reported to affect the defects in the BIF and contribute to V$_{fb}$ [4]. A forming gas anneal (FGA) is known to passivate the Si/SiO$_2$ interfacial defects. The results of Figure 6 are in agreement with these findings where a decrease in the intensity of the 2.9 eV peak is observed after FGA. Figure 7 also identifies an increase in the 2.9 eV amplitude after rapid thermal anneal (RTA), followed by a subtle decrease with FGA, consistent with reports that a higher thermal budget contributes to V$_{fb}$ [4].

CONCLUSION

Evidence is presented correlating optically active dielectric defects with contributions to the observed V$_{fb}$ roll-off phenomenon. These SE results support the recently proposed model suggesting that such defects correspond to oxygen vacancies in the SiO$_2$ interfacial layer, which can be modulated by subsequent high-*k*, metal, and thermal treatments.

REFERENCES

[1]   G. Bersuker *et al.*, ESSDERC, 2008.
[2]   J. Price *et al.*, APL **91**, 061925, 2007.
[3]   G. Bersuker *et al.*, JAPL **100**, 094108, 2006.
[4]   J. K. Schaeffer *et al.*, SSDM, 2007.

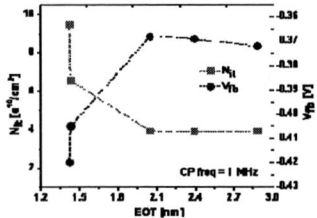

Figure 1: Correlation between V_{fb} and interface trap density (N_{it}) measured on HfO_2/TiN transistors fabricated using terraced oxide structures.

Defect energy	Gate stack location	Origin
2.9 eV	Bottom interface	Intrinsic to Si/SiO$_2$, maybe O$_2$ vacancy
3.6 eV	Bottom interface	Intrinsic to Si/SiO$_2$
3.9 eV	Bottom interface	Intrinsic to Si/SiO$_2$
4.75 eV	Bottom interface	HfO$_2$ specific, maybe O$_2$ vacancy

Table 1: Summary of optically active defects including their energy within the SiO$_2$ BIF bandgap, physical location, and origin. Data taken from reference #2.

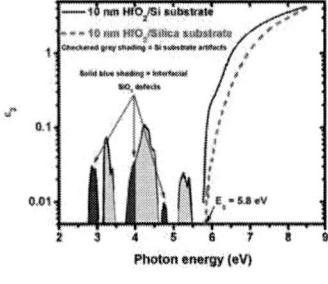

Figure 2: Sub-E$_g$ absorption for two 10 nm thick HfO$_2$ films deposited on Si (black solid line) and Silica (red dashed line). The interfacial specific defects (solid blue shading) are seperated from the Si substrate artifacts (grey checkered shading).

Figure 3: Sub-E$_g$ absorption for a 5 nm intentionally oxygen deficient (red dashed line) and 5 nm control HfO$_2$ film (solid black line). The substantial increase in the 2.9 eV peak as a result of oxygen deficiency suggests that this defect is related to oxygen vacancies.

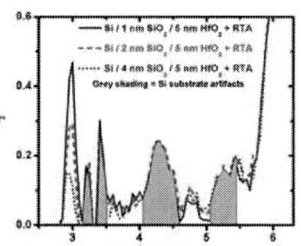

Figure 4: Sub-E$_g$ absorption for 5 nm HfO$_2$ annealed films with different BIF thickness: 1 nm (solid black line), 2 nm (red dashed line), and 4 nm (dotted blue line). A significant increase in the 2.9 eV peak with decreasing thickness indicates a thinner SiO$_2$ BIF has more defects.

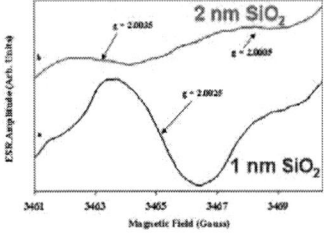

Figure 5: ESR spectra of a 3 nm annealed HfO$_2$ film with 2 nm SiO$_2$ BIF (red line) and 1 nm SiO$_2$ BIF (blue line). A considerable increase in both the P$_b$ and E' features for the thinner BIF suggest more defects are present.

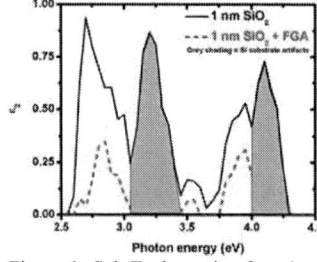

Figure 6: Sub-E$_g$ absorption for a 1 nm SiO$_2$ film with (red dashed line) and without (black solid line) a forming gas anneal (FGA). A pronounced decrease in the 2.9 eV defect peak amplitude indicates a reduction in defects with FGA.

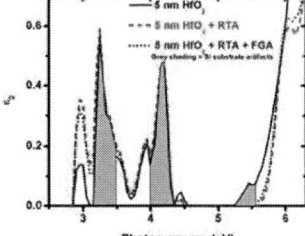

Figure 7: Sub-E$_g$ absorption for a 5 nm HfO$_2$ film (black solid line), followed by an RTA (red dashed line), and a RTA plus FGA (blue dotted line). An increase in the 2.9 eV peak amplitude is observed with high temperature anneal, along with a slight reduction following a FGA.

978-1-4244-2784-0/09 $25.00 © 2009 IEEE

22nm CMOS Approaches by PVD TiN or Ti-Silicide as Metal Gate

C.S. Liu[1*], G. Boccardi[2], H.Y. Wang[1], C.T. Lin[1], J. Petry[2], M. Müller[2], Z. Li[3], C. Zhao[3], C.H. Yu[1]

[1]TSMC, 8, Li-Hsin Rd. 6, Hsinchu Science Park, 300-77 Hsinchu, Taiwan; [2]NXP-TSMC Research Center, [3]IMEC

[1*]Phone: 32-16-287720, Fax: 32-16-281576; E-mail: csliua@tsmc.com

INTRODUCTION

Two cost-effective alternative Vt shift approaches by PVD-TiN and PVD-TiN/Ti metal work function tuning were studied and compared. **Flow-A:** Mid-gap PVD TiN was first adjusted to p-like metal by PVD-TiN film properties for pMOS and then the p-like TiN was adjusted to n-like metal by As ion implantation (I/I) on p-like TiN for nMOS. **Flow-B:** PVD-TiN/Ti was transformed into n-like metal TiN/TiSix for nMOS and the n-like TiN/TiSix was changed into p-like metal by Al I/I in combination with Al diffusion facilitated by snowplow effect during TiSix formation on pMOS. [1] In this paper, we successfully demonstrate the feasibility of two gate first single metal single dielectric (SMSD) manufacturing methods by the changes of PVD TiN process conditions, I/I conditions and/or TiSix formation. Vtn/Vtp= +0.49V/-0.48V were achieved on SMSD gate first planar bulk CMOS devices. The developed SMSD low cost processes meet the equivalent +/-0.2V low Vt target for N22 and beyond fully depleted SOI or FinFET technologies. [2-4]. PVD-TiN metal gate showed 4A lower EOT and hence much better device performance than ALD-TiN one [3]. Lower deposition temperature, less impurity and better film stability in PVD-TiN could be the root cause. Excellent PVD-TiN properties attracted us to further study on varied PVD-TiN films and PVD-TiN/Ti schemes for 22nm fully depleted CMOS device application.

EXPERIMENT

Two CMOS planar SMSD gate stack process flows, flow-A and flow-B, are shown in Fig. 1. For flow-A: After 2nm ALD HfO2 high-k deposition, varied 10-nm-thick TiN films were deposited and followed by 10nm buffer amorphous Si caps and then followed by As I/I on TiN for nMOS. For flow-B: TiN/Ti films were deposited instead and Al I/I was used on TiN/Ti for pMOS. Ti will react with amorphous Si to form TiSix at the subsequent thermal process.

RESULTS AND DISCUSSION

Flow-A: Adjust PVD-TiN to p-like metal

Vt adjustment by TiN process parameters for pMOS and As I/I for nMOS were shown in Fig. 2. Vtn/Vtp can be adjusted to 0.73/-0.48 from 0.64/-0.55 by PVD-TiN process. After further modification by As I/I for nMOS, Vtn/Vtp 0.49/-0.48 was reached. N2 annealing, low oxygen and low deposition temperature after high degas temperature influence the TiN work function of the same TiN thickness. Fig. 3 shows the comparisons of high oxygen, medium oxygen and optimized TiN with low oxygen and lower film deposition temperature. Optimized TiN film has lower oxygen, lower resistivity, smoother film, higher density and finer grain than other conditions. XRD and XRR show the relative grain size and roughness in Fig. 4. Oxygen content versus Vtp was also shown in Fig. 4.

Contrary to general concept that oxygen make metal more p-like along with EOT regrowth [5], more oxygen in well Ti-N-O bonded PVD-TiN didn't increase the work function in current study. From the Fig.3, the N+O composition almost keep the same amounts. The increased oxygen replaced original nitrogen in the TiN. No additional EOT regrowth observed in higher oxygen of TiN supported the mechanism. EOT vs Jg plots are shown in Fig. 5. EOT was reduced to 1.0 from 1.08nm after As I/I on NMOS. Thinner EOT was attributed to the As doping species interaction with interfacial layers. Higher Jg after I/I can be understood due to thinner EOT from the Jg-EOT trend line of non-implanted TiN. No further Vtp shift can be obtained by Al I/I, excess Al I/I only obtained 20mV shift but traded off 3 order higher Jg.

Flow-B: TiN/TiSix and snowplow effect on I/I species

TiN/Ti reacted with amorphous Si to form TiN/TiSix. Vt adjustment by TiN/TiSix for nMOS and Al I/I for pMOS in linear region devices were shown in Fig. 6. Vtn/Vtp was adjusted to 0.52/-0.65 by PVD-TiN/TiSix on nMOS. After modification by Al I/I on pMOS, Vtn 0.52 and Vtp -0.55 Vt were reached. Contrary to Al I/I on PVD-TiN increased 3 order of Jg for small Vt shift, Al I/I on TiN/Ti reduced one order of Jg in combination of 100mV of Vt shift. Lower Jg on with Al I/I than without Al I/I on pMOS was attributed to more uniform of TiSix formation due to Al I/I. More Vt shift was attributed to more Al accumulation at TiN/HfO2 interface due to TiSix snowplow effect. EOT vs Jg plots are shown in Fig. 7. Thinner EOT was attributed to the Al doping species interaction with interfacial layers. Higher Jg on TiN/Ti nMOS and slightly higher Jg on Al I/I TiN/Ti pMOS than on TiN were attributed to thinner EOTs. They followed the TiN Jg-EOT trend lines. Fig. 8 and Fig. 9 show cross-sectional TEM pictures of the TiN/TiSix poly stacks on nMOS capacitor and pMOS transistor. Thicker interfacial layer was observed on Al I/I TiN/TiSix on pMOS. The interaction of Al and original interfacial SiOx contributed to the thicker final interfacial layer AlSiO. This AlSiO resulted in 7.3A only EOT because of its higher k value.

Comparison of flow-A and flow-B

Dit, Nsub and sub-threshold slope are shown in Fig. 10. Fig. 11 shows Id-Vg, CV and g_m curves. Fig. 12 are Vt–Lg and Ion-Ioff behaviors. Both flows achieved high pMOS Ion, 525uA/um@1E-7A/um, without stress booster.

CONCLUSION

Two SMSD gate first planar CMOS devices were demonstrated. Vtn/Vtp= +0.49V/-0.48V were achieved by adjusting TiN to p-like metal and As I/I on nMOS. This enables the equivalent +/-0.2V low Vt target of N22 fully depleted CMOS technologies. Vtn/Vtp= 0.52/-0.55 were obtained by transforming PVD-TiN/Ti into n-like metal TiN/TiSix for nMOS and by Al I/I on TiN/Ti for pMOS. Al diffusion was facilitated by snowplow effect of TiSix formation on pMOS. As low as 7.3A EOT with decent Jg 6.4E-3 A/cm2 @1.1Vwas obtained.

REFERENCES

[1] J. Kedzierski et al, IEDM 2003, p315

[2] ITRS Roadmap

[3] G. Vellianitis et al., IEDM 2007, p681

[4] C. Fenouillet-Beranger et al, IEDM 2007, p267

[5] Z. Li et al, J. Electrochem. Soc., 155, p481, 2008

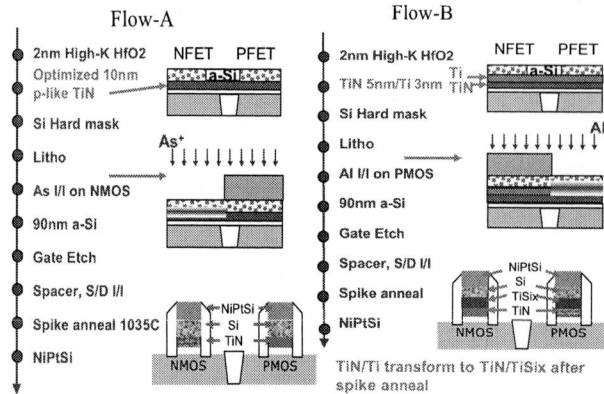

Fig. 1 Schematic diagrams of alternative PVD-TiN (flow-A) and TiN/Ti (flow-B) SMSD process flow

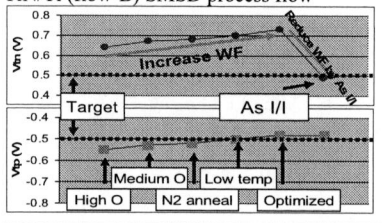

Fig. 2 Vt adjustment by TiN process conditions for pMOS and As I/I on TiN for nMOS in linear region WxL 10x1um devices.

	O (%)	Ti (%)	N (%)	Roughness (A)	Density (g/cm3)	Resistivity (uOhm-cm)	Vtn (V)	Vtp (V)	Grain size
High O	9	39.5	50.9	29.9	4.59	331	0.64	-0.55	Larger grain
Medium O	6.9	40.1	52.2	15.6	4.61	203	0.67	-0.53	Larger grain
Optimized	4.9	40.3	53.9	11.6	4.96	190	0.73	-0.48	Finer grain

Fig. 3 Physical properties comparisons of varied 20nm TiN films

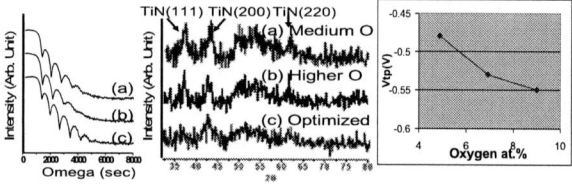

Fig. 4 XRR(left), XRD(middle) and oxygen concentration in TiN vs Vtp (right) of 20nm TiN films. XRR showed smoother TiN surface on optimized TiN. Finer grain was observed on optimized TiN from broader XRD peak. Lower oxygen conc. in TiN has lower |Vtp|.

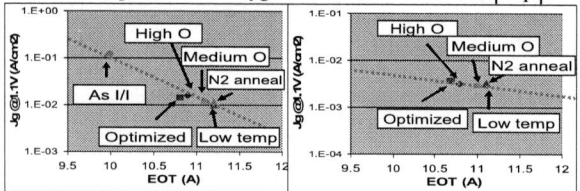

Fig. 5 nMOS(left) EOT vs Jg. As I/I on TiN lay on trend line of the TiN splits. pMOS(right) EOT vs Jg. No trend on oxygen concentration vs EOT.

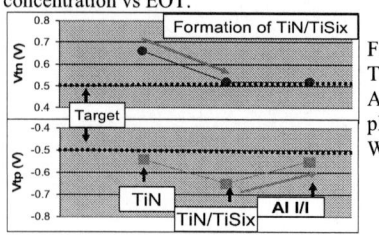

Fig. 6 Vt adjustment by TiN/Ti for nMOS and Al I/I on TiN/Ti for pMOS in linear region WxL 10x1um devices

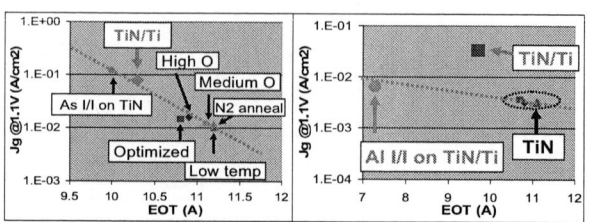

Fig. 7 nMOS(left) EOT vs Jg. TiN/Ti was located on the TiN trend line. pMOS(right) EOT vs Jg. Al I/I on TiN/Ti followed the trend line of TiN. Al I/I reduced Jg of TiN/Ti scheme

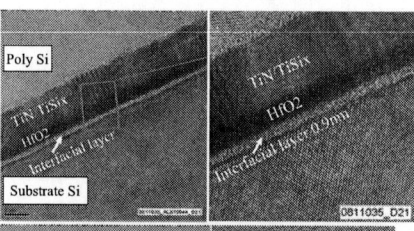

Fig. 8 Cross-sectional TEM of the TiN/TiSix poly stack on nMOS capacitor.

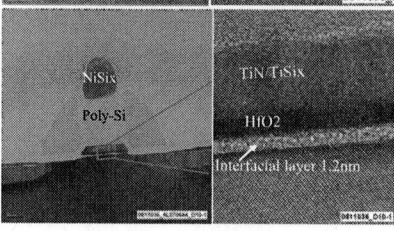

Fig. 9 Cross-sectional TEM of 35nm width TiN/TiSix poly stack with Al I/I on pMOS transistor. Thicker interfacial layer observed because of interaction of Al and interfacial SiOx.

	nMOS Peak Dit (cm⁻²eV⁻¹)	pMOS Peak Dit (cm⁻²eV⁻¹)	nMOS Nsub (cm⁻³)	pMOS Nsub (cm⁻³)	nMOS SS (mV/dec)	pMOS SS (mV/dec)
Flow-A	8.95E+10	1.05E+11	4.60E+17	2.11E+17	66.8	66.2
Flow-B	1.11E+11	8.29E+10	6.29E+17	1.01E+17	66.3	77.8

Fig. 10 Peak Dit, Nsub and Sub-threshold slope(SS) comparison of flow-A and flow-B. Minor effect after I/I.

Fig. 11 Id-Vg, CV and g_m comparison of flow-A and flow-B

Fig. 12 Vtp roll-off (left) and Ion-Ioff(right) comparison of flow-A and flow-B. Both flows achieved high Ion, 525uA/um@1E-7A/um without stress booster.

978-1-4244-2784-0/09 $25.00 © 2009 IEEE

Reliability Assessment of Low |V_t| Metal High-κ Gate Stacks for High Performance Applications

C. D. Young, G. Bersuker, P. Khanal, C.Y. Kang, J. Huang, C.S. Park, P. Kirsch, H.-H. Tseng, and R. Jammy

chadwin.young@SEMATECH.org, 512-356-3612 (o), 512-356-7640 (fax)

SEMATECH, 2706 Montopolis Drive, Austin, TX 78741

INTRODUCTION

With recent significant progress in the performance of high-κ gate/metal gate transistors [1,2], understanding the gate stack breakdown mechanisms becomes a critical factor for reliability assessment. Low |Vt| metal/high-κ MOSFETs with the LaO capping layer for NFET and the Ru-Al metal electrodes for PFET applications [1-6] demonstrated superior performance. However, reliability of La and Ru-Al containing high-κ gate stacks has not yet been thoroughly evaluated [7]. In order to study the governing degradation mechanisms in Hf-based gate stacks with La or Ru-Al incorporation under constant voltage stress (CVS), stress-induced leakage current (SILC) measurements were used to monitor stress-generated traps in these dielectric stacks.

EXPERIMENT

Transistors were fabricated using a gate-first integration flow with the standard source/drain activation. HfO_2 was deposited by atomic layer deposition on the thermally grown SiO_2 films where some samples included an additional LaO_x cap deposited on top of the high-κ dielectric (low V_t NMOS) or Ru-Al electrode (low V_t PMOS). Sample devices in this work had high-κ physical thickness values ranging from 2 nm to 3 nm. Small area transistors were subjected to inversion CVS interspersed with SILC measurements. This measurement sequence was implemented through the "Smart TDDB" approach [8,9], and temperature dependant SILC measurements were performed after breakdown events. Also, electron energy-loss spectroscopy (EELS) analysis was done to determine the chemical composition of the gate stack after high temperature activation.

RESULTS AND DISCUSSION

Low V_t NMOS – In order to capture stress-induced leakage current (SILC) events, a fully-automated "smart" TDDB algorithm, which monitors stress current (I_g) and SILC during CVS, was developed [7,8] with an example shown in Figs. 1 and 2 for a 3 nm HfO_2 stack. During CVS, three phases typically emerge in high-κ gate stacks which contain SiO_2-like interfacial layers (IL): 1) the stress current decreases due to electron trapping in the high-κ film, until 2) a soft BD (SBD) event occurs, followed eventually by 3) progressive breakdown (PBD) and hard breakdown (HBD) (Fig. 2). Fig. 1 shows the resulting SILC data that has been shown to correlate to the corresponding "events" (i.e., stress current changes) that occur during the stress [9,10] which are identified as the relative change in SILC ($\Delta I_g/I_g$) in Fig. 2. There is a steady increase in the stress current after SBD (Fig. 2) that occurs accompanied by the SILC increase indicative of progressive degradation. To better understand and quantify SBD, we calculated differential resistance, $R_{diff}(V_g)$, from the SILC data and the slope of the differential resistance, S_{Rdiff} (Fig. 2) and collected temperature-dependent SILC data (Fig. 3). For pure ohmic conductance, $R_{diff}(V_g)$ = Const (i.e., S_{Rdiff}=0), the slope value $S_{Rdiff} \neq 0$ can be used as a figure of merit for ohmic vs. non-ohmic (tunneling, hopping) conductance. Fig. 2 illustrates a corresponding reduction in S_{Rdiff} with increases in stress current with time. The SILC value extracted at V_g = 0.5 V (corresponding to the maximum of the SILC growth [11]) from each BD regime was used in the slope extraction. As further analysis, the activation energy (E_A) extracted from the SILC temperature dependence (inset, Fig. 3) reduces during post-SBD degradation (Fig. 3). This suggests that the percolation path, which has been formed by the SBD event, becomes more conductive during the progressive breakdown phase. A similar result is obtained for a 2 nm HfO_2 gate stack (Fig. 4). An example of the smart TDDB result for LaO_x capped samples is provided in Fig. 5. A steady increase in SILC during stress on the samples with the LaO_x cap (Fig. 5b) was observed. However, the stress current exhibits a noisy phase followed by a rather abrupt breakdown (Fig. 5a, and inset). After further investigation, the SILC data (Fig. 5b) is also sporadic (i.e., SILC curves temporarily reduce) even though the overall trend increases with time. The reduction of S_{Rdiff} is very abrupt as compared to the control HfO_2 samples also signifying less progressive breakdown (BD) characteristics (Fig. 6). The same approach was invoked to extract the E_A for the LaO_x capped samples during post-SBD degradation (Fig. 7). The E_A results clearly demonstrate sporadic values with no systematic trend in the activation energy values. In an effort to better understand why this is occurring in the La incorporated samples, EELS analysis was performed (Fig. 8). The EELS data clearly shows that La has penetrated into the IL. This La encroachment, which is destructive for the SiO_2 network due to large La size, generates structural defects leading to different degradation characteristics comparing to the 'standard' (no La) high-κ stacks [1,7].

Low V_t PMOS – Fig. 9 shows the increasing SILC results with increasing stress time for a PMOS 3 nm HfO_2 device. Using the same analysis approach as NMOS, Fig. 10 illustrates the correlation between SILC and stress current with corresponding S_{Rdiff}. Results demonstrate a progressive degradation trend since S_{Rdiff} reduces as the SILC gradually increases. The extracted activation energy (E_A) confirms this because the extracted E_A from the SILC temperature dependence reduces during post-SBD degradation for 3 nm (Fig. 11) and 2 nm (Fig. 12) HfO_2. This suggests a similar progressive breakdown phase when compared with the NMOS case. Ru-Al metal gate samples were subjected to S_{Rdiff} extraction and temperature SILC. The S_{Rdiff} slowly reduces with time similar to it PMOS control counterparts (Fig. 13). The extracted E_A from the SILC temperature dependence reduces during post-SBD degradation (Fig. 14) suggesting a percolation path has been formed by the SBD event. Since the results are similar to PMOS control HfO_2 devices, then Ru is most likely not diffusing into the IL. This is confirmed in the EELS data (Fig. 15) which shows no appreciable Ru in the SiO_2 interfacial layer.

SUMMARY

SILC analysis is a powerful tool for the assessment of breakdown characteristics of high-κ devices. By applying the SILC analysis during high field stress, we determined that the degradation mechanism for LaO_x capped devices was drastically different as compared to the conventional Hf-based gate stacks. The La atoms diffused into the interfacial layer disrupting the SiO_2 structure which may affect the reliability of the La-doped stacks. On the other hand, similar analysis applied to the stacks with the Ru-Al bi-layer gate electrode demonstrated that the Al-contained stacks were similar to that of the baseline samples indicating that Al atoms, which preferentially substitute for Si in SiO_2, did not generate defects contributing to SILC.

REFERENCES

[1] J. Huang, et al., *IEDM paper 2.6*, 2008 [2] C.S. Park, et al., *VLSI-TSA*, p.154, 2008 [3] Y. Yamamoto, et al, *Proc. of SSDM*, p.212, 2006 [4] X. P Wang, et al, *IEEE EDL* **27** p. 31 2006 [5] C.Y. Kang, et al., *Proc. of SSDM*, p.250, 2007 [6] M. Takayanagi, et al., *IRPS*, p. 13, 2004 [7] C.Y. Kang et al., *IEDM paper 5.4*, 2008 [8] C. D. Young, et al., *ISAGST*, 2007 [9] C. D. Young, et al., *Micro. Eng.*, to be published [10] G. Bersuker, et al., *IRPS*, p. 49, 2007 [11] R. O'Conner, et al., *IRPS*, p. 324, 2008

978-1-4244-2784-0/09 $25.00 © 2009 IEEE

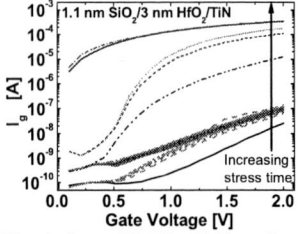

Fig. 1. Smart TDDB SILC results on NMOS where the SILC increases as the increase in stress I_g takes place. Sharp SILC increase changes can be directly correlated with changes in stress I_g.

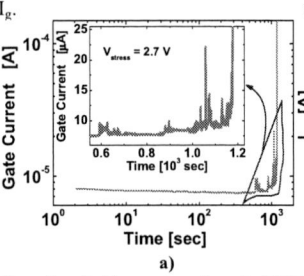

Fig. 2. Correlation of SILC with stress I_g illustrating the evolution of the SILC increase and the slope of the differential resistance, S_{Rdiff}, with CVS stress for a 3 nm HfO_2 NMOS gate stack.

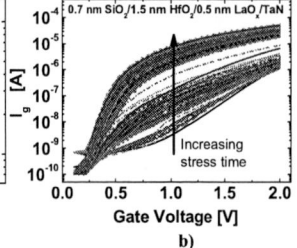

Fig. 3. Extracted E_A from NMOS 3nm HfO_2 SILC temperature dependence (inset) during the breakdown process, where after SBD, a steadily decreasing E_A occurs.

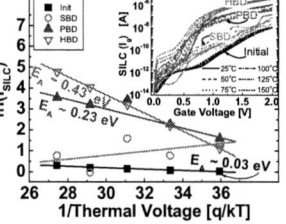

Fig. 4. Extracted E_A from NMOS 2nm HfO_2 SILC temperature dependence (inset) during the breakdown process, where after SBD, a steadily decreasing E_A occurs.

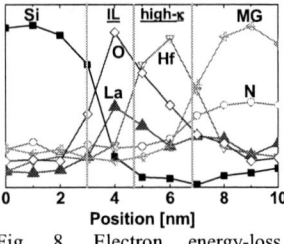

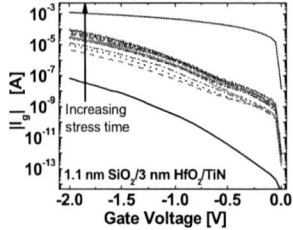

Fig. 5. a) Example: Smart TDDB on HfO_2/LaO_x capping with corresponding b) SILC results. SILC trend appears to have a steady increase throughout the breakdown process; however, the stress I_g exhibits a more "noisy"/less steady increase when compared to HfO_2 only. Further inspection revealed that SILC was also sporadic indicating that a progressive degradation mechanism cannot explain the result.

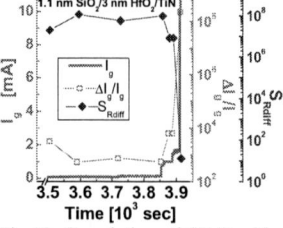

Fig. 6. Evolution of the slope of the differential resistance, S_{Rdiff}, with corresponding stress current during CVS. Stacks with LaO_x capping show abrupt breakdown with less progressive BD characteristics.

Fig. 7. Extracted E_A from the LaO_x cap SILC temperature dependence (inset): sporadic E_A, exhibits non-standard features possibly controlled by impurities in the interfacial layer.

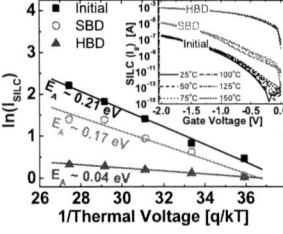

Fig. 8. Electron energy-loss spectroscopy (EELS) profiles indicate that Lanthanum (La) impurities are encroaching into the SiO_2 interfacial layer (IL) thereby possibly creating structural defects.

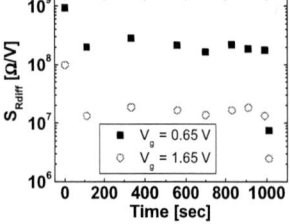

Fig. 9. PMOS SILC results where the SILC increases as the increase (inset) in stress I_g takes place. Sharp SILC increase changes can be directly correlated with changes in stress I_g.

Fig.10. Correlation of SILC with stress I_g illustrating the evolution of the SILC increase and the slope of the differential resistance, S_{Rdiff}, with CVS stress for a 3 nm HfO_2 NMOS gate stack.

Fig. 11. Extracted E_A from PMOS 3 nm HfO_2 SILC temperature dependence (inset) during the breakdown process, where after SBD, a steadily decreasing E_A occurs

Fig. 12. Extracted E_A from PMOS 2 nm HfO_2 SILC temperature dependence (inset) during the breakdown process, where after SBD, a steadily decreasing E_A occurs.

Fig. 13. Evolution of the slope of the differential resistance, S_{Rdiff}, at two different V_g values for Ru-Al devices. S_{Rdiff} of the differential resistance gradually degrades after SBD – similar to baseline samples

Fig. 14. Extracted E_A from the Ru-Al electrode SILC temperature dependence (inset) during the breakdown process, where after SBD, a decreasing E_A occurs. Robust interfacial layer, no signs of impurity-induced degradation

Fig. 15. Electron energy-loss spectroscopy (EELS) profiles indicate no appreciable Ruthenium (Ru) or Aluminum (Al) impurities encroaching in the SiO_2 interfacial layer.

978-1-4244-2784-0/09 $25.00 © 2009 IEEE

High Performance Metal/Insulator/Metal Capacitors Using HfTiO as Dielectric

Hsiao-Hsuan Hsu, Chun-Hu Cheng, and Bing-Yue Tsui

Department of Electronics Engineering & Institute of Electronics, National Chiao Tung University

Room ED641, No.1001, Ta-Hsueh Road, Hsinchu, 300 Taiwan, R.O.C.

Tel: 886-3-5131570   Fax: 886-3-5724361   e-mail: bytsui@mail.nctu.edu.tw

ABSTRACT

Hafnium titanate (HfTiO) film was adapted as the insulator of MIM capacitors for RF/Analog ICs applications. Low leakage current of 3.4×10^{-8} A/cm^2 at -1V and high capacitance density of 17.5fF/μm^2 were obtained. A N$_2$-plasma treatment on HfTiO films can further reduce leakage current by two orders of magnitude and no apparent degradation is observed on the capacitance density and voltage coefficient of capacitance (VCC) properties. Capacitance density of 5.1fF/μm^2, leakage current of 1.3×10^{-9}A/cm^2, and parabolic VCC value of 40ppm/V^2 can be achieved by 51nm thick HfTiO film. These results meet the RF/analog requirements in 2012 predicted by ITRS.

INTRODUCTION

Metal-insulator-metal (MIM) capacitors in silicon analog circuit applications have attracted great attention. To reduce leakage current and capacitor area while increase capacitance density, high dielectric constant (high-κ) dielectrics have been proposed for several years for MIM capacitors. Although TiO$_2$ has a merit of high dielectric constant (κ=50-80), the small band offset is a serious concern [1]. The merit of introducing HfO$_2$ is its medium dielectric constant (κ~25) and large band offset (~ 1.5 eV) which might compensate the leakage issue of TiO$_2$ dielectrics and maintain high dielectric constant simultaneously. Many articles studied the electrical and the structural properties of the mixtures of HfO$_2$ and TiO$_2$ as possible high-κ gate dielectrics [2, 3]. In this works, we demonstrated that the HfTiO dielectric is a good candidate for the insulator of the MIM capacitors.

SAMPLES PREPARATION

Four inches diameter n-type (100) silicon wafers with a 300 nm thick thermally-grown SiO$_2$ were used as substrate. A Ta(250 nm)/ TaN(50 nm) bottom electrode was deposited sequentially by DC sputtering without breaking vacuum. After deposition, NH$_3$ plasma treatment was performed to suppress the formation of interfacial layer. Then HfTiO film was deposited by a dual E-gun evaporation system using the source of TiO$_2$ and HfO$_2$ slits. Post deposition annealing (PDA) was performed at 400°C or 600°C in O$_2$ ambient. Some samples experienced N$_2$ plasma treatment for 20 or 40 sec. Finally, Pt top electrode was formed by E-gun evaporation through a shadow mask.

RESULTS AND DISCUSSIONS

Fig.1 shows the cross-sectional transmission electron microscopy (TEM) image of the Pt/HfTiO/TaN MIM capacitor. The dielectric layer thickness is 19nm including a 5nm thick interfacial layer (IL). Energy dispersion spectroscopy (EDS) analysis reveals that the IL is a TaTiO compound. Inductively-coupled plasma mass-spectroscopy (ICP-MS) analysis identifies that the ratio of TiO$_2$/HfO$_2$ is equal to 1.85 which is also confirmed by the Rutherford backscattering spectroscopy (RBS) analysis. Fig.2 shows the capacitance–voltage (C-V) characteristics of the MIM capacitors with HfTiO and TiO$_2$ as insulator. The capacitance density of the HfTiO sample is 17.6fF/μm^2 at zero bias, which indicates a dielectric constant of 37.

The crystallographes of HfTiO and TiO$_2$ dielectrics experienced different PDA temperatures were investigated by x-ray diffraction (XRD). Fig.3 shows that significant (004) diffraction peak of crystallized-TiO$_2$ with an anatase phase is observed in the TiO$_2$ dielectric after 400°C annealing in O$_2$ atmosphere for 10 min. In sharp contrast, the HfTiO dielectric remains amorphous even after 600°C PDA. Fig.4 shows that the leakage current density of the HfTiO capacitor is lower than the TiO$_2$ capacitor by almost four orders of

magnitude at -1 V at the same capacitance density of ~17.5 fF/μm^2. The low leakage current density of the HfTiO capacitor is attributed to its amorphous structure as well as the high electron barrier height. Positive bias results in higher leakage current due to the low work function of TaN and the formation of the TaTiO IL. Fig.5 shows the current density–voltage (J-V) characteristics of the MIM capacitor measured at temperatures ranging from 25 to 125°C. A small leakage current and weak temperature dependence of 1.74×10^{-8} and 2.62×10^{-7} A/cm^2 at 25 and 125°C are obtained at -1V, respectively. Fig.6 shows that the leakage current mechanism at medium field is Schottky-emission with an effective barrier height of 0.8-0.9eV. At low field (<0.2V), Ohmic conduction dominates the current transport.

The voltage linearity of the MIM capacitor can be depicted as voltage coefficient of capacitance (VCC), which is related to the traps in bulk and interfacial layer. The VCC can be fitted by the following expression: $\Delta C / C_o = \alpha V^2 + \beta V$. The VCC-$\beta$ can be compensated by circuit design while the VCC-α must be reduced by material selection and process technology. Fig.7 shows the variation of $\Delta C(V)/C_o$ as a function of bias voltage. The VCC-α value can be effectively reduced from 8331 to 3730 ppm/V^2 by using HfTiO dielectric to replace TiO$_2$ dielectric.

Fig.8 shows that N$_2$ plasma treatment after HfTiO deposition can reduce leakage current by 50 times at -2V due to surface nitridation. The nirtided layer thickness estimated form the auger electron spectroscope (AES) depth profile is about 6nm as shown in Fig.9. However, Fig.10 shows that the VCC-α is not improved in conjunction with the improvement of leakage current. This observation implies that the VCC-α is related to bulk traps.

By increasing the HfTiO layer thickness to 51nm, the leakage current density at -1V can be further reduced to 1.3nA/cm^2 as shown in Fig.11 while keeping the capacitance density at 5.1fF/μm^2. The VCC-α can be greatly reduced to 40 simultaneously. Fig.12 compares the VCC-α vs. 1/C performance of our results with some published data [4-6]. The exponential decrease of VCC-α with the increase of 1/C is confirmed. The results obtained in this work satisfy the requirement projected to 2012 according to the 2007 ITRS.

CONCLUSIONS

In this work, high performance MIM capacitor with HfTiO as insulator is proposed. Dielectric constant of 37 is demonstrated. The film keeps amorphous after PDA at 600°C. High work function electrode and N$_2$ plasma treatment can reduce the leakage current. The VCC-α is attributed to the bulk traps. A 51nm-thick HfTiO capacitor successfully achieve the ITRS goals of 5fF/μm^2 capacitance density, 1×10^{-8} A/cm^2 leakage current density, and less than 100 ppm/V^2 VCC-α for the RF/analog ICs application in 2012.

ACKNOWLEDGEMENT

This work was supported by the National Science Council, Taiwan, R.O.C. under the contract No.95-2221-E-009-302-MY3.

REFERENCES

[1] J. Robertson, J. Vac. Sci. Tech. B, vol. 18, p.1785, 2000.
[2] A. Paskaleva, et al., J. Appl. Phys., vol. 95, p.5583, 2004.
[3] M. Li, et al., J. Appl. Phys., vol. 98, p.054506, 2005.
[4] T. Ishikawa, et al., in IEDM Tech. Dig., p.940, 2002.
[5] H. Hu, et al., in IEDM Tech. Dig., p.879, 2003.
[6] S. J. Kim, et al., in Proc. of Symp. on VLSI Tech., p.77, 2003.

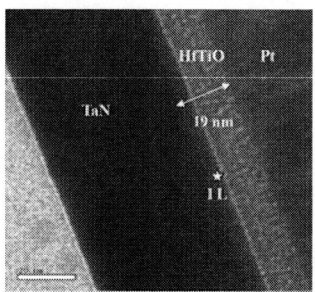

Fig.1 Cross-section TEM micrograph of the Pt/HfTiO/TaN MIM capacitor. The bottom interfacial layer is TaTiO.

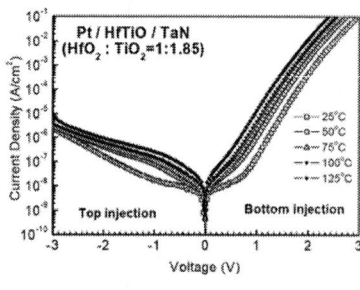

Fig.5 Current density - Voltage (J-V) characteristics of the Pt/HfTiO/TaN MIM capacitor measured at temperatures from 25 to 125°C.

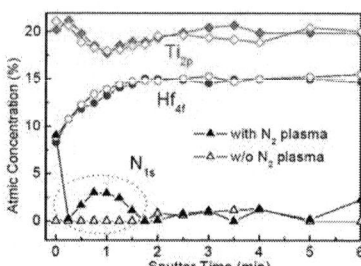

Fig.9 AES depth profile of the HfTiO dielectric with and without N_2 plasma treatment. Nitrogen accumulates at the top 6nm surface.

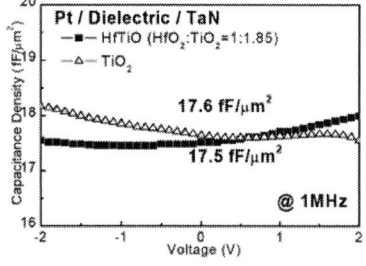

Fig.3 Capacitance density - Voltage (C-V) characteristics of the MIM capacitors using HfTiO or TiO_2 as a insulator.

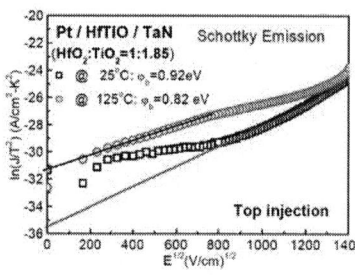

Fig.6 Current transport mechanism examination of the Pt/HfTiO/TaN MIM capacitors at 25 and 125°C. Schottky emission seems the dominant mechanism at medium bias.

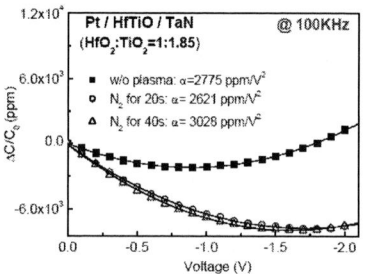

Fig.10 Voltage linearity characteristics of the MIM capacitors with various post HfTiO deposition N_2 plasma treatment. The very close VCC-α values indicate that bulk traps dominate the voltage linearity.

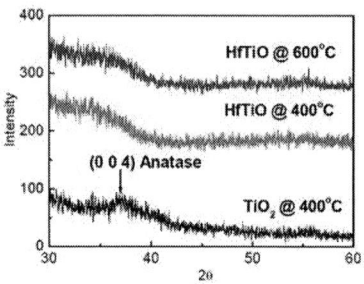

Fig.2 XRD patterns of the TiO_2 dielectric annealed at 400°C and the HfTiO dielectric annealed at 400 and 600°C.

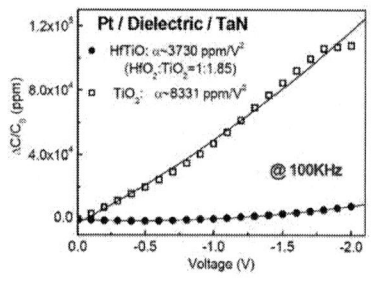

Fig.7 Voltage linearity characteristics of the MIM capacitors with HfTiO and TiO_2 as insulator. The HfTiO capacitor exhibit much lower VCC-α than the TiO_2 one.

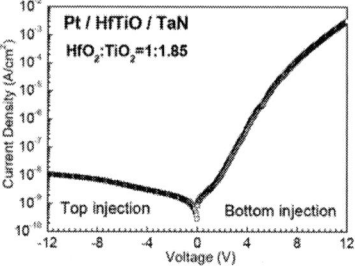

Fig.11 Current density - Voltage (J-V) characteristics of the Pt/HfTiO(51nm)/TaN MIM capacitor. The leakage current density is only 1.3nA/cm² at -1V.

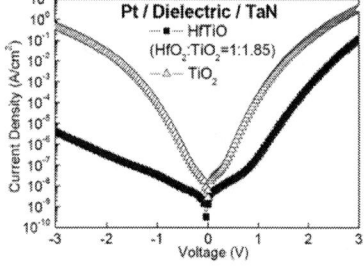

Fig.4 Current density - Voltage (J-V) characteristics of the MIM capacitors using HfTiO or TiO_2 as insulator. The HfTiO sample exhibits 5 orders of magnitude lower leakage current at -2V.

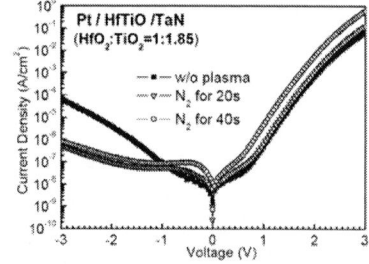

Fig.8 Current density - Voltage (J-V) characteristics of the MIM capacitors with various post HfTiO deposition N_2 plasma treatment. Fifty times improvement is obtained at -2V.

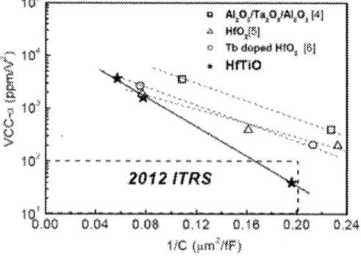

Fig.12 VCC-α verse 1/C plot for various high-κ MIM capacitors. Our results satisfy the requirement of RF/analog ICs in 2012. The exponential decrease with increasing 1/C is important to design capacitors for different applications.

978-1-4244-2784-0/09 $25.00 © 2009 IEEE

High-K/ Metal-Gate Stack Work-function Tuning by Rare-Earth Capping Layers: Interface Dipole or Bulk Charge?

H.Y. Yu[1], S.Z. Chang[2], M. Aoulaiche[3], B. Kaczer[3], P. Absil[3], C. Adelmann[3], T.Hoffmann[3], S. Biesemans[3], C.Wann[2], Y.J. Mii[2]

[1]School of EEE, Nanyang Technological University, 50 Nanyang Avenue, Singapore, 639798, email: hyyu@ntu.edu.sg;

[2]TSMC, 8, Li-Shin Rd., 6, Hsinchu Science Park, Hsinchu, Taiwan, R.O.C; [3]IMEC, Kapeldreef 75, B-3001 Leuven, Belgium

Abstract

The transistor V_T tuning mechanism in metal-gate/high-k (MG/HK) gate stack doped with rare-earth elements (Dysprosium or Dy in this work) is studied in transistors fabricated by either a gate-first or a gate-last approach. Except the commonly believed interface dipole, this work provides additional evidence that the bulk trapping charges can also play an important role in determining the device V_T for above-mentioned gate stacks. It is thus suggested that careful design of capping layer thickness as well as the thermal budget for intermixing the capping layer with host dielectrics are necessary to eliminate the impact from bulk trapping charges to the device performance.

Introduction

A gate-first approach with La-family (Dy, La, and so on) elements in gate stacks successfully demonstrates low-V_T metal-gate/high-κ n-FETs [1,2]. It is commonly believed that the work function tuning is due to the interface-dipole formations stemming from the addition of La-family element [3,4]. Recently, bulk trapping induced transient charging effect is reported to lead to the abnormal V_T shifts during the bias-temperature-instability (BTI) stress [5] in the La (or Dy)-containing MG/HK gate stacks. In this work, by fabricating the n-FETs using both the gate-first and the gate-last approaches, the transistor V_T tunability in terms of interfacial dipoles and bulk trapping charges is investigated, which can also link to the "turnaround" abnormalities we observed in both the transistor V_T and the BTI induced ΔV_T. It is thus suggested that careful design of capping layer thickness as well as the thermal budget for intermixing the capping layer with host dielectrics are necessary to eliminate the impact from bulk trapping charges to the device performance.

Experimental

N-FETs with NiSi-FUSI gate and DyO/SiON dielectric are prepared with 0.5, 1.0, or 2.0nm DyO cap on 1.8nm-thick SiON (dielectrics-first, which undergoes S/D annealing). TiN/HfSiO p-FETs with DyO cap are also prepared in a replacement–gate (RPG) manner (i.e. DyO does NOT receive S/D annealing). 1030°C spike anneal is employed for both FUSI and RPG devices. The gate-stack intermixing is studied by X-ray reflectivity (XRR) analysis. The interface defect density D_{it} is extracted through charging pumping current measurement. PBTI is performed at 110°C by using sense-and-measure technique [6]. ΔV_T relaxation during the recovery period (100s) after each BTI stress is also considered [5,7].

Results and Discussion

Fig.1 shows the I_D-V_G of Ni-FUSI n-FETs with either SiON or SiON capped with 0.5-2.0nm DyO. Significant (~500mV) V_T lowering (i.e. *reduction of the effective work function of NiSi*) is observed after adding 0.5nm DyO cap. However, V_T turnarounds appear and tunability cannot be maintained with DyO-cap thickness > 0.5nm. **Fig.2** shows that the D_{it} is ~ 6E10 cm^{-2} on SiON, and increases to 1-3E11 cm^{-2} after adding 0.5-2.0nm DyO

caps. D_{it} seems to be saturated at this point. Meanwhile, obvious I_D-V_G hysteresis and crossover are observed on 2.0nm DyO-capped devices, by sweeping V_G back and forth from 0 to 2V (**Fig.3**). The XRR spectra and schematics of best fitting models of Poly-Si/DyO/SiO$_2$/Si stack before/after annealing are shown in **Fig.4**. No significant reaction between DyO and poly-Si electrode (before FUSI electrode formation) is observed, and it is seen ~2nm DySiO$_x$ is formed by consuming the respective DyO and SiO$_2$ each of 1nm after 1035°C annealing.

Fig.5 summarizes the PBTI induced V_T shift vs. stress time under different V_G-V_T bias. Both negative ΔV_T (abnormal PBTI [5]) and ΔV_T turnaround are observed, depending on cap thickness, stress field, as well as the stress time. Negative ΔV_T of ~ −25 and −100mV after 2000s stress are observed on the 1.0 and 2.0nm DyO-capped n-FETs at low stress fields (V_G-V_T = 0.8V). ΔV_T turns back to positive values when the stress field/stress time increases. Meanwhile, ΔV_T relaxation during the stress recovery period is also examined (**Fig.6**). Negative ΔV_T in the PBTI stress phase is believed to correlate with electrons de-trapping from the pre-existing negatively charged defects, while positive ΔV_T in the recovery phase is related to electrons trapping-back [5].

Fig.7 shows the conceptual integration flow/TEM of the RPG (gate-last) process. DyO (1.0nm) and TiN are deposited on existing HfSiO after Poly-Si gate removal. P-FET's I_D-V_G and C-V curves in **Fig.8** show that the adding of DyO-cap with low thermal budgets (much less than the required one for intermixing) results in ~4A increase T_{inv} and 160mV increase in V_{fb} (or equivalently *an increase of TiN EWF* by 160meV), similar to the thick DyO-cap FUSI case (**Fig.1**). Combining **Fig.1** to **Fig.8**, for Dy containing gate stack, we believe that the V_T tuning (e.g. 0.5nm-cap case in **Fig 1**) is related to the interface dipoles triggered by the thermally induced dielectric intermixing. While in DyO itself (before intermixing with host dielectrics), bulk negatively trapping charges (**Fig.5**) could exist. If with limited thermal budgets applied (i.e. DyO is not fully converted to Dy-silicate), the bulk charge effects from DyO become visible. As a result, V_T/V_{fb} shift back to p-type band edges (**Fig.1 and 8**), and trapping/de-trapping occurs in the BTI measurement. Hence the optimization in thermal budget and cap thickness is critical for low-V_T device applications. **Fig.9** illustrates the phenomena.

Conclusion

This work suggests both interface dipoles and bulk trapping charges contribute to the V_T for Dy-containing gate stacks. The optimization in both thermal budget and cap layer (i.e. DyO, LaO) thickness determines V_T and BTI characteristics of MG/HK FETs.

Acknowledgement: Yu HY acknowledges support from NAP-SUG of NTU.

References: [1] S. Kubicek et al., IEDM, p.49 (2007); [2] V. Narayanan et al., Tech. VLSI, 22.2 (2006); [3] P. Sivasubramani et al., Tech. VLSI, p.68 (2007); [4] K. Iwamoto et al., Tech. VLSI, p.70 (2007); [5] S.Z. Chang et al., Sym VLSI Tech, pp.7.1 (2008); [6]. B. Kaczer et al., Tech. IRPS, p. 381 (2005); [7] T. Grasser et al., IEDM, p.801 (2007)

978-1-4244-2784-0/09 $25.00 © 2009 IEEE

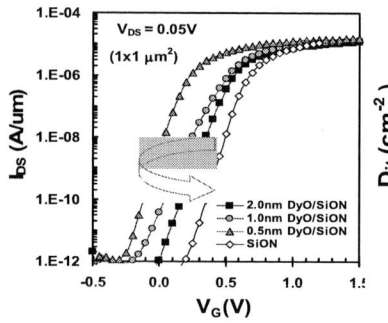

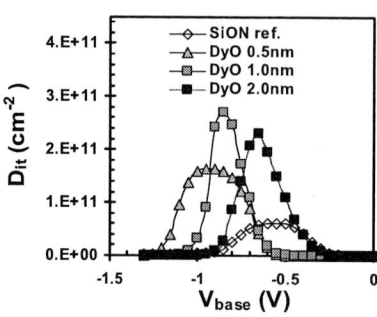

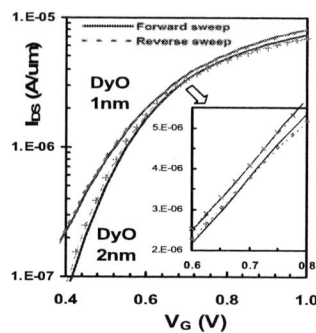

Fig.1 I_D-V_G characteristics for the Ni-FUSI n-FETs with either SiON or SiON capped 0.5, 1.0 and 2.0nm DyO. V_T turnaround happens as the capping thickness increases to 1.0 and 2.0nm.

Fig.2 D_{it} obtained from charge-pumping current measurements on DyO-capped n-FETs (W/L=10/1um^2) at 1MHz. V_{amp} keeps at 1V during the measurement. D_{it} seems to be saturated at DyO >1nm.

Fig.3 Hysteresis and crossover in I_D-V_G are observed on both the 1.0 and 2.0nm DyO-capped n-FETs. Negative Vt-shift at lower V_G range while positive Vt-shift at higher V_G range, respectively.

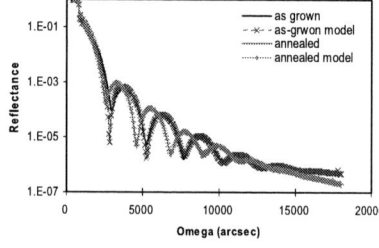

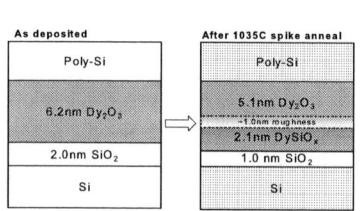

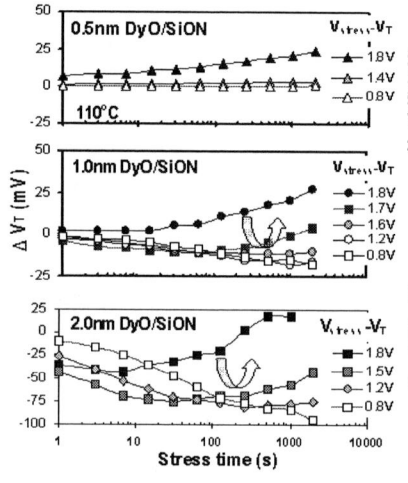

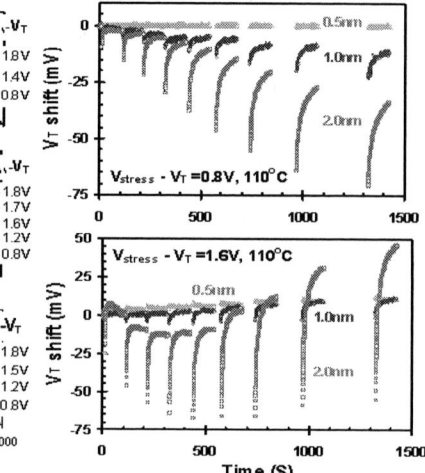

Fig.4 XRR measuring and fitting spectra for Poly-Si/DyO (6nm)/SiO$_2$ (2nm)/Si gate stacks before and after 1035°C spike anneal. Poly-Si gate then transfer to NiSi during the silicidation step. Schematics of best fitting models are shown. There is no Poly-Si/DyO intermixing after annealing (no interface broadening)

Fig.5 PBTI V_T shift vs. stress time under different V_G stress conditions for the Ni-FUSI n-MOSFETs (W/L=10/1um) with 0.5, 1.0 and 2.0nm DyO cap layers on SiON. ΔV_T turnaround at higher stress filed or longer stress time occurs when DyO cap > 1.0nm.

Fig.6 V_T shift vs. time during the recovery periods (100s) after each PBTI stress, for Ni-FUSI n-FETs with DyO cap 0.5-2nm. V_G=V_{stress}, V_D= V_S = 0V during the stress period; V_G=V_T, V_D= 50mV, and Vs= 0V during the recovery period.

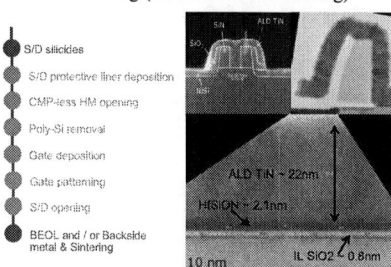

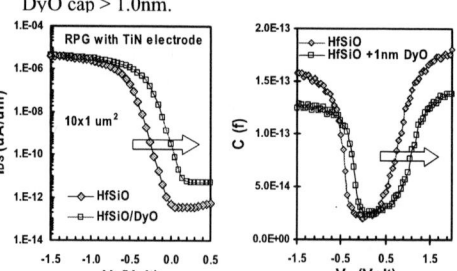

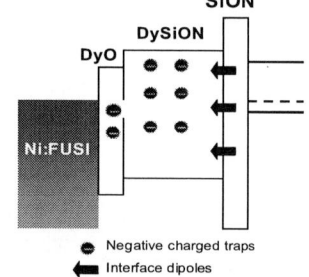

Fig. 7 Simplified RPG flow with CMP-less self-aligned gate opening. The poly-Si removal stops on already present HfSiO. XSEM (XTEM) of the fabricated device with good trench fills and 20nm thick TiN (ALD). In some cases a DyO capping layer was deposited prior to the metal deposition.

Fig.8 Both I_D-V_G and C-V curves on RPG p-FETs with DyO (1.0nm) cap layers deposited on already present HfSiO. T_{inv} and V_{fb} increases ~4A and ~160mV, respectively, under low thermal budget.

Fig.9 Schematic diagram illustrates that both interface dipoles and bulk trapping charges exist in a non-optimized Ni-FUSI/DyO/SiON stack, and the bulk charges shift V_T back to high values.

978-1-4244-2784-0/09 $25.00 © 2009 IEEE

Scaling Challenges of MOSFET for 32nm Node and Beyond

Yasuo Nara

Fujitsu Microelectronics Ltd.

1500, Mizono, Tado, Kuwana, Mie 511-0192

Japan

ABSTRACT

Scaling challenges for MOSFET fabrication process with design rule of 32nm and below will be reviewed. This paper will especially focus on the scaling issues of conventional planar bulk CMOS technology and discuss about multiple stress engineering, junction engineering and high-k/metal gate stack as key technology boosters to enhance CMOS performance with scaled dimensions.

INTRODUCTION

In order to realize high-performance, low-power, and low-cost CMOS devices, shrinkage of device size is essential but not sufficient, because the simple scaling is reaching the limit in terms of CMOS performance improvement. Therefore, in addition to the scaling of device size, introduction of new technologies such as strained channel, metal gate, high-k gate dielectrics, thin body SOI, and multi-gate transistor (FinFET) is extensively studied. From recent reports on scaled CMOS for 32nm node [1-4], conventional planar bulk technology has been still utilized and may continue to be the most probable integration approach for next generation especially for low-cost application. Thus, in this presentation, process technology boosters on planar bulk technology, such as stress engineering, shallow junction technology and high-k/metal gate stack, is focused and recent topics including our results are reviewed.

MULTIPLE STRESS ENGINEERING

In order to boost carrier mobility and hence MOSFET performance, stress introduction is believed to be one of the most efficient technique. Embedded SiGe (eSiGe) formation in source/drain region of pMOSFET, tensile and compressive SiN stress liner (tCESL and cCESL) formation for n- and pMOSFET, and stress from poly-Si gate (poly-Si stress: PSG or stress memorization technique: SMT) are recognized as standard method. Recently, combination of these mobility enhancement technologies (multiple stress engineering) was introduced to further enhance the CMOS performance. Figure 1 shows the CMOS structure with such multiple stress engineering [5, 6]. Two kinds of PSG were applied to nMOSFET's fabrication process; one is after offset spacer formation (OS-PGS) and the other is after source/drain ion-implantation (SD-PGS). By optimizing these PGS process, 23% improvement of nMOSFET's Ion-Ioff performance was successfully obtained as shown in Fig. 2. Id-Vd characteristics with multiple stress engineering is shown in Fig. 3, showing competitive high performance nMOSFET and pMOSFET drive currents of 1.22/0.765 mA/μm at 100 nA/μm off-current.

ULTRA SHALLOW JUNCTION FORMATION

The demand for ultra-shallow junction (USJ) formation with high dopant activation, high profile abruptness, and low contact resistance between doped-Si and NiSi has been greatly increasing with the scaling of CMOS.

For the activation of ion-implanted dopants, various kinds of annealing methods including millisecond annealing have been proposed. Among them, flash lamp annealing (FLA) is a promising candidate for future annealing techniques because of its diffusion-less and high-dopant-activation features. It has already been confirmed that sheet resistance vs. junction depth characteristics by FLA were superior to the conventional spike RTA or solid phase epitaxial regrowth (SPER) [7]. It is, however, reported that application of FLA to high-k/metal gate device degrades carrier mobilities and NBTI (negative bias temperature instability) reliability, probably due to the increase of electron trapping site in high-k during FLA treatment. In order to recover from such degradation, additional thermal annealing (recovery annealing) is necessary, but this degrades shallow junction characteristics. Thus, new FLA called flexibly-shaped pulse FLA (FSP-FLA) is being considered [8]. FSP-FLA has a feature with new pulse shape in which high-power short pulse (0.8ms) and following low-power long pulse (approx. 10ms) are combined. By using FSP-FLA, electron mobility and NBTI degradation were improved to the level with conventional FLA and recovery anneal without deteriorating shallow junction characteristics.

Reduction of contact resistance between shallow junction and silicide (NiSi) is also one of the concerns for source and drain formation of scaled CMOS. One possible approach for this issue is to control schottky barrier height between doped-Si of source/drain region and NiSi, and this was realized by ultra shallow Al implant into NiSi layer (Fig. 4) [9]. Figure 5 shows the increase of on-current of nMOSFET using such Al implant into NiSi. Ion gain was significant especially for short gate length and more than 50% improvement has been obtained with Lg=45nm. It seems to be caused by the decrease of parasitic contact resistance between n$^+$ source/drain region and NiSi due to reduced schottky barrier height.

HIGH-K/METAL GATE CMOS

Metal gate/high-k stack is required for scaled MOSFETs in order to achieve low gate leakage current and high current drivability at the same time. One of its biggest problems is threshold voltage (Vth) control. Dual metal gates are proposed so far [10, 11], but they still suffer from insufficient Vth control after high temperature annealing required in a standard gate-first fabrication. Here, integration scheme with single metal/dual high-k (SMDH) gate stack for aggressively scaled gate-first CMOSFETs is discussed. The basic device concept is to control the Vth by two different high-k, but to use a single metal gate for precise control of the gate profile as shown in Fig. 6. Individual Vth control for n- or pMOSFET by high-k is reported [12-13], but our focus is on controlling both with a common metal gate for scaled CMOSFETs [14]. SMDH CMOSFET was fabricated with Al_2O_3-containing high-k for pMOSFET, MgO-containing high-k for nMOSFET and common stack formation with TiN metal gate. NMOSFETs with PVD-TiN/PVD-MgO/ALD-HfSiO were fabricated in a standard gate-first flow with a spike annealing at 1000°C. Flat-band voltage (Vfb) – equivalent oxide thickness (EOT) trend extracted by CV characteristics showed a sufficient negative Vfb shift by MgO incorporation. For pMOSFET, it is clarified that the Al_2O_3 insertion position is key to effectively control Vth and found that Al_2O_3 under HfSiO could shift Vth about 0.15V. By taking advantage of above dual high-k with a common TiN gate, low Vth operations for n and pMOSFETs with Lg=100nm are successfully demonstrated.

978-1-4244-2784-0/09 $25.00 © 2009 IEEE

SUMMARY

Key process technologies for scaling planar CMOS for 32nm and beyond have been discussed. Recent progress for multiple stress engineering, shallow junction technology, and high-k/metal gate stack have been reviewed. Combination of these technologies will boost the planar CMOS performance even for 32nm node and beyond.

Acknowledgement

The author would like to express thanks to Mr. T. Miyashita and Dr. H. Fukutome in Fujitsu Microelectronics Ltd., and members of Semiconductor Leading Edge Technologies (Selete) for their support for this presentation.

REFERENCES

[1] C. D. Diaz et al., Technical Digest of IEDM, p.629, 2008.
[2] F. Arnaud et al., Technical Digest of IEDM, p.633, 2008.
[3] S. Hasegawa et al., Technical Digest of IEDM, p.938, 2008.
[4] S. Natarajan et al., Technical Digest of IEDM, p.941, 2008.
[5] T. Miyashita et al., Technical Digest of IEDM, p.251, 2007.
[6] T. Miyashita et al., Technical Digest of IEDM, p.55, 2008.
[7] S. Kato et al., Extended Abstract of IWJT, p.143, 2007.
[8] T. Onizawa, et al., Symposium on VLSI Tech., p.110, 2008.
[9] H. Fukutome et al., Technical Digest of IEDM, p.59, 2008.
[10] Z. B. Zhang et al., Symposium on VLSI Tech., p.50, 2005.
[11] F. Ootsuka et al, Extended Abstract of SSDM, p. 1116, 2006.
[12] V. Narayanan et al., Symposium on VLSI Tech., p.224, 2006.
[13] H.N. Alshareef et al., Symposium on VLSI Tech., p.10, 2006.
[14] N. Mise et al., Technical Digest of IEDM, p.527, 2007.

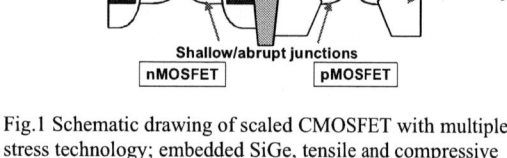

Fig.1 Schematic drawing of scaled CMOSFET with multiple stress technology; embedded SiGe, tensile and compressive SiN stress liner (tCESL and cCESL), and multiple poly-Si stress.

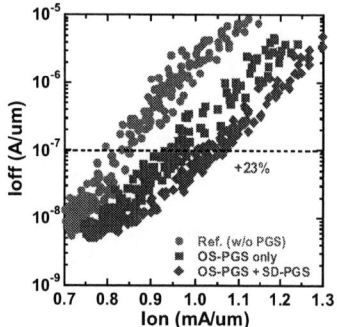

Fig.2 NMOS Ion-Ioff characteristics using multiple PGS.

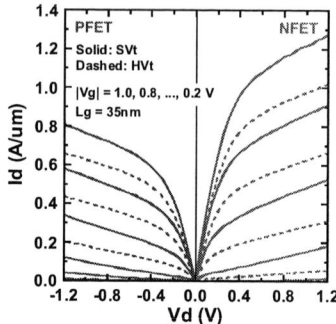

Fig.3 Id-Vg characteristics of CMOS using multiple stress technology.

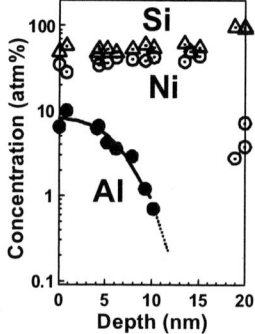

Fig.4 Al depth profile with ultra shallow Al implant into NiSi layer.

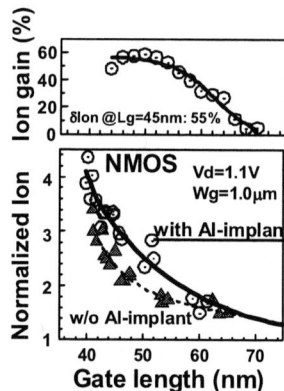

Fig.5 Ion gain vs. gate length. Effect of reducing parasitic resistance is dominant for short gate length.

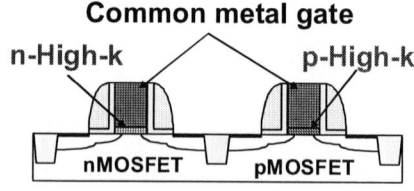

Fig.6 Concept of single metal/dual high-k (SMDH) CMOS for low Vth high-k/metal gate integration.

978-1-4244-2784-0/09 $25.00 © 2009 IEEE

P-FinFETs with Al Segregated NiSi/p^+-Si Source/Drain Contact Junction for Series Resistance Reduction

Mantavya Sinha, Rinus T. P. Lee, Sivasubramaniam Nandini Devi, Guo-Qiang Lo*, Eng Fong Chor, and Yee-Chia Yeo.

Department of Electrical and Computer Engineering, National University of Singapore (NUS), 117576 Singapore.
* Institute of Microelectronics, 11 Science Park Rd, 117685 Singapore.
Phone: +65 6516-2298, Fax: +65 6779-1103, E-mail: yeo@ieee.org.

ABSTRACT

This paper demonstrates the integration of Al segregated NiSi/p^+-Si S/D contact junction in p-FinFETs for parasitic series resistance reduction. Al is introduced by ion implant into p^+ S/D region followed by nickel deposition and silicidation. Drive current enhancement of ~15 % is achieved without any degradation of short channel effects. This is attributed to the lowering of $\Phi_B{}^p$ of NiSi on p-Si from 0.4 eV to 0.12 eV with low Al dose of 2×10^{14} atoms-cm^{-2}, leading to lowering of contact resistance at NiSi/p^+-Si S/D junction.

INTRODUCTION

FinFETs (or Multiple gate transistors) are promising for possible adoption beyond the 22 nm technology node due to excellent control of short-channel effects (SCEs) [1,2]. The primary integration challenge in FinFETs is the formation of narrow Si-fins with low parasitic series resistance (PSR) [3]. For FinFETs, the NiSi/(heavily doped Si) S/D region contact resistance is the biggest contributor to PSR, which gets further amplified at shorter fin-widths [3]. Recently we have reported the tuning of the effective Schottky barrier height of holes ($\Phi_B{}^p$) for NiSi on p-Si using aluminum (Al) implant [4], which is promising for drive current enhancement in p-FinFETs. While S and Se implant have been shown to be effective in tuning the effective Schottky barrier height of electrons ($\Phi_B{}^n$) for NiSi on n-type Si and have been demonstrated successfully in n-FETs [5], there have been no reports of an analogous integration on p-FETs.

In Part I of this paper, we report the extensive material results for the modulation of the $\Phi_B{}^p$ of NiSi on p-Si using Al implant into p-Si, and its segregation at NiSi/p-Si interface after Ni deposition and silicidation. Optimum process conditions are extracted for integration of the Al segregated NiSi/p^+-Si S/D contact junction into S/D of p-FinFETs, to yield significant drive current enhancement. Part II of the paper discusses the electrical device results of the p-FinFETs with and without Al implant.

EXPERIMENT

To prepare NiSi/p-Si junctions, 200 mm p-type Si (100) wafers were patterned using thermal oxide, and then implanted with Al at an energy of 10 keV with dose ranging from $2\times10^{14} - 2\times10^{15}$ atoms-cm^{-2}. 10 nm of Ni was deposited and one step silicidation was done at 500 °C for 30 s to form NiSi. The unreacted Ni was removed using sulphuric-peroxide mixture (SPM) at 120 °C for 60 s.

RESULTS AND DISCUSSIONS

I. Material Characterization

SIMS depth profile in Fig. 1 shows that after nickel silicidation Al segregates at the NiSi/p-Si interface and lowers the $\Phi_B{}^p$ of NiSi on p-Si, as seen by the ohmic I-V characteristics of NiSi/p-Si junctions (inset of Fig. 1). $\Phi_B{}^p$ is extracted using low temperature I-V

measurements (Fig. 2) to be 0.12 eV at Al dose of 2×10^{14} atoms/cm^2 (inset of Fig. 2), a drop of ~70 % from the reference sample without Al implant. This is extremely promising for p-FinFETs device performance. With increase in Al implant dose, $\Phi_B{}^p$ saturates at slightly higher value 0.14 eV. Thus, the low Al dose of 2×10^{14} atoms-cm^{-2} is used for integration into p-FinFETs. Al is known to introduce negatively charged acceptor type trap levels in Si bandgap near the valence band which will generate an electric dipole at the NiSi/p-Si interface [4] (Fig. 3). This leads to the lowering of $\Phi_B{}^p$ due to enhanced hole tunneling through thinner Schottky barriers.

II. Device Integration and Electrical Results

Fig. 4 details the process flow for p-FinFETs with Al segregated NiSi/p^+-Si S/D contact junction. 200-mm SOI substrates with 40 nm thick Si and 140 nm thick BOX were used in the device fabrication process. Fin width (W_{FIN}) down to 55 nm were defined using 248 nm lithography, resist trimming and reactive ion etching. After gate stack (poly-Si on 3.0 nm SiO$_2$) and spacer formation, and deep S/D implant and activation, Al was implanted at dose of 2×10^{14} atoms-cm^{-2}, followed by Ni salicidation. Cross-sectional TEM image of the fabricated p-FinFETs is shown in Fig. 5.

I_{DS}-V_{DS} characteristics of a pair of p-FinFETs (having gate length L_G of 170 nm and W_{FIN} of 55nm) is shown to have an I_{DSAT} enhancement of 14 % with Al implant. (Fig. 6) The two devices are closely matched with similar OFF-current, DIBL (0.03V/V), and subthreshold swing SS (100 mV/decade), as extracted from their I_{DS}-V_{GS} characteristic (inset of Fig. 6). On average, the drive current enhancement obtained with Al implant at an I_{OFF} of 100 nA/μm is 15 % as shown in Fig. 7. Device performance compared at a fixed DIBL of 100 mV/V and at a fixed SS of 120 mV/decade also show I_{DSAT} enhancement of 14 % (Fig. 8), and 13 % (Fig. 9), respectively, with Al implant.

Statistical data plots in Fig. 10 show that Al implant does not affect device V_{TSAT}, DIBL and SS. Shorter L_G FinFETs are shown to have an higher drive current enhancement, reaching ~20 % for L_G of 170 nm (Fig. 11). This increase in drive current is credited to the lowering of parasitic series resistance from 931 Ω-μm to 814 Ω-μm with Al implant, yielding a drop of 13 % (Fig. 12), which in turn is attributed to the lowering of $\Phi_B{}^p$.

CONCLUSION

We have demonstrated a detailed material study to lower the $\Phi_B{}^p$ at the NiSi/p-Si junction by Al implant to an extremely low value of 0.12 eV, and have integrated it into p-FinFETs to yield a drive current enhancement of ~15 %.

REFERENCES

[1] P. Verheyen *et al.*, *Symp. VLSI Tech.*, p. 194, 2005.
[2] K.-M. Tan *et al.*, *IEEE Elect. Dev. Lett.* **28**, p. 905, 2007.
[3] A. Dixit *et al.*, *IEEE TED* **52**, p. 1132, 2005.
[4] M. Sinha *et al.*, *Appl. Phys. Lett.* **92**, 222114, 2008.
[5] H.-S. Wong *et al.*, *IEEE Elect. Dev. Lett.* **29**, p. 841, 2008.

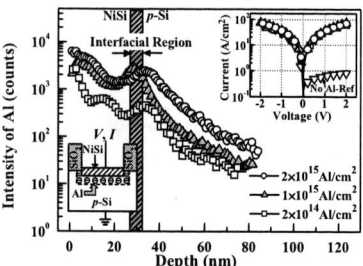

Fig. 1. SIMS analysis reveals Al segregation at the NiSi/*p*-Si interface. Inset on top-right shows ohmic *I-V* characteristics of NiSi/*p*-Si junctions with Al implant.

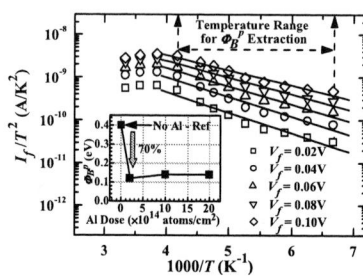

Fig. 2. Activation energy plot to extract Φ_B^p of NiSi on *p*-Si (with Al dose = 2×10^{15} atoms/cm²). Inset shows lowest Φ_B^p of 0.12 eV at Al implant dose = 2×10^{14} atoms/cm².

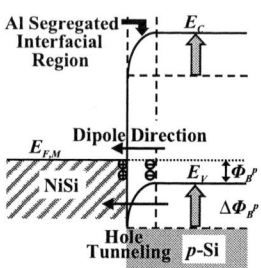

Fig. 3. Schematic of interface dipole generated by Al (owing to acceptor-type trap levels) at the NiSi/*p*-Si interface, leading to lowering of Φ_B^p.

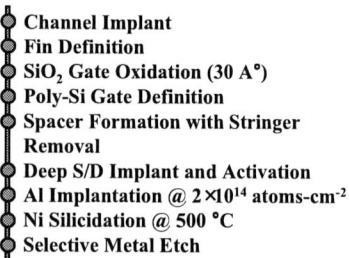

Fig. 4. Key process steps for fabrication of *p*-FinFETs with Al implant at a dose of 2×10^{14} atoms/cm². Control device did not receive any Al implant.

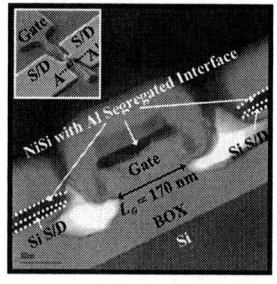

Fig. 5. x-TEM of a *p*-FinFET with 170 nm L_G, after Al implant and Ni silicidation. FIB cut along A-A' (shown in the tilt-SEM in the inset of the figure) overestimates L_G.

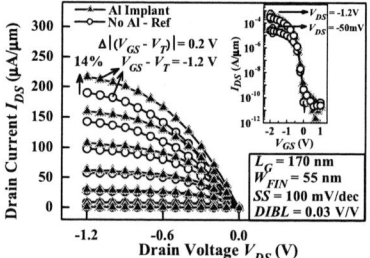

Fig. 6. I_{DS}-V_{DS} characteristics of a *p*-FinFET with Al implant exhibits 14 % higher I_{DSAT} over the control FinFETs. I_{DS}-V_{GS} (inset) shows that the pair of devices have comparable *SS*, DIBL, V_T and OFF-current.

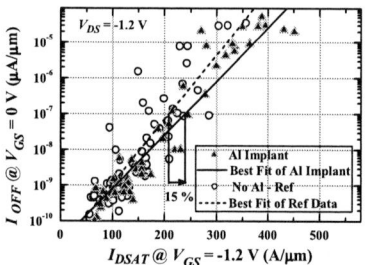

Fig. 7. A 15 % increase in I_{DSAT} is achieved for *p*-FinFETs with Al implant (over the control FinFETs), at I_{OFF} = 100 nA/μm.

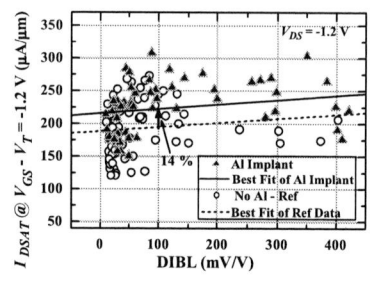

Fig. 8. At a fixed DIBL = 100 mV/V, *p*-FinFETs with Al implant show an I_{DSAT} enhancement of 14 % over control FinFETs.

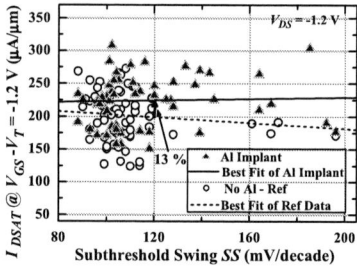

Fig. 9. At a fixed *SS* = 120 mV/decade, *p*-FinFETs with Al implant show an I_{DSAT} enhancement of 13 % over control FinFETs.

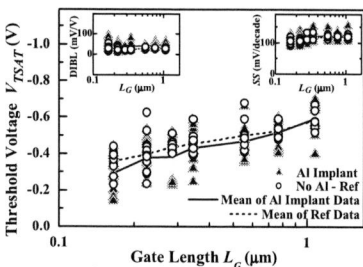

Fig. 10. *p*-FinFETs with and without Al implant have comparable SCEs, e.g. similar V_{TSAT}, V_{TSAT} roll-off, DIBL (inset @ top-left) and *SS* (inset @ top-right).

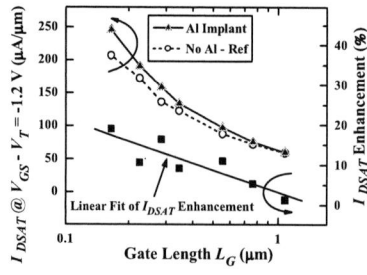

Fig. 11. Mean I_{DSAT} versus L_G showing that I_{DSAT} enhancement with Al implant (over control FinFETs) increases with smaller L_G, reaching ~20 % for L_G = 170 nm.

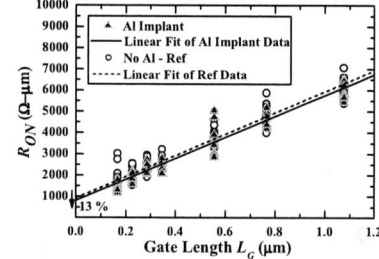

Fig. 12. The parasitic series resistance extracted at zero gate length shows a drop of 13 % for p-FinFETs with Al implant (over the control FinFETs).

978-1-4244-2784-0/09 $25.00 © 2009 IEEE

Impacts of NBTI on SRAM Array with Power Gating Structure

Hao-I Yang, Ching-Te Chuang, and Wei Hwang
Department of Electronics Engineering & Institute of Electronics,
National Chiao-Tung University, Hsinchu, Taiwan, R. O. C.

INTRODUCTION

This paper presents a detailed study on impacts of NBTI on power-gated SRAM arrays based on PTM 32nm CMOS technology model [1]. NBTI effects become more significant with technology scaling, causing PMOS threshold voltage (V_T) drift with stress time as shown in Fig. 1 [1, 2]. Previous papers have reported that SRAM Read-SNM (RSNM) degraded with NBTI, while SRAM Write-Margin (WM) improved [3]. These works, however, focused only on the standard SRAM and cells. Many of the state-of-the-art low-power SRAM designs have employed power-gating techniques to suppress leakages. As such, it is crucial to understand the NBTI degradation of the power-gating structure, in addition to the cell, and the resulting impacts on the power, performance, margin, and wake-up time, etc.. Furthermore, PMOS-type pre-charge circuits are also susceptible to NBTI, thus degrading the bit-line pre-charge level or pre-charge time. Finally, judicious choice of sense amplifier design is shown to be an important consideration in minimizing the NBTI impacts.

NBTI MODEL

The V_T drift of PMOS due to NBTI can be described by DC Reaction-Diffusion (RD) framework when the stress signal doesn't change (i.e. Static stress). If the stress signal changes with time (i.e. Alternating stress), the DC RD model is modified to AC RD model:

$$\triangle V_T(t) = \alpha (S, f) \times K_{DC} \times t^{0.25} \qquad (1)$$

where the prefactor, α, accounts for the signal (stress) probability and recovery mechanism, S, and frequency, f. K_{DC} is a technology dependent parameter and t is time. However, according to experimental results of [5], the impact of the signal frequency on V_T drift is relatively insignificant. Thus, we neglect the effect of signal frequency, and analyze cases with various signal (stress) probabilities. Notice also that NBTI induced V_T drift depends strongly on the V_{GS} bias and temperature, but barely on V_{DS}.

SRAM POWER-GATING STRUCTURE

Fig. 2(a) shows the schematic of a column-based power-gated SRAM, where M3 is the PMOS power switch for Standby leakage reduction and M6 is the clamping device to bias VVDD to proper level for data retention in Standby mode. However, switching of M3 induces virtual supply bounce because large wake-up currents flow through parasitical capacitance, inductance, and resistance of the package (Fig. 2(b)) [4]. The following sections present detailed simulation results based on BSIM Predictive Model PTM 32nm [1], with the AC RD model based NBTI induced V_T drift calibrated against published data [2]. Table I shows 7 simulation cases with different PMOS stress probability of Fig. 2, where nodes N1 and N2 are assumed to be at "logic 0" and "logic 1" respectively. In power-gated SRAM arrays, the power switch should provide sufficient supply voltage and current for SRAM cells to maintain adequate margin and performance during Read and Write operations. As the power switch is constantly under NBTI stress (V_{GS} stress) in Active mode, it has the highest probability of being stressed. Hence, we assume the worst case scenario that the power switch is always stressed. In contrast, the clamping device (diode) is shunted and shorted by the power switch and experiences no NBTI stressing during Active mode. In Standby or Sleep Mode, the stressing voltage (V_{GS}) of the clamping device is about one V_T (MOS diode voltage), thus the NBTI effect on the clamping device is negligible.

SRAM READ/WRITE OPERATION

The RSNM (Read SNM) degradation is dominated by V_T mismatch in a cell induced by NBTI (Fig. 3). Worse RSNM appears with larger V_T drift resulting from longer stress time. Furthermore, RSNM is also impacted by V_T drift of the power switch M3 and signal probability of unselected cells. The reason is as follows: The active-mode VVDD is determined by M3 equivalent resistance and SRAM array equivalent resistance. It varies with V_T drift (stress time) as shown in Fig. 4. If V_T drifts of unselected (loading) cells' PMOS transistor pairs are significant to offset the V_T drift of power switch M3, the active-mode VVDD would remain relatively unchanged. On the other hand, for Case 1, only the power switch M3 suffers NBTI (thus higher V_T and higher resistance), and active-mode VVDD would decrease. Fig. 5 shows the degradation of RSNM after 3 years of usage versus VDD. Notice that SRAM VDD_{MIN} would rise with time of usage with larger NBTI induced V_T drift. Fig. 6 shows that NBTI effects on Read bit-line discharge latency is negligible. This is due to the fact that there is no V_T drift for the access NMOS and driver NMOS transistor of the selected cell. Fig. 7 shows that Read power of SRAM decreases with larger NBTI induced V_T drift. The reason is that leakages through M3 decrease with larger V_T drift.

The Write-Margin (WM) varies with NBTI stress time as shown in Fig. 8. The V_T drift of power switch M3 reduces active-mode VVDD, thus easing writing of the cell. The signal probability and V_T drift of unselected cells also affect active-mode VVDD, thus influencing writing of the cell as well. For the selected cell, when the cell signal (stress) probability is not 100% (0%), both PMOS loading transistors become weaker. Hence, during Write operation, the storage node which originally stores "logic 1" can be easily pulled down to "logic 0". Consequently, Write Margin (WM) improves. However, when the cell signal (stress) probability is 100% (0%), only one PMOS loading transistor becomes weaker. For the worst case pattern, the PMOS holding the original "logic 1" storage node is not stressed/weakened, so the pull down of the "logic 1" storage node is not getting easier. The PMOS corresponding to the original "logic 0" storage node, however, would be fully stressed/weakened, and thus slowing down the charging of its storage node to "logic 1" during Write operation. As a result, the WM degrades. This worst case only slightly degrades the WM as can be seen in Fig. 8 as its effect is compensated by the reduction of the active-mode VVDD. Write delay decreases with larger V_T drift due to better WM of the selected cell (Fig. 9). Moreover, the write power decreases with larger V_T drift of M3 (Fig. 10).

SRAM WAKE-UP TRANSITION

Fig. 11 shows the supply bounce during SRAM mode transition. It is dominated by V_T drift of the power switch, and decreases with larger V_T drift as wake-up currents decrease with larger V_T drift. Virtual supply bounce is slightly affected by signal probability of cell array. The wake-up time increases with larger V_T drift of the power switch as shown in Fig. 12. This figure also implies that the decrease of wake-up current is more significant when the unselected (loading) transistor pairs are also stressed.

PRE-CHARGE CIRCUIT AND SENSING AMPLIFIER

PMOS-type pre-charge circuits (Fig. 2(a)) are impacted by NBTI. BL pairs may be pre-charged to a lower level in a fixed pre-charge phase when V_T drifts are large, leading to amplifier sensing error. According to Fig. 13, pre-charge circuits with longer pre-charge phase, although facing longer NBTI stress and potentially larger V_T drifts, have better ability to tolerate performance degradation induced by NBTI.

Finally, we compare two basic sense amplifier structures. These two structures would have similar performance were it not for the NBTI effects. PMOS pairs of AMP_A (Fig. 14(a)) are under NBTI stress all the time; while PMOS pairs of AMP_B (Fig. 14(b)) are only stressed when it senses data. Therefore, NBTI induced V_T drifts on AMP_A is significantly larger than AMP_B as shown in Fig. 15.

CONCLUSIONS

We have analyzed impacts of NBTI on power-gated SRAM arrays in terms of RSNM, WM, power, performance, and wake-up time. We also studied PMOS-type pre-charge circuit degradation, and compared two basic sensing amplifier structures when they were under NBTI stress.

Our results indicated that V_T drift of power switch degraded RSNM but improved WM in power-gated SRAM. Signal probability of unselected cells also impacted SRAM RSNM and WM. The leakage currents and virtual supply bounce were reduced, but wake-up time became longer. Longer pre-charge phase and judicious choice of sense amplifier structure would improve the tolerance to NBTI effects.

ACKNOWLEDGEMENTS

This research is supported by National Science Council, R.O.C., under project NSC 96-2220-E-009-027. This work is also supported by Ministry of Education, R.O.C., under the project 5Yr5B. The authors would like to thank ITRI, TSMC, and Ministry of Economic Affairs for their support.

REFERENCES

[1] http://www.eas.asu.edu/~ptm/.
[2] S. Zafar, et al., *Symp. VLSI Tech.*, pp. 23-25, 2006.
[3] K. Kang, et al., *IEEE TCAD*, pp1770-1781, Oct. 2007.
[4] S. Kim, et al., *ISLPED*, pp. 22-25.
[5] R. Fernandez, et al., *IEDM*, Dec. 2006.

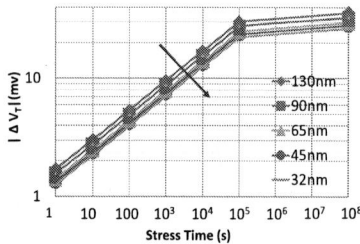

Fig. 1. V_T drifts induced by NBTI in different technology nodes based on Reaction-Diffusion (RD) model calibrated with published data [1].

Table I Stress Prob. of PMOSs

	Selected Cell		Power Gate	Unselected Cells	
Stress Prob.	M1	M2	M3	M4	M5
Case1	0	0	1	0	0
Case2	0.25	0.75	1	0.25	0.75
Case3	0.75	0.25	1	0.25	0.75
Case4	0.5	0.5	1	0.25	0.75
Case5	0.25	0.75	1	0.5	0.5
Case6	0.75	0.25	1	0.5	0.5
Case7	0.5	0.5	1	0.5	0.5

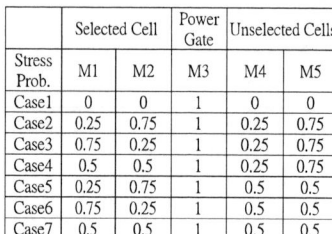
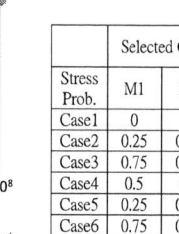
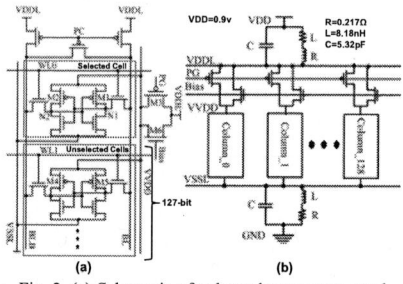

Fig. 2. (a) Schematic of column-base power-gated SRAM, and (b) SRAM and package model.

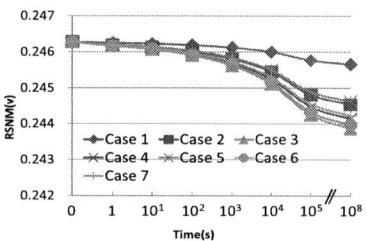

Fig. 3. RSNM decrease with NBTI stress time.

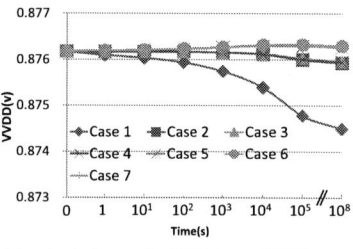

Fig. 4. Active-mode VVDD vs. NBTI stress time.

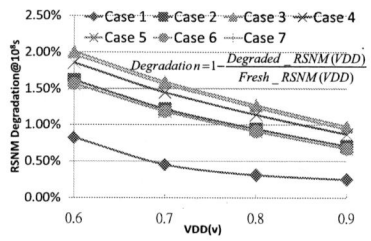

Fig. 5. RSNM degradation after 3 years vs. VDD.

$$Degradation = 1 - \frac{Degraded\_RSNM(VDD)}{Fresh\_RSNM(VDD)}$$

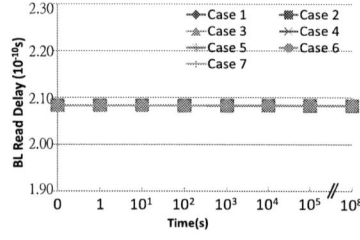

Fig. 6. BL Read delay relatively insensitive to NBTI stress.

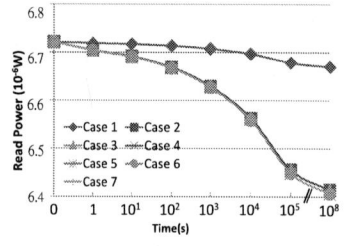

Fig. 7. Read power decrease with NBTI stress time.

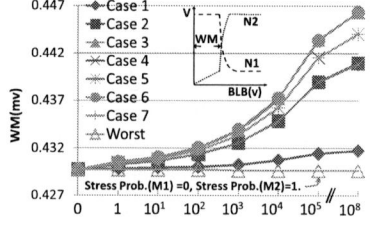

Fig. 8. WM vs. NBTI stress time.

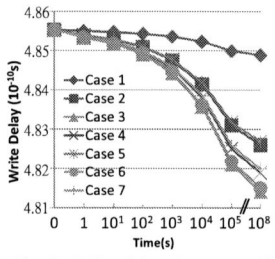

Fig. 9. Write delay decrease with NBTI stress time.

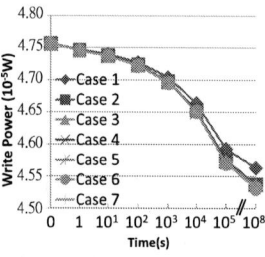

Fig. 10. Write power decrease with NBTI stress time.

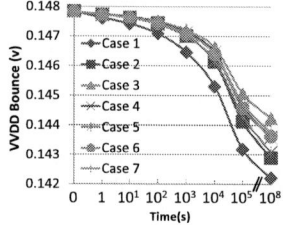

Fig. 11. VVDD bounce during wake-up transition decrease with NBTI stress time.

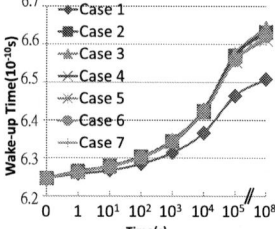

Fig. 12. Wake-up time increase with NBTI stress time.

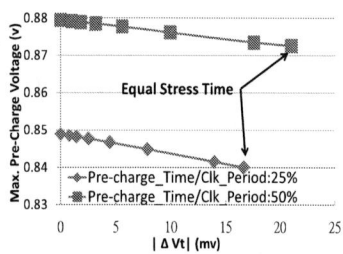

Fig. 13. Bit-line pre-charge voltage level decrease with NBTI stress time.

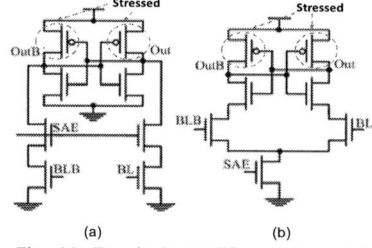

Fig. 14. Two basic amplifier structures (a) AMP_A, and (b) AMP_B.

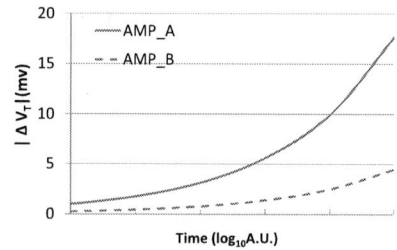

Fig. 15. V_T drifts of AMP_A is more serious than AMP_B with NBTI stress time.

978-1-4244-2784-0/09 $25.00 © 2009 IEEE

CMOS Technology Roadmap Projection Including Parasitic Effects

Lan Wei[*], Frédéric Boeuf[+], Thomas Skotnicki[+], H.-S. Philip Wong[*]

[*] {lanw, hspwong}@stanford.edu, Stanford University, Stanford, California, USA
[+] {frederic.boeuf, thomas.skotnicki}@st.com, STMicroelectronics, Crolles, France

ABSTRACT

In this paper, we revisit the Si CMOS roadmap projection by taking into consideration the parasitic capacitances, which significantly affect the device performance beyond 32nm technology. Capacitance components are analytically modeled and different design rules are examined.

INTRODUCTION

Historically, the delay of CMOS devices is benchmarked by the intrinsic delay $C_{gc}V/I$, where C_{gc} is the intrinsic gate-to-channel capacitance, scaled with the gate length[1]. With device scaling down to 32nm technology and beyond, parasitic capacitances, which are also charged/discharged during the device switching, significantly affect the actual device delay [2][3]. Thus, it is necessary to include the parasitic capacitance and modify the delay expression to $C_{tot}V/I$, where C_{tot} is the total load capacitance including both intrinsic and parasitic components. Analytical capacitance models for bulk CMOS, fully-depleted Silicon-on-Insulator devices ("FDSOI") and double gate devices ("DG") are introduced at the device level. The capacitance models are incorporated into the I-V calculator, MASTAR [4], for the estimation of FO1 and FO3 delay of the inverter chains. The efficacy of various device structure engineering is also evaluated including the parasitics.

Capacitance Modeling

Fig 1 shows the three types of devices studied in this paper: bulk CMOS, FDSOI and DG. These are the most promising candidates proposed by ITRS Roadmap [1]. The models for different capacitance components (Fig 2), are summarized in Table 1. Briefly, the parallel plate capacitance model is used to model charge-sheet capacitances, such as the gate-to-channel capacitance (C_{gc}) and overlap capacitance (C_{ov}). Conformal mapping is applied for fringe type of capacitance, such as outer-fringe capacitance (C_{of}), inner-fringe capacitance (C_{if}) and corner capacitance (C_{corner}) [5][6]. All device-related data are taken from the ITRS 2007 edition, and we used the ITRS device parameters as a starting point (Table 2) [1].

Inverter Chain Load Capacitance

FO1 and FO3 delays of inverter chains are widely adopted figures of merit for device delay benchmarking [4]. It is necessary to be aware of which capacitance components are affecting the delay. In Fig 3, the driving stage switches the loading stage, during the 50% input level to 50% output level. The current charges/discharges the capacitances looking into the drain node (C_d) of the driving stage, and those looking into the gate node of the loading stage, which switches between on and off (C_{g_on} and C_{g_off}). Averaging over the fall-to-rise and rise-to-fall delays, the effective load capacitance is expressed by (1). Here, M is the Miller effect coefficient. Interconnect capacitance is not included. By implementing (1) within MASTAR program, the delay for nominal cases with parasitic capacitances is illustrated in Fig 4. The intrinsic CV/I is way below the actual FO1 or FO3 delay, however, it reasonably reveals the speed-up per generation. This is because for the nominal cases, the device is proportionally scaled in all three dimensions. By switching to DG at 16nm and 11nm, the parasitic effects are relieved.

Selectively Scaling of Device Dimensions

For the same device structure, scaling down is less and less effective to speed up the devices. As shown by Fig 5(a), as devices are scaled down, the delay curve of the bulk devices gets flattened out the fastest, followed by FDSOI, and then DG devices. This is mainly due to the short channel effect and DIBL which degrades the

$$C_{tot} = C_d + (0.25C_{g\_off} + 0.75C_{g\_on}) \cdot FO \qquad (1)$$

$$= (C_{ov} + C_{of} + C_{pcca} + C_{corner}) \cdot M + C_j + \left[\begin{array}{l} 0.25(C_{gb\_off} + 2C_{ov} + 2C_{if} + 2C_{of} + 2C_{pcca} + 2C_{corner}) \\ +0.75(C_{gc} + 2C_{ov} + 2C_{of} + 2C_{pcca} + 2C_{corner}) \end{array} \right] \cdot FO$$

on-current. By relaxing the gate length (as in Table 2), the flatten-out trends retard, giving a delay penalty of less than 10% (Fig 5). While relaxing the gate length increases the intrinsic gate-to-channel capacitance, the short channel effect and DIBL effect are improved, without introducing more parasitic capacitance to the first order. The devices with relaxed gate length are actually faster than or comparable to the nominal devices at 16nm and 11nm for bulk and FDSOI devices. Fig 6 shows the delay dependence on the poly gate edge to contact edge distance (PCCA) at each technology node. A smaller PCCA introduces larger C_{pcca}, with a slight benefit of reducing C_{of}. Reducing PCCA by half increases the delay by 8% - 19%, while increasing PCCA to 1.5x only reduces the delay by <6%. The penalty from reducing PCCA can potentially be compensated at the circuit level, because a smaller device pitch enables shorter interconnect, thus a smaller interconnect RC delay [2]. Gate height (t_{poly}) dependence is revealed in Fig 7. Devices with a lower gate height suffer less from parasitics, because of smaller C_{pcca}, C_{of}, and C_{corner}. Lowering the gate height by half of the nominal case speeds up the devices by 16%, for all technologies from 45nm to 11nm. Traditionally, the device's width statistical distribution is divided into two main families: "large" width (~0.5μm) in logic cells and small width (<0.1μm) in SRAM cells. As logic cell size is scaled down, the average "large" width is becoming smaller and is now close to 0.1μm. As a consequence, the C_{corner} becomes significant. As shown in Fig 8, wide devices, where C_{corner} is negligible, is 35% (45nm) – 58% (11nm) faster than the one with minimum widths. Conventionally, for static circuit design, the ratio of the width of PMOS over NMOS is about 2. There has been intense effort to boost PMOS driving capability [8-9], which can potentially reduce the PMOS device width and thus reduce the total capacitance. Fig 9 shows that 12% (11nm) – 20% (45nm) delay reduction is obtained by boosting PMOS driving capability by 2x, (i.e. same driving capability as NMOS). Boosting PMOS driving capability by 1.5x achieves 2/3 of the improvement of 2x, (7% (11nm) – 14% (45nm)). This improvement diminishes towards more advanced technology because of the disproportionally scaled parasitic capacitances.

Proposed Roadmap

A proposed roadmap is shown in Fig 10. By introducing lower gate heights and better PMOS, the device delay performance is improved by 26% (11nm) – 34% (45nm). Without sacrificing too much of device speed (<8%), relaxing the gate length alleviates variability issues and improves the short channel effect [10]; and reducing the PCCA shortens the interconnect length. A shorter interconnect length improves speed at the circuit level.

CONCLUSION

Starting from the ITRS 2007 edition, we use the MASTAR model to study the device delay including the parasitic capacitances for technology beyond 32nm, where the intrinsic delay $C_{gc}V/I$, which relies mainly on L_{gate} and does not capture the relevant design parameters necessary for MOSFET optimization, is inadequate for device performance evaluation. By selectively engineering the device dimensions (gate height, PCCA, device width, gate length, etc.) and the P/N ratio, the device performance can be boosted by reducing parasitics and preserving electrostatic integrity. A possible improvement of the roadmap is proposed.

978-1-4244-2784-0/09 $25.00 © 2009 IEEE

ACKNOWLEDGEMENT

The authors would like to thank G. Bidal and K. Herve for useful discussion and programming help. L. Wei and H.-S. P. Wong are supported in part by the Focus Center Research Program (FCRP) Center for Circuit and System Solutions (C2S2), and the member companies of the Stanford Initiative for Nanoscale Materials and Processes (INMP). L. Wei is additionally supported by the Stanford Graduate Fellowship.

REFERENCES

[1] ITRS http://www.itrs.net/reports.html
[2] J. Deng et al, p. 159, VLSI-TSA, 2008
[3] A. Khakifirooz, et al, p.1391, TED2008
[4] MASTAR http://www.itrs.net/models.htm
[5] K. Suzuki, p1895, TED1999
[6] A. Bansal et al, p.256, TED2005
[7] J. Deng et al, p.2377, TED2007
[8] B.Yang, et al, p.126, VLSI2007
[9] H.C.-H. Wang, p.67, IEDM2006
[10] F. Boeuf , et al, p. 1433, TED 2008.

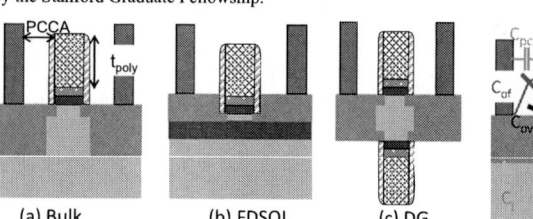

Figure 1 Three structures studied in this paper. (a) Bulk device (b) FDSOI (c) Double gate device.

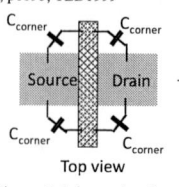

Figure 2 Schematic of the capacitance components.

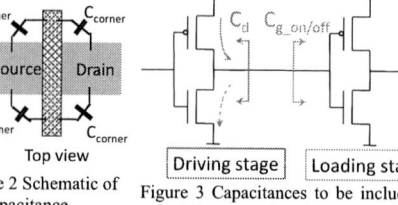

Figure 3 Capacitances to be included for delay calculation are the drain capacitance of the driving stage and the gate capacitance of the loading stage.

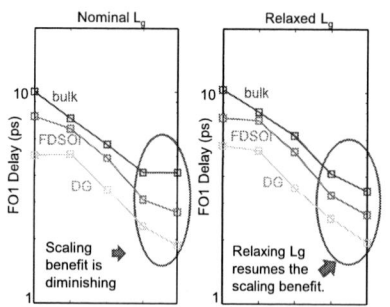

Figure 4 The delay calculated by MASTAR program. τ_{1fF} is the delay with a fixed 1fF load. The actual FO1 delay is way above the intrinsic CV/I because of the proper use of all extrinsic and intrinsic capacitances. But the intrinsic CV/I can still predict the scaling trends reasonably well.

Figure 5 Device scaling is less and less effective due to short channel effect and DIBL. Relaxing the gate length (as in Table 2) can resume the scaling return without too much speed penalty.

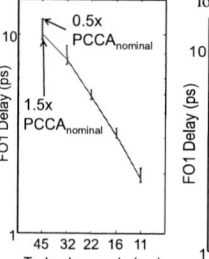

Figure 6 Delay dependence on PCCA distance. Upper and lower bound show the delay with PCCA to be 0.5x and 1.5x of the nominal case, respectively.

Figure 7 Delay dependence on gate height. Lower bound shows the delay with gate height to be half of the nominal case.

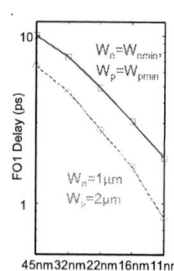

Figure 8 C_{corner} is an important parasitc for narrow devices. Wide devices are much faster, especially at very advanced technology.

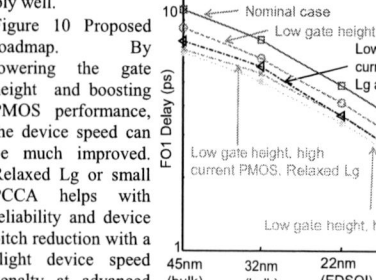

Figure 9. Delay dependence on N/P ratio. Boosting PMOS current driving capability.

Figure 10 Proposed roadmap. By lowering the gate height and boosting PMOS performance, the device speed can be much improved. Relaxed Lg or small PCCA helps with reliability and device pitch reduction with a slight device speed penalty at advanced technology nodes.

Table 2 Key parameters for nominal cases (nm)

Technology	45nm	32nm	22nm	16 nm	11nm
Device	Bulk	Bulk	FDSOI	FDSOI	DG
L_g	25	18	13	9	6
Relaxed L_g	32	25	18	14	10
EOT	1.1	0.65	0.5	0.5	0.5
t_{poly}	50	36	26	18	12
PCCA	17	12	9	6	4
$W_{n\ min}$	65	45	32	22	16
$W_{p\ min}$	130	90	64	44	32

Table 3 Parameter Definition

L_g	gate length	ΔL	gate overlap length
t_{ox}	physical oxide thickness	$t_{ox\ el}$	electrical oxide thickness
t_{poly}	gate height	$t_{rS/D}$	raised S/D height
t_{dep}	Si depletion width	PCCA	distance between gate edge to contact edge
$W_{n\ min}$	minimum NMOS width	$W_{p\ min}$	minimum PMOS width
W_{ext}	gate overhang per side	L_{pitch}	device pitch
ε_{ox}	dielectric permittivity	ε_{spacer}	spacer permittivity
ε_{box}	box layer permittivity	ε_{cap}	cap layer permittivity
N_a	channel doping	N_d	S/D doping

Table 1 Capacitance models

gate-to-channel capacitance	$C_{gc\_bulk} = C_{gc\_SOI} = \dfrac{\varepsilon_{ox}}{t_{ox\_el}}\left(L_g - \Delta L\right)W,\ C_{gc\_DG} = 2C_{gc\_bulk}.$
overlap capacitance	$C_{ov\_bulk} = C_{ov\_SOI} = 0.5\dfrac{\varepsilon_{ox}}{t_{ox\_el}}\Delta L \cdot W,\ C_{ov\_DG} = 2C_{ov\_bulk}.$
gate-to-substrate off-state capacitance	$C_{gb\_bulk} = series\left[\dfrac{\varepsilon_{ox}}{t_{ox}}(L - \Delta L)W, \dfrac{\varepsilon_{si}}{t_{dep}}(L - \Delta L)W\right],\ C_{gb\_DG} = 0,$ $C_{gb\_SOI} = series\left[\dfrac{\varepsilon_{ox}}{t_{ox}}\left(L_g - \Delta L\right)W, \dfrac{\varepsilon_{si}}{t_{si}}\left(L_g - \Delta L\right)W, \dfrac{\varepsilon_{box}}{t_{box}}\left(L_g - \Delta L\right)W\right]$
inner-fringe capacitance	$C_{if\_bulk,SOI} = \dfrac{2\varepsilon_{si}}{\pi}W\ln\left(1 + \dfrac{\min(0.5L_g - t_{ox}, t_{eff})}{t_{ox}}\right),\ C_{if\_DG} = \dfrac{4\varepsilon_{si}}{\pi}W\ln\left(1 + \dfrac{\min(0.5L_g - t_{ox}, t_{eff})}{t_{ox}}\right)$
outer-fringe capacitance	$C_{of\_bulk} = C_{of\_SOI} = 0.5C_{of\_DG}$ $= \dfrac{\varepsilon_{cap}W}{\pi}\ln\left(\dfrac{PCCA + \sqrt{t_{ox}^2 + PCCA^2}}{t_{ox}}\right) + 0.35\dfrac{\varepsilon_{cap}W}{\pi}\ln\left(\dfrac{\pi W}{t_{ox}}\right)$
gate-to-contace capacitance	$C_{PCCA} = \dfrac{0.5\pi\varepsilon_{cap}W}{\ln\left[\dfrac{2\pi\left(PCCA + L_g\right)}{2L_g + \tau_{bk}\cdot t_{poly\_eff}}\right]} + \dfrac{\varepsilon_{cap}Wt_{poly\_eff}}{PCCA} + \dfrac{\varepsilon_{spacer}Wt_{rSD\_eff}}{t_{spacer}}$ $\tau_{bk} = \exp\left[2 - 2\sqrt{1 + \dfrac{2\left(t_{poly\_eff} + L_g\right)}{PCCA}}\right]$
corner capacitance	$C_{corner\_bulk} = C_{corner\_SOI} = \dfrac{2\pi\varepsilon_{cap}\left(t_{poly\_eff} + 2W_{ext}\right)}{\cosh^{-1}\left(\dfrac{t_{eff} + 2t_{ox}}{t_{eff}}\right)},\ C_{corner\_DG} = 2C_{corner\_bulk}$
junction capacitance	$C_{j\_bulk} = W\left[x_j + 0.5\left(L_{pitch} - L_g\right)\right]\sqrt{\dfrac{q\varepsilon_{si}}{2V_{bi}}\dfrac{N_aN_d}{N_a + N_d}}\Bigg/\left(1 + \dfrac{V}{V_{bi}}\right)^{0.33}$ $C_{j\_SOI} = \dfrac{\varepsilon_{box}}{t_{box}}W \cdot 0.5\left(L_{pitch} - L_g\right),\ C_{j\_DG} = 0.$ $t_{eff} = x_j(bulk), t_{si}(SOI)$ or $0.5t_{si}(DG)$ $t_{rSD\_eff} = 0(bulk, DG)$ or $t_{rSD}(SOI)$ $t_{poly\_eff} = t_{poly}(bulk, DG)$ or $(t_{poly} - t_{rSD})(SOI)$

978-1-4244-2784-0/09 $25.00 © 2009 IEEE

THE DEVICE ARCHITECTURE DILEMMA FOR CMOS TECHNOLOGIES: OPPORTUNITIES & CHALLENGES OF FINFET OVER PLANAR MOSFET

B. Parvais, A. Mercha, N. Collaert, R. Rooyackers, I. Ferain, M. Jurczak, V. Subramanian, A. De Keersgieter, T. Chiarella, C. Kerner, L. Witters, S. Biesemans and T. Hoffman

IMEC, Kapeldreef 75, 3001 Leuven, Belgium
Phone: +32 16 28 83 18, Fax: +32 16 281844, E-mail : parvais@imec.be

ABSTRACT

Despite their excellent control of short channel effects, FinFETs suffer from different trade-offs in the mixed-signal domain, with respect to planar devices. For the first time, we report a complete and comprehensive comparative analysis showing that these trade-offs can be alleviated in advanced FinFET technology. As such, higher voltage gain and transconductance than planar MOSFETs are reached at the same time. V_T mismatch smaller than 3mV.µm is obtained for narrow (10nm) fins. Reduced speed sensitivity to gate pitch scaling and invertor delay reduced below 10 ps will be demonstrated.

INTRODUCTION

Multi-gate MOSFETs, such as FinFETs, are the most promising approach to address Short Channel Effects (SCE) and leakage issues in deeply scaled CMOS. In addition, ultra-thin body devices eliminate the need for channel doping, thereby reducing parametric spread due to dopant fluctuations and reduced junction leakage due to high electric fields. In this contribution, the challenges and opportunities of FinFETs are benchmarked to planar MOSFETs, regarding the integration complexity, as well as the DC and HF performances.

HIGH-K/METAL GATE INTEGRATION CHALLENGES

Both experimentally and through TCAD simulations, we previously demonstrated for planar devices that the optimum workfunction (WF) of the gate electrode is a function of the V_T (or leakage) targeted [1]. Fig. 1 shows that for a given I_{OFF} target of 100nA/µm (low-V_T applications) the optimum WF separation between nMOS and pMOS can be greatly relaxed with FinFET as compared to planar, thanks to intrinsically superior SCE control with no/low channel doping. This enables a significant simplification of the High-K/Metal gate integration scheme: as an example, Fig. 2 shows that to achieve similar V_T's as FinFET, using a single metal for nMOS and pMOS, a dual High-K capping layer approach is required (LaO and AlO for nMOS and pMOS, respectively)[2]. Moreover, the lower level of Coulomb scattering reached in undoped FinFETs results in a higher mobility (Fig. 3). However, the length dependence of mobility points to additional (remote) scattering mechanisms, which are more pronounced in undoped fins.

DC PERFORMANCE: FINFET OPPORTUNITIES

Even if intrinsic fin doping is advantageous for intra-die variability, increased mismatch was reported for narrow fin devices [3]. We demonstrate here that with well optimized FinFETs [4], the mismatch coefficient only marginally increases when the fin is downscaled to 10 nm (Fig. 4) and the performance is very good as compared to planar transistors [3, 5]. With A_{VT}<3 mV.µm, fin scaling is not a showstopper for this technology in applications that require good matching, like SRAMs for which the enhanced FinFET density is another asset [4]. Furthermore, the inherent excellent electrostatic control of the channel by the gate is beneficial in the analog domain. FinFETs can achieve high output resistance and peak transconductance efficiency (Fig. 5). However, nFinFETs suffer from low transconductance with respect to planar devices. The reasons for this are: (i) enhanced parasitic S/D resistance, (ii) reduced low-field mobility along (100) crystalline plane, and at high-field levels (iii), reduction of velocity saturation (Fig. 6), which is related to the different dominant surface orientations used in planar and FinFET and to the lower volume available for current flow beyond pinch-off in case of FinFET. Normalized to the gate leakage, the planar n-MOSFET (p-MOSFET) drive current is therefore increased (decreased) w.r.t. n-FinFETs (p-FinFETs), as seen in Fig. 8. However, with an optimized process, FinFETs offer at the same time a higher voltage gain and transconductance than planar MOSFETs (Fig. 7), corresponding to a reduced trade-off between I_{ON} and DIBL. This is a key advantage for both digital and high-precision analog circuits.

HIGH FREQUENCY CHALLENGES

Despite excellent intrinsic behavior, FinFETs are known to suffer from a high level of parasitics, which limit their speed [6]. If the series resistance can be reduced by epitaxial growth and layout optimization, the parasitic capacitance depends on the architecture. However, the additional capacitance penalty of FinFETs is limited, when compared to planar transistors (Fig. 9). Using [7], the FinFET parasitic capacitance is shown to be dominated by the overlap and the fringing capacitance between the gate and fin extensions (~60% of the parasitics). Even with a speed reduction due to excess of parasitics, the 11ps FinFET invertor stage delay is comparable with the performance of planar transistors (Fig. 10), as a result of the high g_m values shown in Fig.7. This delay can be further decreased down to 7.25ps at V_{DD}=1.6V, at the cost of increased consumption (Fig. 11). For speed optimization, the width ratio of n- and p-transistors is around 1 for FinFET and ~2 for planar, related to their respective I_{ON} levels [4]. Similarly, the f_T reaches~120 GHz for W_{fin}=20nm for both transistor polarities (Fig. 12). Note that only the pad parasitics are deembedded from RF measurements, but not the device interconnect. Even if planar devices outperform the FinFETs (Fig. 13) for the above mentioned reasons, SOI FinFETs exhibit a better robustness to pitch reduction and BEOL scaling (Fig.18), which is attributed to the absence of STI induced stress that is detrimental for the g_m of n-transistors.

CONCLUSIONS

Beneficiating from high voltage gain, FinFET technology is appealing in high precision analog applications. We showed here that at the same time, FinFETs can achieve high values of transconductance, which translates into low inverter delay in ring oscillators (~10 ps). However, thanks to a lower level of parasitics, planar bulk transistors reach higher f_T values. Finally, the excellent matching performance obtained on narrow fin devices is a key advantage of FinFET in high density memory applications.

978-1-4244-2784-0/09 $25.00 © 2009 IEEE

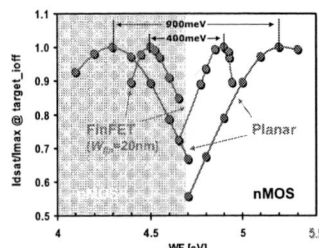

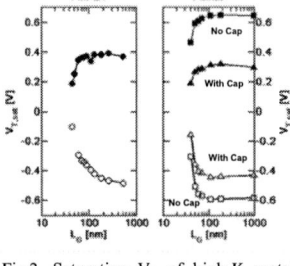

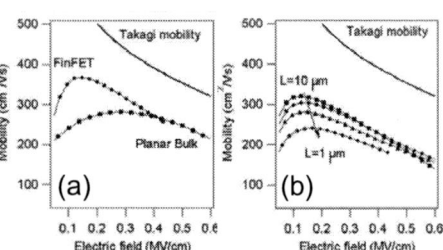

Fig.1. TCAD simulations of the saturation current @I_{ON}=100pA/μm as a function of the workfunction, comparing FinFET devices with 20nm fin width to planar devices.

Fig.2. Saturation V_T of high-K metal gates: (left) FinFET with TiN/HfSiON; (right) Planar with and without LaO and AlO cap for n- and p-MOS, respectively.

Fig.3. (a) Increased n-FinFET mobility w.r.t. n-Planar due to low doping. (b) Gate length dependence of n-FinFET mobility (W_{fin}=25nm), resulting from remote Coulomb and phonon scattering in undoped metal gates high K.

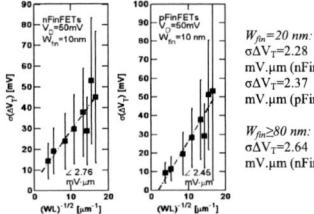

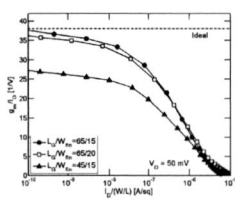

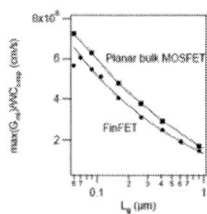

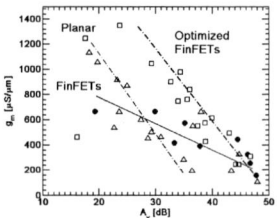

Fig.4. Pelgrom plot of 10nm wide SOI n-FinFETs (left) and p-FinFETs (right) with TiN/HfSiON gate stack.

Fig.5. Transconductance efficiency (g_m/I_D) and SCE control of FinFETs depends on the L_G/W_{fin} ratio.

Fig.6. Velocity saturation (measured by $g_m/(W.C_{ox})$) of planar outperforms FinFETs.

Fig.7. Optimized FinFETs for high g_m and voltage gain; V_{DS} = 1.1 V; V_G=V_T+0.2V and V_G+0.6V.

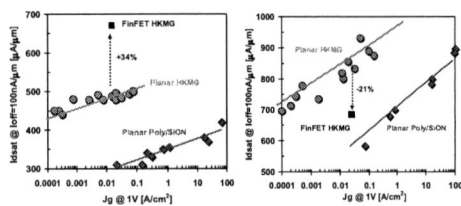

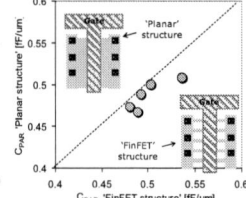

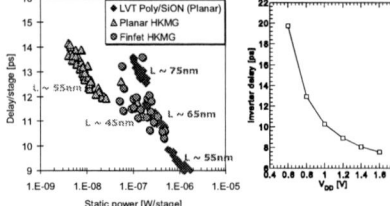

Fig.8. I_{ON} at target I_{OFF} versus gate leakage is higher for nFinFETs (left) and lower for pFinFETs, (right) w.r.t. planar devices. V_D = 1 V.

Fig.9. Parasitic capacitance of FinFETs is slightly higher than the planar counterpart; SOI substrate.

Fig.10. Power/speed tradeoff in ring-oscillators illustrates that FinFET can achieve a comparable performance as planar; V_{DD}=1V.

Fig.11. FinFET inverter delay decreases down to 7.25ps when V_{DD}=1.6 V.

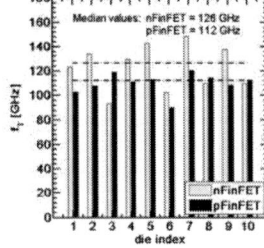

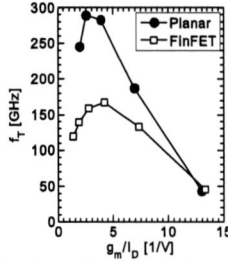

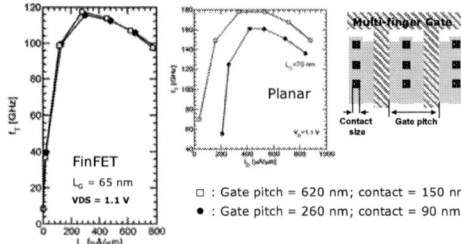

Fig.12. Distribution of the cut-off frequency of the current gain of n- and p-FinFETs across the wafer.

Fig.13. Planar transistors (Poly/SiON, L_G=40nm) outperform the best FinFETs (TiN/HfSiON, L_G=40nm) in terms of speed; V_{DD}=1.1V.

Fig.14. When compared to planar devices (polySi/SiON), the speed of SOI FinFETs is less affected by gate pitch reduction and BEOL scaling.

REFERENCES

[1] T. Hoffmann *et al.*, *Proc. IEDM,* 2006, pp. 1-4, Dec. 2006; [2] S. Kubicek *et al.*, *Proc. IEDM,* 2007, pp. 49-52, Dec. 2007; [3] C. Gustin *et al.*, *Proc. VLSI-TSA,* 2007. pp. 168-169, April 2007; [4] N. Collaert *et al.*, *ICICDT 2008*, pp. 59-62, June 2008; [5] J. Johnson *et al.*, *IEEE Electr. Dev. Let.*, July 2008, pp. 802-804; [6] B. Parvais *et al.*, *IEEE SOI Conf. 2007*, paper 3-5, Oct. 2007; [7] W. Wu *et al.*, *IEEE Trans. Electr. Dev,* April 2007, pp. 692-698.

978-1-4244-2784-0/09 $25.00 © 2009 IEEE

Investigation of Low Frequency Noise in Uniaxial Strained PMOSFETs

Jack J.-Y. Kuo, William P.-N. Chen, and Pin Su

Department of Electronics Engineering, National Chiao Tung University, Hsinchu, Taiwan
E-mail: pinsu@mail.nctu.edu.tw

Abstract

We have investigated the low frequency noise characteristics for uniaxial strained PMOSFETs. In the low $|V_{gst}|$ regime, the 1/f noise is dominated by the carrier-number-fluctuations and the S_{Id}/I_d^2 is increased by the enhanced g_m/I_d for the strained device. Nevertheless, the S_{Id}/I_d^2 of the strained device is almost the same as the unstrained one at a given g_m/I_d. Furthermore, with the application of uniaxial compressive strain, the attenuation length λ is reduced because of the increased out-on-plane effective mass and tunneling barrier height. The reduced λ may result in a smaller S_{Vg}. In the high $|V_{gst}|$ regime, the 1/f noise is dominated by the mobility-fluctuations and the S_{Id}/I_d^2 is increased due to the larger Hooge parameter for the strained device.

Introduction

Low frequency noise is becoming a concern for continuously scaled down CMOS devices because it may limit the functionality of analog, mixed signal, and RF circuits. As strained-silicon is widely used in state-of-the-art CMOS technologies to enable the mobility scaling [1-2], the low frequency noise performance of these strained devices is especially important. Although Giusi et al. [3] have reported that the low frequency noise of strained PMOSFETs could be degraded due to the worse gate dielectric quality when processing the SiGe source/drain, Simoen et al. [4] and Ueno et al. [5] reported that the gate dielectric quality of the strained device may not be degraded and the low frequency noise performance may be preserved. Whether there exists an intrinsic stress effect on low frequency noise characteristics is still not clear and merits investigation. Through a comparison between co-processed strained and unstrained devices, this work investigates the low frequency noise performance for uniaxial strained PMOSFETs.

Devices and Measurement

The strained PFETs used in this study were fabricated by state-of-the-art process-induced uniaxial strained-silicon technology featuring SiGe source/drain and compressive Contact Etch Stop Layer (CESL) [6]. For the transistors with gate length L_{gate} = 65 nm, the saturation drain current of the strained device is improved more than 100% as compared with its control counterpart. Low frequency noise measurements were carried out using the BTA9812 measurement system.

Results and Discussion

The drain current noise spectral densities (S_{Id}) for the strained and unstrained devices biased at gate overdrive $|V_{gst}|$=0.2V and 0.8 V are shown in Fig. 1. The spectra show typical $1/f^\gamma$ noise type with the frequency index γ close to one. Fig. 2 shows the measured S_{Id} versus I_d for the strained and unstrained devices at f=10Hz. It can be seen that the S_{Id} of the strained device is larger than that of the unstrained one. Moreover, the increasing S_{id} with I_d^2 (for $I_d < \sim 10^{-5}$A) indicates a carrier-number-fluctuations dominated 1/f noise.

The carrier-number-fluctuations origin is further illustrated in Fig. 3, which shows the normalized noise spectral density (S_{Id}/I_d^2) as well as $(g_m/I_d)^2$ versus I_d for strained and unstrained devices. In Fig. 3, S_{Id}/I_d^2 shows a fairly good proportionality with $(g_m/I_d)^2$, which also points to a carrier-number-fluctuations origin. In addition, it implies that the discrepancies in S_{Id}/I_d^2 between the strained and unstrained devices can be attributed to $(g_m/I_d)^2$. Fig. 4 compares the g_m/I_d versus $|V_{gst}|$ for strained and unstrained devices in the low $|V_{gst}|$ regime. The g_m/I_d for the strained device is higher than its control counterpart because of the higher V_g sensitivity of carrier mobility present in the strained device (Fig. 5). Fig. 6 further shows that for a given g_m/I_d (in the low $|V_{gst}|$ regime), the S_{Id}/I_d^2 of the strained device is almost the same as that of the unstrained one.

Fig. 7 shows the input-referred voltage spectral density ($S_{Vg}=S_{Id}/g_m^2$) at f = 10Hz as a function of V_{gst} taken from the average of 10 devices. From Fig. 7, it can be seen that the S_{Vg} for strained and unstrained devices are nearly the same. The gate bias independent S_{Vg} for $|V_{gst}|$ below 0.2V indicates a carrier-number-fluctuations origin of S_{Vg}, which stems from oxide traps and can be expressed as [7]:

$$S_{Vg} = \frac{kTq^2}{fWL_{gate}C_{OX}^2} \lambda \times N_t \qquad (1)$$

Where λ is the tunneling attenuation length, N_t is the number of occupied traps per unit area, kT is the thermal energy, W is the gate width, L_{gate} is the gate length, and C_{OX} is the gate capacitance. The nearly identical S_{Vg} implies that the $\lambda \times N_t$ product of the strained and unstrained devices are almost the same. To investigate the impact on strain on λ, the N_t is extracted by the charge pumping method [8] (Fig. 8). It is found that the N_t of the strained device is 9% larger than its control counterpart. In other words, λ of the strained device is reduced. λ can be further expressed as:

$$\lambda = \sqrt{\hbar^2/2m^*\varphi_B} \qquad (2)$$

where m^* is the out-of-plane effective mass and φ_B is the tunneling barrier height. The impact of strain on m^* and φ_B is illustrated in Fig. 9. The increased m^* and φ_B by strain [9, 10] results in a smaller λ for the strained device.

In the high $|V_{gst}|$ regime, the strained device also shows larger S_{Id}/I_d^2 than its control counterpart (Fig. 10). Fig. 11 shows the extracted Hooge parameter (α_H) [11] versus $|V_{gst}|$ from the average of 10 devices. It can be seen that α_H shows weak V_{gst} dependence, which indicates the mobility-fluctuations origin of the 1/f noise.

Acknowledgement

This work was supported in part by the National Science Council of Taiwan under Contract NSC97-2221-E-009-162, and in part by the Ministry of Education in Taiwan under ATU Program.

References

[1] Y. Tateshita et al., *IEDM Tech. Dig.*, 2006.
[2] S.E. Thompson et al., *IEEE Trans. Electron Devices*, vol. 53, no. 5, p. 1010, May 2006.
[3] G. Giusi et al., *IEEE Electron Device Lett.*, vol. 27, no. 6, p.508, Jun. 2006.

[4] E. Simoen et al., *IEEE Electron Device Lett.*, vol. 28, no. 11, p.987-989, Nov. 2007.
[5] T. Ueno et al. *VLSI Symp. Tech. Dig., 2006*, pp.104-105.
[6] J. Kuo et al., *Semicond. Sci. Technol.* vol. 22, no. 4, p. 404, Apr. 2007.
[7] K.K. Hung et al., *IEEE Trans. Electron Devices*, vol. 37, no. 3,p. 654, Mar. 1990.
[8] S.S. Chung et al., *VLSI Symp. Tech. Dig., 2002*, pp.74-75.
[9] S.E. Thompson et al., *IEDM Tech. Dig.*, 2004, pp.221-224.
[10] X. Yang et al., *IEDST* 2007, pp.149-152.
[11] F. Hooge, *Physica B*, vol. 83, pp. 14, 1976.

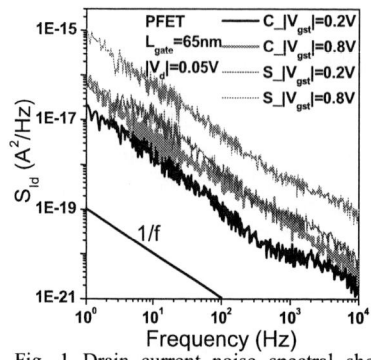

Fig. 1 Drain current noise spectral showing larger S_{Id} for the strained device at $|V_{gst}| = 0.2$V and 0.8V.

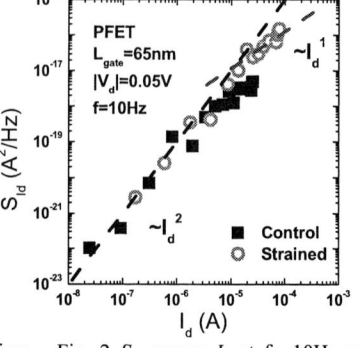

Fig. 2 S_{Id} versus I_d at f =10Hz and $|V_d|$=0.05V.

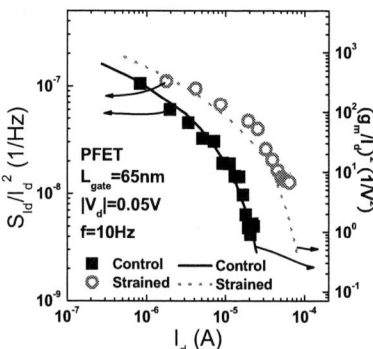

Fig. 3 S_{Id}/I_d^2 and $(g_m/I_d)^2$ versus I_d at $|V_d|$=0.05V.

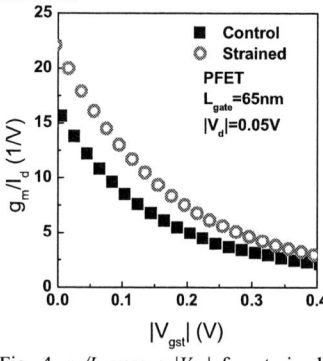

Fig. 4 g_m/I_d versus $|V_{gst}|$ for strained and unstrained devices with L_{gate} =65nm at $|V_d|$=0.05V.

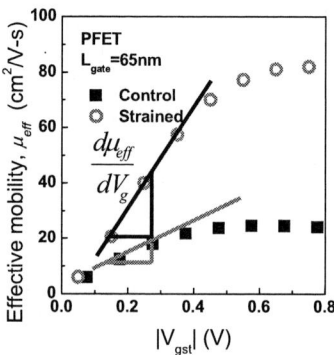

Fig. 5 The extracted effective hole mobility versus $|V_{gst}|$ for strained and unstrained devices with L_{gate} =65nm.

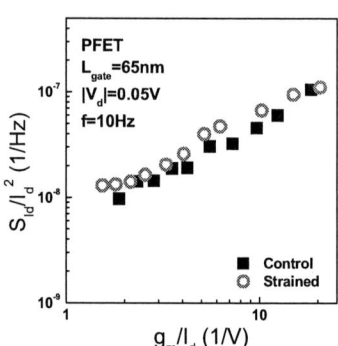

Fig. 6 Normalized drain current noise spectral density S_{Id}/I_d^2 versus g_m/I_d for devices with L_{gate} =65nm and $|V_d|$=0.05V.

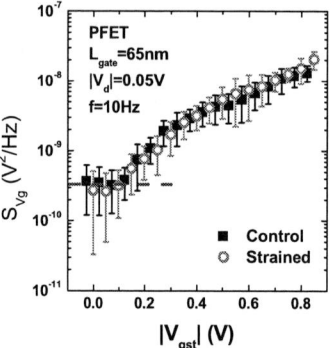

Fig. 7 S_{Vg} versus $|V_{gst}|$ for devices with L_{gate} =65nm at f=10Hz and $|V_d|$=0.05V.

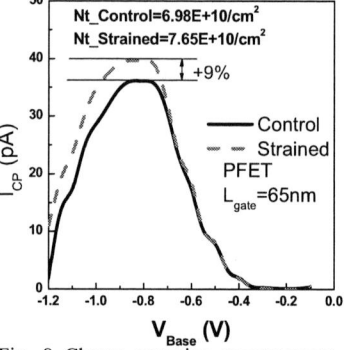

Fig. 8 Charge pumping measurement for strained and unstrained devices with L_{gate} =65nm.

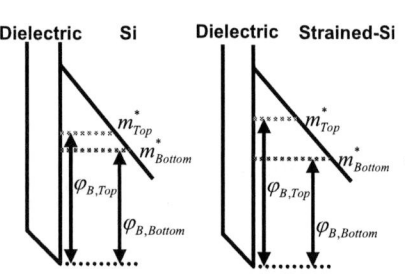

Fig. 9 The band diagram for strained and unstrained devices. For the unstrained device, holes populate in the top two bands. When the strain is applied, the bands split and more holes populate in the top band.

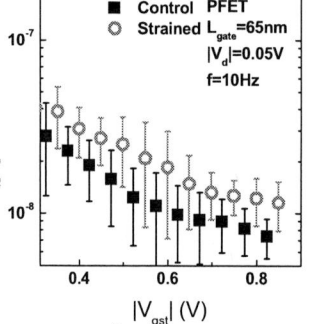

Fig. 10 S_{Id}/I_d^2 versus $|V_{gst}|$ (the average of 10 devices) at $|V_d|$=0.05V and f =10Hz.

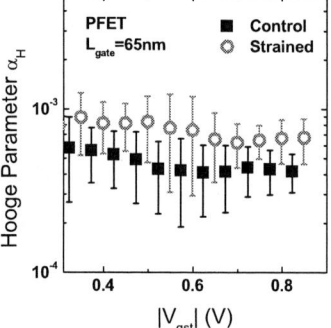

Fig. 11 Hooge parameter versus $|V_{gst}|$ for devices with L_{gate} =65nm.

978-1-4244-2784-0/09 $25.00 © 2009 IEEE

FDSOI CMOS with Dual Backgate Control for Performance and Power Modulation

Jeng-Bang Yau, Jin Cai, Leathen Shi, Robert. H. Dennard, Arvind Kumar, Katherine L. Saenger, Alexander Reznicek, Paul. M. Solomon, Qiqing Ouyang, Steven Koester, and Wilfried E. Haensch

IBM Semiconductor Research and Development Center, T.J. Watson Research Center, Yorktown Heights, NY 10598, USA
Tel: 914-945-2279, Fax: 914-945-4184, E-mail: jyau@us.ibm.com

Abstract

We demonstrate, for the first time, modulation of power-performance of a ring oscillator fabricated on thin-BOX (buried oxide) FD (fully-depleted) SOI using independent backgate controls for nFET and pFET. The thin BOX facilitates an effective modulation of ring characteristics with small (1–2V) independent backgate voltages. Leakage current per stage can be reduced by more than 100× with 30% increase of inverter delay. In addition, the inverter delay can be improved by 15% with 2× increase of the stand-by current. Compatible with conventional CMOS process, our results suggest the backgate technology, an additional knob for power/performance optimization and variability control, is attractive for continued CMOS scaling.

Introduction

FDSOI transistors have gained extensive attention for their improved scaling potential over their PDSOI and bulk Si counterparts. As Si thickness (T_{Si}) in FDSOI is scaled down to enable shorter channel length, however, it encounters V_T variability issues arising from within-wafer and wafer-to-wafer T_{Si} variations. Another difficulty with FDSOI technology is to satisfy different V_T requirements for SRAM and logic devices. The concept of backgate has been proposed [1, 2] for V_T control and the simplest implementation consists of a common backgate for nFET and pFET, which allows rebalancing CMOS for best power/performance trade-off [3–6]. While a common backgate inadvertently shifts V_T of nFET and pFET in opposite directions, it is beneficial to decouple the backgate for better controllability and flexibility. In this work we present experiments that investigate the V_T tunability of FD FET by independent backgates and show how the operation of a ring oscillator can be modulated for either a performance or a power objective.

Device Design and Fabrication

Fig. 1(a) shows a schematic backgated FDSOI CMOS. Thin-BOX SOI with BOX thickness ranging from 5 to 30nm, with negligible bonding defect and edge chip, was made by surface oxidization of an SOI wafer and a bulk Si wafer, followed by bonding the oxidized surfaces to form a BOX layer. The bonded pair was grinded from the SOI side to produce a thin-BOX SOI. The process flow and a typical device for the backgated SOI CMOS are shown in Figs. 1(b) and 1(c), respectively. The fabrication of backgated SOI CMOS is similar to traditional CMOS process, with modifications tailored for backgated FDSOI: After the active area is patterned, shallow trenches are formed to isolate both the backgate wells and FETs. Prior to STI fill, a layer of silicon nitride is deposited to protect the STI in subsequent oxide-consuming steps. Figures 2(a) and (b) shows the STI structures in wafers without and with the silicon nitride liner after aggressive clean steps (120sec 100:1 DHF and 80sec 50:1 BOE etches), where the latter shows improvements of STI recess and significant suppression of junction leakage current measured between the nFET and pFET backgate wells (Figure 2(c)).

Device Results

Figures 3(a)–(c) show tuning of threshold voltages by the backgate in our FDSOI CMOS. Figure 4(a) shows that by symmetrically reverse-biasing the nFET and pFET backgates ($V_{NBG}= -1V, -2V$ and $V_{PBG}= 1V, 2V$. Note: V_{PBG} is referenced to the pFET source terminal, which is connected to V_{DD} in a ring oscillator), the quiescent leakage current per stage (Iddq) of a ring oscillator is reduced by 10×–30× with corresponding 44%–70% increases of ring delay. One major advantage of our backgated SOI CMOS is independent backgate controls for nFET and pFET, which is particularly useful in performance tuning for a device with unmatched V_T. We then focus on the chip circled in Fig. 4(a), where the pFET has a lower V_T than NFET (Fig. 4(b)). Figure 5 shows that by keeping $V_{NBG} = 0V$ and reverse-biasing V_{PBG}, Iddq is reduced by over 100× with only a 30% increase of delay, which achieves better leakage reduction compared to data measured for the chip circled in Fig. 4(a) with unmatched V_T (open triangles in Fig. 5). On the other hand, the ring performance can be improved by 15% with ~2× increase of Iddq by forward-biasing V_{NBG} and keeping V_{PBG} constant. Iddq vs. delay is also measured with common backgate bias (V_{BG}=0.25, 0.5, 1.0, 1.5, and 2.0 V) and included in Fig. 5 (open diamonds) as a comparison to the independent backgate. While common backgate achieves delay time comparable to that in independent backgate as V_{BG} increases, it produces higher Iddq due to limited threshold tunability. This further illustrates the advantage of independent backgate control in performance optimization. The impact of backgate controls on the transport properties in our FDSOI CMOS is also investigated. Figures 6(a) and (b) show V_T tuning and backgate sensitivity ($\Delta V_{Tsat}/\Delta V_{BG}$) vs. V_{BG}, respectively. The increasing $\Delta V_{Tsat}/\Delta V_{BG}$ for forward biased V_{BG} can be attributed to the inversion charge being attracted towards the SOI/BOX interface while decreasing $\Delta V_{Tsat}/\Delta V_{BG}$ for reverse biased V_{BG} indicates the inversion charge is pushed toward the front gate oxide interface. This is corroborated by improved DIBL (drain-induced barrier lowering) and subthreshold slope with reverse biased V_{BG}, as shown in Fig. 7. This improvement comes from enhanced vertical confinement field in the channel, which pushes the charges toward the front gate oxide interface, leading to better short channel effects. The drive current for $|V_G-V_T|$=0.7V, however, decreases with reverse-biased V_{BG}, as shown in Fig. 8. This is likely due to mobility degradation under stronger vertical field. This suggests a trade-off between inversion charge confinement and mobility degradation when the backgate is used to tune the short channel effect.

Conclusion

We have demonstrated CMOS on thin-BOX FDSOI with independent backgate controls that can achieve power-performance modulation of a ring oscillator. Our results suggest that the backgate SOI CMOS technology is a good candidate for high-performance and energy-efficient applications.

Acknowledgments

The authors would like to acknowledge the managerial support from G. Shahidi. We would also like to thank A. Majumdar and D. Greenberg for discussions and the Watson Microelectronics Research Laboratory for the technical support.

References

[1] I. Y. Yang, et al., IEEE Trans. Elect. Dev., p. 822 (1997).
[2] H. Oh, et al., Jpn. J. Appl. Phys., v43, p. 2140 (2004).
[3] R. Tsuchiya et al., IEDM Tech. Dig., p. 631 (2004).
[4] H. –Y. Chen et al., Symp. VLSI Tech., p. 16 (2005).
[5] R. Tsuchiya et al., IEDM Tech. Dig., p. 475 (2007).
[6] Y. Morita, et al., Symp. VLSI Tech., p. 166 (2008).

978-1-4244-2784-0/09 $25.00 © 2009 IEEE

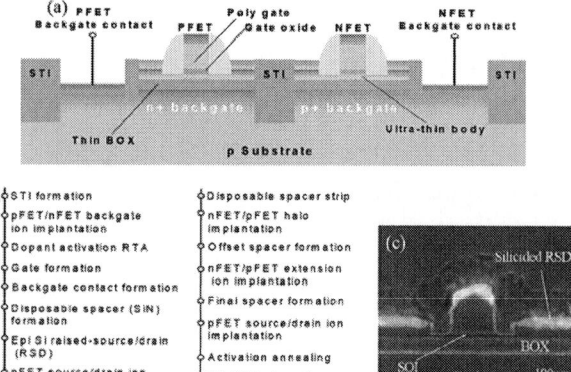

FIG. 1 (a) Schematic structure, (b) process flow, and (c) Cross-sectional SEM image of the backgated FDSOI CMOS device.

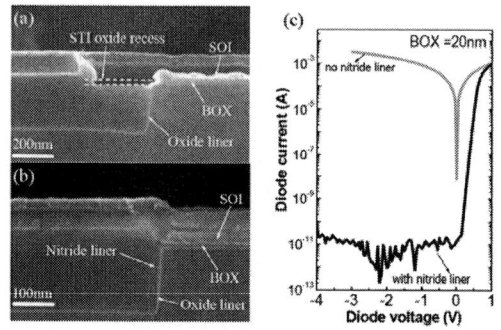

FIG. 2 Cross-sectional images of STI for wafers (a) without nitride liner and (b) with nitride liner after aggressive cleans. (c) Current measured between the nFET and pFET backgate wells in backgated SOI with and without nitride liner.

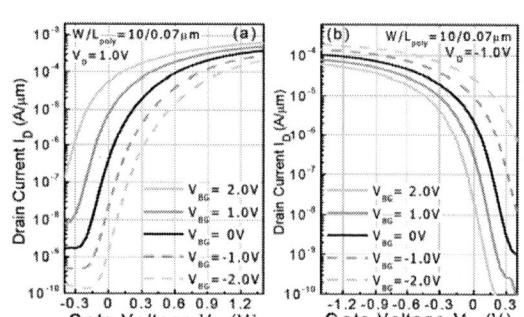

FIG. 3 (a) and (b): Modulation of the transfer curves of nFET and pFET, respectively, by V_{BG}.

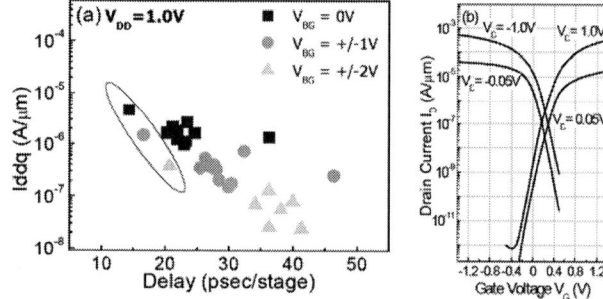

FIG. 4 (a) Iddq vs. delay of ring oscillators with symmetric reverse backgate biases applied to nFET and pFET. (b) I_D-V_G of the device circled in (a), which exhibits unmatched V_T between nFET and pFET.

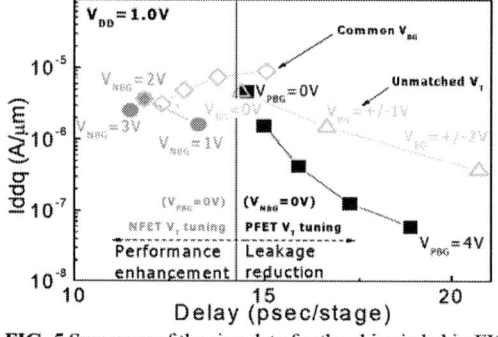

FIG. 5 Summary of the ring data for the chip circled in FIG. 4 with various backgate bias schemes, demonstrating the advantages of the independent backgate controls.

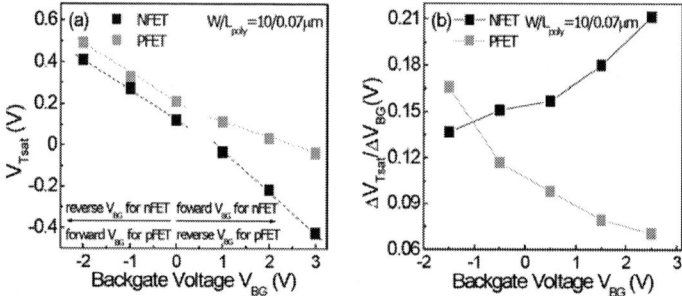

FIG. 6 (a) Tuning of V_T for nFET and pFET by V_{BG}, showing differences in slopes between forward and reverse backgate biases. (b) Backgate sensitivity ($\Delta V_{Tsat}/\Delta V_{BG}$) vs. V_{BG}.

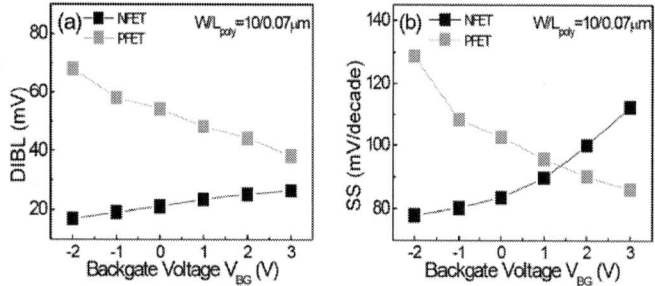

FIG. 7 Modulation of: (a) DIBL and (b) Subthreshold slope by V_{BG} in thin-BOX SOI.

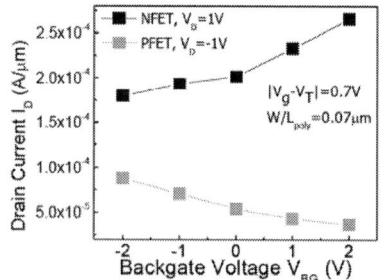

FIG. 8 I_D at $|V_G$-$V_T|$=0.7V for V_{BG}= −2 to 2V.

978-1-4244-2784-0/09 $25.00 © 2009 IEEE

Sub-32nm CMOS Technology Enhancement For Low Power Applications

R. M. Huang, P. W. Liu, E. C. Liu, W. T. Chiang, S. H. Tsai, Jonas Tsai, Tzermin Shen, C. H. Tsai, C. T. Tsai and G. H. Ma

United Microelectronics Corporation (UMC), No.18 Nanke 2nd Rd., Tainan 741, Taiwan R.O.C

email: Rai_Min_Huang@umc.com

ABSTRACT

In this paper, we have systematically investigated the factors for performance enhancement on sub-32nm CMOS technology. We report that PMOS gains the drive current by slim spacer, S/D silicide resistance reduction by e-SiGe, and compressive CESL. The three factors improve the PMOS performance by 7%, 10% and 25% respectively. Combined with the three factors can gain the device drive current 30%. In addition, the optimized integration scheme can reduce NMOS extension resistance. The main cause is that post e-SiGe clean processes would loss the extension dopant and increases the extension resistance. We successfully reduce the NMOS total resistance 22% compared to control without compromise PMOS device performance.

INTRODUCTION

Strain engineering has been the critical keys to boost CMOS device performance while scaling and meet ITRS road map in the past few years. For instance, utilizing selective epitaxially SiGe in source/drain regions, uniaxially compressive stress was introduced into PMOS transistors and the carrier mobility was enhanced significantly, resulting in higher drive current. The integration of e-SiGe into source/drain regions as channel stressors would be the candidate to boost PMOS device performance for sub-32 nm technology [1]. Source/drain e-SiGe with low resistivity by well silicidation becomes important while source/drain resistance increases with CMOS device scaling. However, their interaction makes such integration challenge [2]. In this paper, the performance factor resulted from integration of strained SiGe were systemically investigated. For certain implant conditions, either before or after SiGe deposition, new process scheme was also proposed.

EXPERIMENTAL

Fig.1 lists key process steps used in the sub-32nm technology fabrication sequence. PMOS transistors with EOT=12A SiON and Lg=30nm were fabricated by utilizing strained e-SiGe (Ge~20 to 30 at. %) source/drain structures on 300mm (001) Si wafers. After active area definition by shallow trench isolation, channel implants were used to determine device threshold voltage (Vth). Then, poly-Si were deposited and patterned, followed by slim spacer1 formation and LDD engineering. After main spacer formation, various S/D junction engineering, and nickel silicidation are formed subsequently. To enhance the mobility, dual stressed contact etch stop layers (DSL) were deposited [3]. **Table I** shows the wafer sets with different split conditions. Transmission electron microscopy (TEM) cross-sectional image is shown in **Fig.2** for a completed PMOS transistor with 30nm gate length. To achieve the minimum 6T SRAM area (0.122um² cell), advanced double patterning immersion technology and double etch is used. **Fig.3** shows the tilt SEM view image on SRAM region [4].

RESULT AND DISCUSSION

From the sub-32nm technology process steps (**Fig.1**), there are three critical key steps, which will have much influence on device performance. (1) **Slim spacer**: slim spacer will make the e-SiGe strain effect closer to channel and enhance the hole mobility more obviously. **Fig.4** shows the slim spacer case (Sample A) gains the drive current 7% compared to counterpart. However, slim spacer needs to be optimized or make the short channel effect more serious. (2) **S/D silicide resistance reduction by e-SiGe**: e-SiGe is an indispensable PMOS performance enhancement factor on sub-32nm technology. But when Ge percentage in SiGe increases, the silicidation condition above SiGe would be poorer. **Fig.5** shows the S/D silicide resistance reduction optimized case (Sample B) gains the drive current 10% compared to counterpart. (3) **DSL**: stress liner also plays very important role on sub-32nm technology. Compressive strain liner improves PMOS hole mobility more effectively on scaling device. Form **Fig.6**, compressive strain liner case (Sample C) gains the drive current 25% compared to counterpart. Combined with these three optimized cases, Sample D demonstrates 30% drive current enhancement compared to baseline case (**Fig.7**). **Fig.8** summarizes the different factors improvement percentage. It can be seen that combined improvement factor is not linear. Sample D combines with all optimized factor cases, so its Gm,max shows 80% enhancement compared to baseline case (**Fig.9**). **Fig.10** shows the 80% performance enhancement from sample D and baseline case under the same level DIBL value. **Fig.11** also shows 80% enhancement in drive current under the same level subthreshold swing value.

The influence of dopant loss caused by e-SiGe clean processes would increase the extension resistance. **Fig.12** shows the schematic of integration scheme, which NMOS implantation is before or after e-SiGe process. The SiGe first scheme **Fig.12(a)** can avoid the NMOS extension implantation to suffer the dopant loss caused by post e-SiGe clean processes other than SiGe last scheme **Fig.12(b)**. The comparison of NMOS extension resistance can be seen in **Fig.13**. Obviously, SiGe first scheme can reduce the total resistance 22% and gain the device performance. While SiGe first scheme reduces the NMOS total resistance, it also has no impact on PMOS device extension resistance. **Fig.14** shows similar behavior on the PMOS total resistance from SiGe first and SiGe last scheme.

CONCLUSION

Through the analysis of device electrical behavior, PMOS device shows drive current gain 7% with slim spacer, 10% with S/D silicide resistance reduction, and 25% with cCESL. Combined with these three factors would result in 30% drive current gain. By using SiGe first integration scheme, NMOS total resistance can be further reduced 22% compared to SiGe last scheme. In this work, we successfully demonstrate sub-32nm node device performance boost featuring with double patterning, immersion technology and gate SiON for low power applications.

REFERENCES

[1] Yu, M.H. et. al., IEDM Tech. Dig. p. 1, 2006.
[2] Umezawa, K. et. al., ISSM Tech. Dig. p. 1, 2007.
[3] Y. C. Liu et al., IEDM, p. 836, 2005.
[4] Shien-Yang Wu et al., IEDM Tech. Dig. p. 263, 2007.

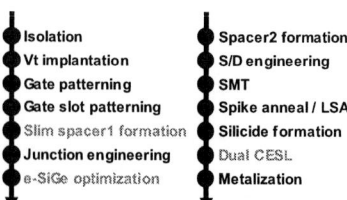

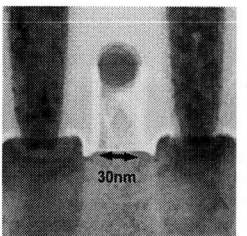

Split condition	Slim Spacer1	S/D Silicide Resistance Reduction by e-SiGe	DSL
Sample A	✓	-	-
Sample B	-	✓	-
Sample C	-	-	✓
Sample D	✓	✓	✓

Fig.1 Low power sub-32nm process flow illustration. (featuring with gate SiON, double patterning, immersion technology, multi-stressor..)

Fig.2 TEM picture of PMOS transistor with 30nm poly-Si gate length.

Fig.3 Tilt SEM view of 0.122um² SRAM region. SRAM poly definition uses double pattern and double etch to make good profile.

Table.I Wafer sets with different processing conditions, such as slim spacer1, S/D silicide resistance reduction and dual stress liner.

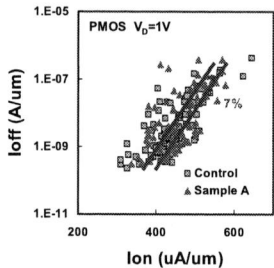

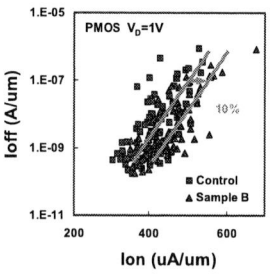

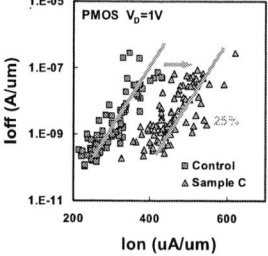

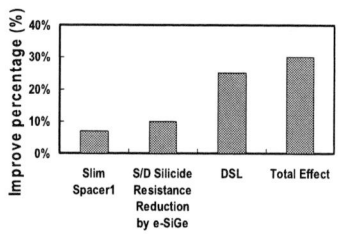

Fig.4 PMOS Sample A current gains 7% from counterpart due to slim spacer1, which makes e-SiGe strain more closer to channel.

Fig.5 PMOS Sample B current gains 10% from counterpart due to better NiSi formation above e-SiGe, which reduces the S/D silicide resistance.

Fig.6 PMOS Sample C current gains 25% from counterpart due to compressive CESL (cCESL), which improves PMOS carrier mobility.

Fig.7 The sum of the improvement factors, such as slim spacer1, S/D silicide reduction and DSL, gains 30% of drive current totally.

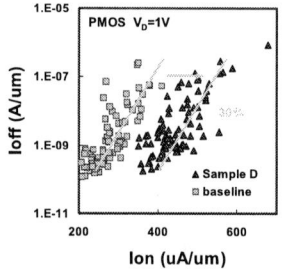

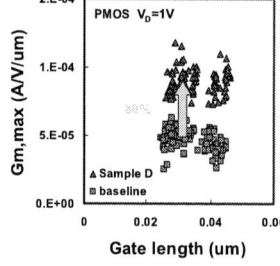

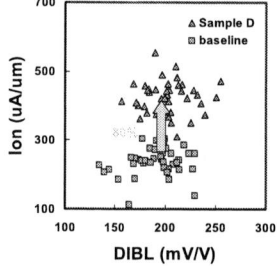

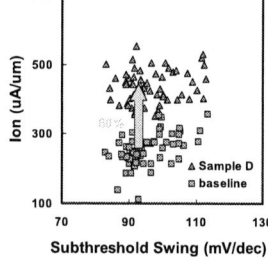

Fig.8 PMOS Sample D current gains 46% from baseline with all applied enhancement factors.

Fig.9 PMOS Sample D shows 80% gain on transconductance from baseline.

Fig.10 PMOS Sample D shows 80% current gain from baseline under the same DIBL level.

Fig.11 PMOS Sample D shows 80% current gain from baseline under the same subthreshold swing level.

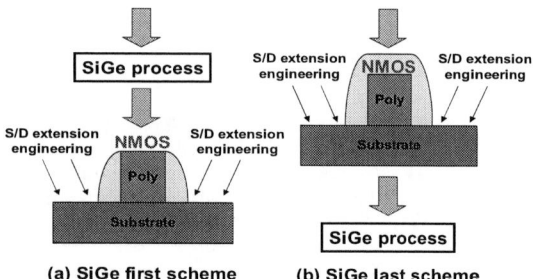

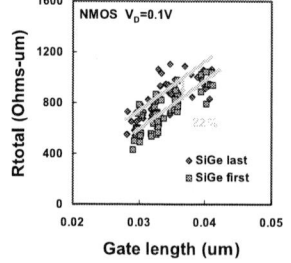

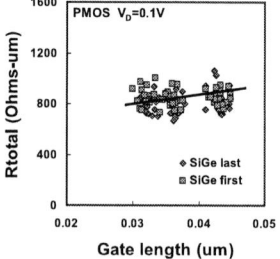

Fig.12 For NMOS improvement, it is found the clean processes after SiGe epitaxy influence the Si loss of device. (a) SiGe first scheme avoids the NMOS S/D extension dopant loss (b) SiGe last scheme will suffer the dopant loss.

Fig.13 NMOS total resistance versus gate length. SiGe first scheme shows 22% improvement of total resistance from SiGe last scheme.

Fig.14 PMOS total resistance versus gate length. SiGe first scheme shows no degradation impact from SiGe last scheme.

978-1-4244-2784-0/09 $25.00 © 2009 IEEE

Highly performant FDSOI pMOSFETs with metallic source/drain

T. Poiroux, M. Vinet, F. Nemouchi, V. Carron, Y. Morand [1], B. Previtali, S. Descombes [1], L. Tosti, O. Cueto, L. Baud, V. Balan, M. Rivoire [1], S. Deleonibus, O. Faynot

CEA-LETI/Minatec – 17, rue des Martyrs – 38054 Grenoble Cedex 9 – France

[1] STMicroelectronics – 850, rue Jean Monnet – 38926 Crolles – France

ABSTRACT

We report in this paper the fabrication and the characterization of FDSOI pMOSFETs with metallic source and drain exhibiting the best performance obtained so far on metallic source/drain devices, with Ion=345µA/µm and Ioff=30nA/µm at -1V for a 50nm gate length device. These results have been achieved thanks to a careful optimization of the source/drain to channel contacts, which can allow specific contact resistivities as low as $0.1~\Omega.\mu m^2$.

INTRODUCTION

While the CMOS transistor dimensions are scaled down to the 32 and 22nm technology nodes, the reduction of parasitics becomes a crucial issue in order to fully benefit from the performance advantages of such device shrinking. In particular, it is now well established that access resistance to the channel will constitute a first order parameter to be optimized in future technologies [1]. With that respect, several studies have been carried out on metallic source and drain reservoirs, either with Schottky [2,3] or modified Schottky [4,5] contacts to the channel. It turns out from these experiments that lowering the Schottky barrier at the channel entrance and using dopant segregation or narrow doped regions between the metallic electrode and the channel to reduce the tunnel width is mandatory to achieve high MOSFET performance [6]. In the first part of this paper, we analyze the requirements in terms of metal to semiconductor contact resistivity in order to enhance the device performance with respect to that of conventional MOSFETs. Then we show that contacts with very low resistivity, in the required range, can be achieved between PtSi and p-doped silicon. Finally, we describe the integration of such optimized contacts in a state of the art FDSOI process, and we present the electrical characterization of highly performant FDSOI pMOSFETs with W/TiN/Ti/PtSi source/drain electrodes, exhibiting Ion=345µA/µm and Ioff=30nA/µm at -1V.

METALLIC SOURCE/DRAIN OPTIMIZATION

Contact resistivity requirements

In order to determine the source/drain to channel contact resistivity (ρ_c) range required to improve the transistor access resistance with metallic source/drain, we have extracted from electrical simulations the resistance between the contact plug and the channel entrance (not accounting for the current crowding part) of a FDSOI transistor as a function of the remaining unsilicided silicon thickness in the source/drain regions for various silicide/silicon contact resistivities (Fig. 1). These results, obtained for a resistivity of the doped source/drain layer and of the silicide equal to $5~\Omega.\mu m$ and $0.2~\Omega.\mu m$ respectively, indicate that a full salicidation of the source/drain regions translates into a degradation of the overall resistance as long as ρ_c exceeds a few tenths of $\Omega.\mu m^2$, as previously shown in [7]. For values below this limit, the full silicidation is beneficial and the access resistance can be further improved by a controlled penetration of the silicide below the spacers (negative values of the x-axis on figure 1 indicate the silicide penetration in the LDD region).

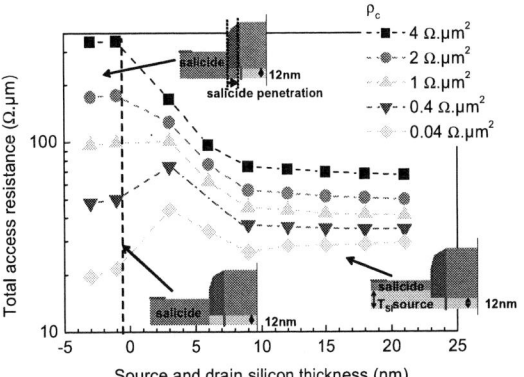

Figure 1. Simulated access resistance between the contact plug and the channel entrance of a FDSOI transistor (with a body thickness of 12nm) as a function of the unsilicided source/drain thickness for different silicide/silicon contact resistivities. Negative values on the x-axis indicate the silicide penetration in the LDD region.

Metallic source/drain module development

To characterize accurately the metal to channel contact resistance for pMOSFETs, we have fabricated specific 4 probes resistances on SOI substrates, composed of a 20nm thin boron-doped silicon film, laterally salicided (PtSi) and contacted with the source/drain metallization (Fig. 2). Thanks to various geometries (L and W) of the doped silicon active area, the resistance of these lateral contacts can be extracted separately from that of the active area.

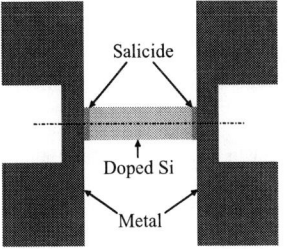 

Figure 2. Top view (left) and TEM cross section (right) of the 4-probes resistance structure dedicated to contact resistance optimization.

We have studied in particular the impact of the barrier deposited onto the silicide (TiN versus TiN/Ti), of an additional dopant diffusion annealing (DA), and of an additional lateral BF₂ implant (DSP) prior to salicidation. Table 1 summarizes the measured contact resistivities.

We notice that ρ_c can be efficiently reduced by an additional lateral BF₂ implant and by an additional diffusion annealing step.

However, the most significant ρ_c improvement is obtained when the TiN barrier is replaced by a TiN/Ti bilayer, which is related to a different species distribution at the barrier/PtSi interface [8]. Combining the TiN/Ti barrier together with a dopant diffusion annealing and a pre-silicidation lateral BF$_2$ implant leads to a mean ρ_c value of 0.6 $\Omega.\mu m^2$, with measured values as low as 0.1 $\Omega.\mu m^2$.

Barrier	DA	DSP	Contact resistivity ($\Omega.\mu m^2$)		
			Min	Mean	Max
TiN	No	No	7.7	11.6	29.0
TiN	Yes	No	7.4	8.0	22.0
TiN	No	Yes	7.0	7.6	20.5
Ti/TiN	No	No	0.4	1.7	3.7
Ti/TiN	**Yes**	**Yes**	**0.1**	**0.6**	**1.4**

Table 1. Measured contact resistivities for various source/drain metallurgies, with/wo additional diffusion anneal (DA), with/wo additional BF$_2$ lateral implant (DSP)

Finally, in order to integrate this source/drain metallization in very short transistors, the width of the lateral salicidation has to be finely controlled. On figure 3, we show that the silicide lateral penetration in a thin silicon film can be efficiently reduced to 10nm by decreasing the deposited platinum thickness down to 6nm and by reducing the salicidation temperature down to 450°C. With such a process, contact resistivities of about 2 $\Omega.\mu m^2$ are achieved.

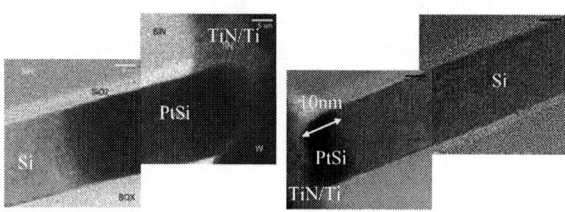

Figure 3. TEM cross section showing the width of the lateral silicide formed with 16nm (left) and 6nm (right) of deposited Pt.

DEVICE FABRICATION AND CHARACTERIZATION

Device fabrication

Starting from a state of the art FDSOI process [9] with 12nm film thickness, 3nm HfO$_2$ dielectrics, TiN metal gate and p-doped LDD, we have etched away the silicon from the source/drain region after spacer formation. Then, an additional BF$_2$ tilted implantation of the channel edges has been performed on some of the wafers prior to lateral salicidation with 6nm of platinum. The source/drain metallization, composed of a W/TiN/Ti stack, has been deposited together with an oxide capping layer, and planarized in a Damascene way. Figure 4 presents a 50nm gate length transistor. Notice the reduced penetration of the silicide below the spacers.

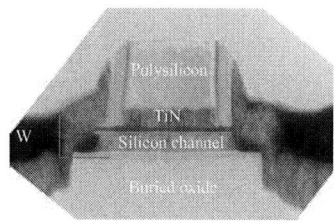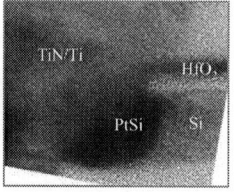

Figure 4. Left: TEM cross section of a 50nm gate length FDSOI transistor. Right: detail of the source to channel contact.

Electrical results

Figure 5 shows the Ion-Ioff cloud obtained at -1V. The best compromise is obtained with an additional lateral implant before salicidation, which is coherent to the measured contact resistivities on test structures. Good performance are then achieved, with Ion=345$\mu A/\mu m$ and Ioff=30nA/μm at -1V (Ion=480$\mu A/\mu m$, Ioff=47nA/μm at -1.2V) for a 50nm gate length (Fig. 5, right). These performances are the best results obtained so far on metallic source/drain transistors, and could be further optimized to benefit from the sub-1 $\Omega.\mu m^2$ contact resistivities demonstrated above.

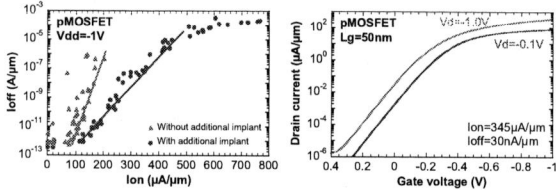

Figure 5. Left: Ion-Ioff results obtained on pMOSFETs at a -1V supply voltage with (circles) and without (triangles) additional BF$_2$ implant. Right: Transfer characteristic of a 50nm gate length device, exhibiting Ion=345$\mu A/\mu m$ Ioff=30nA/μm. The ambipolar conduction generally observed in Schottky transistors [6] is suppressed thanks to the presence of the LDD beneath the silicide.

Figure 6 shows the output characteristics of the 50nm gate length device and its output conductance as a function of the drain voltage. We notice that a perfectly ohmic contact is achieved, with a monotonous decrease of the conductance even at low drain voltage.

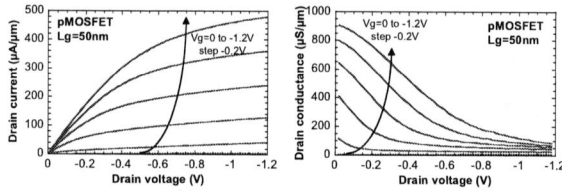

Figure 6. Drain current (left) and drain conductance (right) of a 50nm gate length pMOSFET as a function of the drain voltage.

CONCLUSION

We have fabricated performant FDSOI pMOSFETs with metallic source and drain, thanks to a careful optimization of the source/drain metallurgy, which can allow contact resistivities to the channel as low as 0.1 $\Omega.\mu m^2$. To our knowledge, these results are the best obtained on transistors with metallic source/drain, with Ion=345$\mu A/\mu m$ for Ioff=30nA/μm at a supply voltage of -1V.

Acknowledgements- This work has been carried out in the frame of the LETI-STMicroelectronics collaboration and partly funded by the UE Pullnano project IST-026828.

REFERENCES

[1] International Technology Roadmap for Semiconductors, 2007.
[2] A. Itoh et al., Jpn. J. Appl. Phys. **39**, pp. 4757-4758, 2000.
[3] J. Kezierski et al., IEDM Tech. Dig., pp. 57-60, 2000.
[4] C. Wang et al., Appl. Phys. Lett. **74**, pp. 1174-1176, 1999.
[5] A. Kinoshita et al., Symp. on VLSI Tech., pp. 168-169, 2004.
[6] J. Knoch et al., Appl. Phys. A **87**, pp. 351-357, 2007.
[7] L.T. Su et al., IEEE Elec. Dev. Lett. **15**, no 9, pp. 363-365, 1994.
[8] V. Carron et al., Mat. for Adv. Metal. Conf. (MAM), 2008.
[9] F. Andrieu et al., IEDM Tech. Dig., #23.7, 2006.

High-κ/Metal Gate Low Power Bulk Technology - Performance Evaluation of Standard CMOS Logic Circuits, Microprocessor Critical Path Replicas, and SRAM for 45nm and beyond

D.-G. Park, K. Stein[1], K. Schruefer[2], Y. Lee[3], J.-P. Han[4], W. Li[1], H. Yin[1], C. Pacha[2], N. Kim[3], M. Ostermayr[4], M. Eller[4], S. Kim[4], K. Han[4], S. Han[5], K. von Arnim[2], N. Moumen[1], M. Hatzistergos[1], T. Tang[3], R. Loesing[1], X. Chen[1], D. Jaeger[1], H. Zhuang[4], J. Chen[3], W. Yan[1], T. Kanarsky[1], M. Chowdhury[6], Jens Haetty[1], D. Schepis[1], M. Chudzik[1], V.-Y. Theon[6], S. Samavedam[6], V. Narayanan, M. Sherony[1], R. Lindsay[2], A. Steegen[1], R. Divakaruni[1], and M. Khare

IBM T.J. Watson Research Center, Yorktown Heights, NY 10598, USA, [1]IBM Microelectronics Division, Semiconductor Research and Development Center (SRDC), Hopewell Junction, [2]Infineon Technologies, 81726 Munich, Germany, [3]Chartered Semiconductor, [4]Infineon Technologies, [5]Samsung LSI logic, [6]Freescale Semiconductor, Inc., 2070 Route 52 Hopewell Jct, NY 12533, USA, Phone: 845-894-9409, FAX: 845-892-6483, E-mail: dpark@us.ibm.com

Abstract This paper presents performance evaluation of high-κ/metal gate (HK/MG) process on an industry standard 45nm low power microprocessor built on bulk substrate. CMOS devices built with HK/MG demonstrate 50% improvement in NFET and 65% improvement in PFET drive current when compared with industry standard 45nm Poly/SiON devices. No additional stress elements were used for this performance gain. The critical path circuits of this low power microprocessor built with HK/MG show dynamic performance gain over 50% at same supply voltage and 36% lower dynamic energy at same performance. Superior SRAM minimum operating voltage characteristics are achieved due to V_t variability reduction from HK/MG. Analog circuit functionality is demonstrated by a fully integrated PLL circuitry without any modification to process.

Introduction Despite many roadblocks [1], thermally stable HK/MG CMOS devices were demonstrated in gate-first processing [2-5] with dual work-function HK/MG in high-performance SOI devices [4] and 32nm node bulk Si CMOS device [5]. Aggressive T_{inv} scaling and bend-edge (BE) work function control enabled gate length scaling with good short-channel control, resulting in highly performing HK/MG devices [4,5]. This work presents performance benefit of 45nm Low Power technology with gate-first HK/MG simple CMOS devices fabricated on 300mm wafers for all relevant system building blocks of an industry standard microprocessor. This includes standard CMOS logic circuits, microprocessor critical path circuit replicas, PLL and SRAM.

Device Fabrication A schematic process flow of gate-first simple CMOS HK/MG device fabrication with bulk-Si substrate is described in Fig. 1. Thermally stable BE HK/MG devices were formed with Hf-based dielectric/interface control layer and a thin single metal layer deposition, followed by polycrystalline-Si (poly) capping over HK/MG gate stack. Following a lithographic patterning and gate stack etch process, a conventional self-aligned implant flow, single dielectric stress liner, and conventional MOL and BEOL were used. Fig. 2 shows a cross-sectional SEM image of SRAM array with full-build Cu BEOL up to M7, where the gate-first CMOS HK/MG stack is highlighted in the inset.

Logic Device Performance and Critical Path Circuit Validation
DC and AC device characteristics: Figures 3-6 show DC transfer characteristics of HK/MG, exhibiting superior drive current enhancement and short channel control as compared to 45nm-node LP 18Å SiON/Poly control. Nominal DC Idsat of 940μA/μm for NFET and 540μA/μm for PFET at Ioff = 0.45nA/μm and V_{dd}=1.1V are attained. This is superior drive current enhancement, i.e., +50% I_{dsat} for NFET and +65% for PFET, as compared to Poly/SiON control. No additional stress techniques were utilized to achieve these performance gains. To our knowledge, this is the best device performance demonstrated so far in 45nm industry standard low power device technology. Excellent short channel control with BE V_t as manifested by improved V_t roll-off (Figs. 4,6) and DIBL as low as 120mV are notable using gate-first CMOS HK/MG. Fig. 7 exhibits ring oscillator (RO) delay vs. standby power characteristics of HK/MG CMOS, showing 40% faster RO for same V_{dd} (1.1V) than control with significant standby power reduction. Library test chip (LTC) results are shown in Figs 8-9, exhibiting outstanding performance gain of 47% and 52% from complex circuitry of

NAND and NOR, respectively.

Critical Path Circuit Validation: Energy-performance trade-offs are assessed using critical path replicas of an industry standard microprocessor built with HK/MG and Poly/SiON. The critical path replicas consist of a typical mix of Inverter, NAND, NOR gates, and a synchronizing flip-flop to enable an energy and performance analysis on product-like level. Fig. 10 shows the performance of HK/MG critical path circuits, yielding a performance gain of +50% at same V_{dd} =1.1V (85°C) and +20% for 100mV lower V_{dd} compared to Poly/SiON. A strongly improved low V_{dd} gain of +70% is observed at V_{dd}=0.8V (25°C). In addition, the dynamic energy-performance trade-off (Fig.11) as function of V_{dd} demonstrates 36% lower energy at same frequency f_{CLK} and 44% higher f_{CLK} at same energy for HK/MG compared to Poly/SiON circuits. This also indicates that the +9% higher dynamic circuit energy of HK/MG due to the thinner T_{inv} can be compensated on product level by dynamic V_{dd}. Note that the speed and energy improvements of simple CMOS HK/MG circuits over Poly/SiON control were achieved without design optimization. Furthermore, high performance analog circuit functionality has been demonstrated by integrating a PLL, showing full functionality at target frequencies with high yield.

SRAM Functionality and Yield Characteristics of SRAM functionality are displayed in Figs.12-15. Butterfly curve (Fig. 12) shows a SNM of ~230mV from 0.249 μm² cell and Pelgrom plot (Fig. 13) exhibits Vt-mismatch ($V_{t\text{-}mm}$) improvement by > 40% over Poly/SiON due to HK/MG T_{inv} scaling. In Fig. 14, SRAM standby leakage (I_{stby}) as a function of V_{dd} is plotted, depicting lower I_{stby} from HK/MG over Poly/SiON. Notably, I_{stby} is further lowered at low V_{dd} region due to the gate leakage reduction benefit with HK/MG device, that can reduce standby power in the retention mode at V_{dd} < 0.8V. Fig. 15 represents the functionality of 1Mb SRAM cell, displaying no disturb fails at 0.7V/0.8V and high V_{min} yield equivalent to Poly/SiON down to 0.9V. It is noteworthy to mention that the high-temperature operating life assessment of 8Mb SRAM modules met the reliability criteria.

Conclusions Performance benefit of gate-first HK/MG process on an industry standard 45nm low power microprocessor built on bulk substrate has been demonstrated. Significant RO (> 40%) performance improvement was shown when compared to standard 45nm Poly/SiON low power technology. Critical paths circuit replicas of an industry standard microprocessor built with this technology shows over 50% performance gain at same V_{dd} and 36% lower dynamic energy in the critical path due to improved voltage scaling behavior. Analog circuit functionality has been also realized without any design modifications. SRAM functionality, with reduced V_t variability and excellent yield at low V_{dd}, has been demonstrated with HK/MG. This technology offers an easy, low cost and scalable solution for the low power applications space towards 45nm and beyond.

Acknowledgment: This work was performed at the IBM Microelectronics division, Semiconductor Research and Development Center, Hopewell Junction, NY 12533. The authors thank to Drs. T. Chen, G. Patton, G. Shahidi, P. Gilbert, R. Jaganathan, R. Mahnkopf, J. Winnerl, and F. Neppl for their support.

References

[1] E. Gusev, V. Narayanan, M. Frank, IBM J. Res. & Dev. Vol. 50, pp. 387-410 Jul-Sep. 2006.
[2] D.-G. Park et al., VLSI Tech. Dig., p.186, 2004,
[3] V. Narayanan et al., VLSI Tech. Dig., p.224, 2006,
[4] M Chudzik et al. VLSI Tech. Dig., p.194, 2007.
[5] X. Chen et al., 2008 VLSI Tech. Dig., p.88, 2008.
[6] E. Josse et al., IEDM Tech. Dig., p.679, 2006.

Fig 1: A schematic process flow of HK/MG simple CMOS integration.

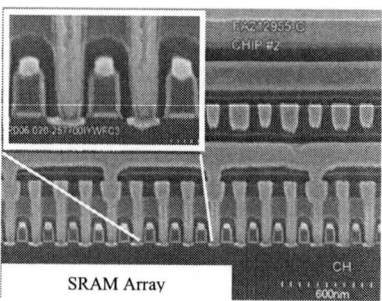

Fig. 2. Cross-sectional SEM image of gate-first HK/MG simple CMOS full-built for large SRAM array.

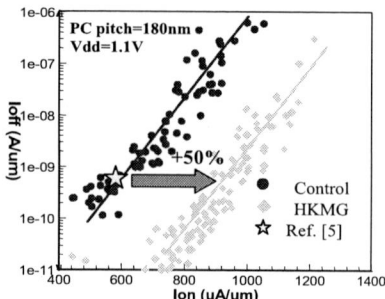

Fig. 3: NFET performance enhanced by ~50% using HK/MG as compared to Poly/SiON control.

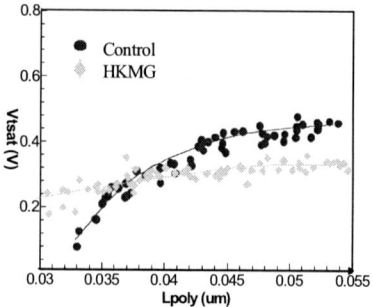

Fig 4: NFET Vtsat as a function of gate length showing significant Vt roll-off improvement.

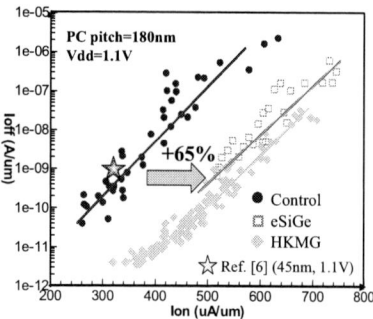

Fig.5: PFET performance improved by ~65% with HK/MG over control, by 8% over eSiGe at Vdd=1.1V.

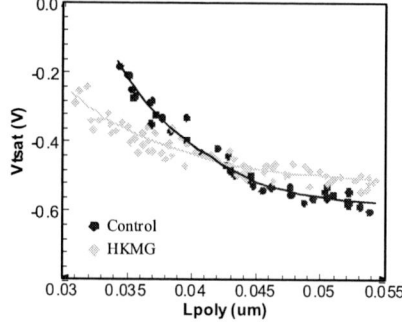

Fig 6: PFET Vtsat as a function of gate length showing significant Vt roll-off improvement with HK/MG.

978-1-4244-2784-0/09 $25.00 © 2009 IEEE

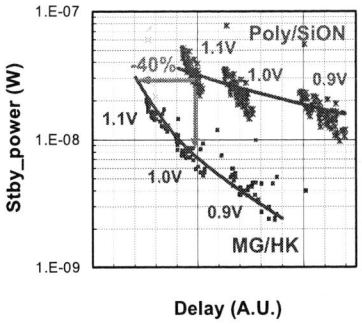

Fig. 7: Inverter (RO) delay vs. standby power characteristics of HK/MG as compared to poly/SiON at various Vdd.

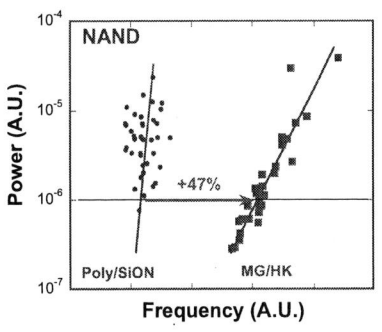

Fig.8: LTC NAND frequency vs. power of HK/MG as compared to Poly/SiON.

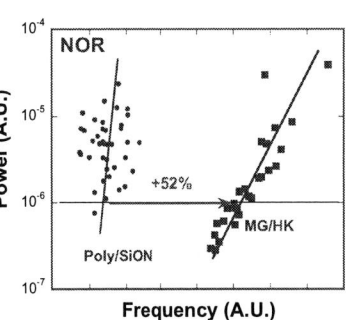

Fig. 9: LTC NOR frequency vs. power of HK/MG as compared to Poly/SiON.

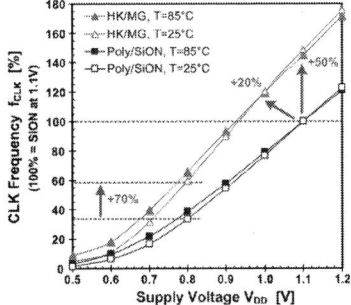

Fig.10: Performance comparison of HK/MG critical path circuit compared to poly/SiON as a function of Vdd.

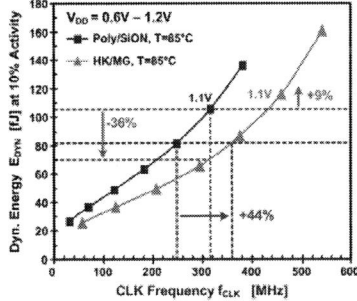

Fig.11: Dynamic energy versus performance of HK/MG critical path circuit compared to poly/SiON as a function of Vdd.

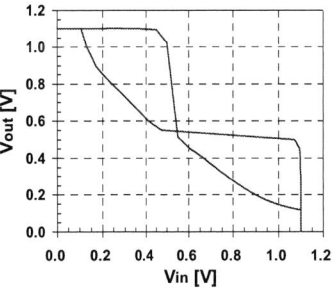

Fig.12. SRAM Butterfly curve of HK/MG simple CMOS devices, showing SNM ~230mV in 0.249 μm2 cell.

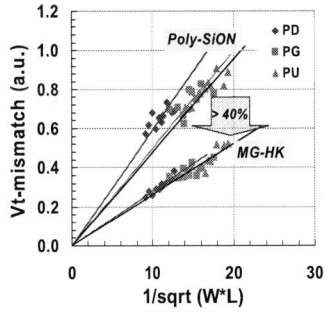

Fig.13. Vt-mismatch (Vt-mm) of HK/MG SRAM as a function of $1/(W*L)^{1/2}$, showing > 40% reduced Vt-mm as compared to Poly/SiON.

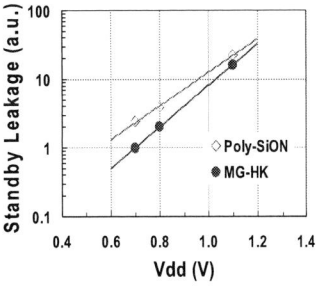

Fig.14: SRAM Istby leakage as a function of Vdd, comparing cell leakage between HK/MG and Poly-SiON.

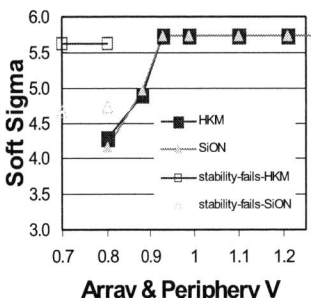

Fig.15: SRAM cell functionality, displaying no stability fails at 0.7V/0.8V for HK/MG and equivalent Vmin yield to poly/SiON. Note that a soft sigma value of 5.74 corresponds to the perfect yielding SRAM.

978-1-4244-2784-0/09 $25.00 © 2009 IEEE   85

NGL Overview

Burn Lin

TSMC, Taiwan

ArF water-immersion lithography supports 1.35 NA or slightly higher but cannot reach the theoretical limit of 1.44 NA. It is increasing difficult to resolve half pitches below $k_1 = (HP/\lambda)*NA = 0.28$, i.e. 40-nm half pitch.

Double Patterning Interactions with Wafer Processing, OPC and Physical Design Flows

Kevin Lucas

Synopsys, Inc., USA

In this work we study interactions of double patterning technology (DPT) with lithography, masks synthesis and physical design flows for the 22nm device node. DPT methods decompose the original design intent into two individual masking layers which are each patterned using single exposures and existing 193nm lithography tools. Double exposure and etch patterning steps create complexity for both process and design flows. DPT decomposition is a critical software step which will be performed in physical design and also in mask synthesis. Decomposition includes cutting (splitting) of original design intent polygons into multiple polygons where required; and coloring of the resulting polygons. We evaluate the ability to meet key physical design goals such as: reduce circuit area; minimize rework; ensure DPT compliance; guarantee patterning robustness on individual layer targets; ensure symmetric wafer results; and create uniform wafer density for the individual patterning layers.

Multiple electron beam maskless lithography for high-volume manufacturing

Jack J.H. Chen, S.J. Lin, T.Y. Fang, S.M. Chang, Faruk Krecinic, and Burn J. Lin
Taiwan Semiconductor Manufacturing Company
8, Li-Hsin 6th Road, Hsinchu 30077, Taiwan, R.O.C.

INTRODUCTION

The steeply increasing price and difficulty of the masks makes the mask-based optical lithography, such as ArF immersion lithography and extreme ultra-violet lithography (EUVL), unaffordable when going beyond the 32-nm half-pitch (HP) node[1]. The electron beam direct writing (EBDW), or called maskless lithography provides an ultimate resolution without the troubles from mask but the extremely low productivity of traditional single beam systems made it impossible for mass manufacturing after over 3 decades of development.

Though electron beam lithography has been long used for mask writing, it is yet very slow and typically takes almost a day to complete a high-end mask. Of course direct writing on a 300-mm wafer with around 100 fields would take much longer. To make it production-worthy, its throughput should achieve at least >10 wafers per hour (WPH) from a single chamber or >100 WPH by a cluster within a scanner footprint while maintaining high resolution. A more than 3-order of improvement in throughput is required. Increasing the beam current in the conventional single beam system is apparently not enough. Using the maturing MEMS technology and electronic control technology to make more than ten thousand electron beamlets write in parallel is the most possible and straightforward solution. Several groups[2][3][4][5] have proposed different multiple electron beam maskless lithography (MEBML2) approaches, with either Gaussian beams, variable shape beams or cell projections, to increase the throughput. In this paper, the challenges and possible solutions to achieve production-worthy throughput of MEBML2 for the 32-nm HP node and beyond are discussed.

THROUGHPUT

In an EBDW system, the average throughput in WPH can be simply calculated as,

$$WPH = \frac{3600}{t} = \frac{3600}{t_w + t_m + t_o},$$

Where t is the average total time to process one wafer, t_w is the writing time, t_m is the total time for movements from shot to shot including turnarounds, and t_o is the overhead time for load/unload the wafer, alignment, calibration and else activities before writing. Reducing any of these times directly results in higher throughput. The writing time can be simply calculated from the required dosage over the beam current as

$$t_w = \frac{Q}{I} = \frac{Dose \times Area}{n \times i},$$

Where Q is the total electric charge deposited on wafer; I, the total beam current; n, the number of parallel beamlets; and i, the beam current of each beamlet. A lower dosage or a higher total beam current may directly reduce the writing time. But considering the shot-noise-effect induced line width roughness (LWR), the lowest dosage to facilitate LWR < 1 nm should be around or higher than 30 $\mu C/cm^2$. The 3-order throughput increment shall then be overcome by increasing the beamlet counts to over ten thousands as the single beam current i is limited by the source brightness and spot size. The

parallelism of electron beams can be multiple beams in a column, and multiple columns in a chamber. However, due to physical limitation of electron source brightness or Coulomb effect at the crossover inside the optics, the throughput from a single column of each multiple beam approach is yet limited up to 1-20 WPH. A MEBML2 tool from MAPPER technology contains 13,000 e-beams focused on the wafer by MEMS electrostatic lens array, writing with high-speed optical-switched blanker array, as illustrated in Figure 1. Each beamlet has its own optical column to avoid the central crossover and hence secures the possibility of > 10-wph throughput at the 32-nm HP node. Clustering multiple chambers can then further multiply the throughput.

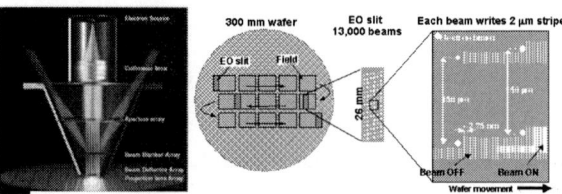

Figure 1. Principle of MAPPER Technology.

Removing mask restrictions, the exposure scan route can omit the unwanted turnarounds and last scanning at constant speed to cross the whole wafer diameter, field-by-field, then turn back. Therefore the turnaround counts are minimized. As a result, the stage only requires a below 40-mm/s scan speed to achieve a 20-WPH throughput, comparing with 600 mm/s scan speed for >145 WPH in the latest optical scanner. Consequently, the footprint of a chamber including an e-beam column and a 300-mm wafer stage can be encapsulated to within 1×1 m^2, whose cost and footprint can be minimized for realizing the cluster concept. To achieve a >100-WPH throughput, a cluster contains 6 chamber for 20-WPH per chamber, or 10 chambers if only 10-WPH per chamber, and hence only occupy a footprint of 14.4 or 23.4 m^2 respectively, which is comparable to

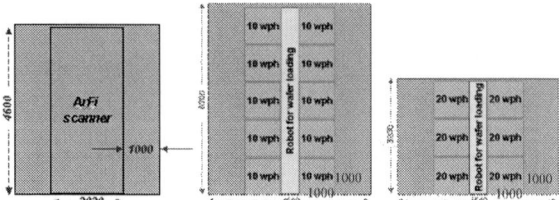

Figure 2. Footprint of an ArF immersion scanner, and MEBML2 clusters of 10x10WPH and 6x 20WPH.

the optical scanner's 18.5 m^2 and much smaller than 48.7 m^2 of a >100-WPH EUV HVM. This clustered MEBDW system can then interfaced to an existing track with conventional clean room layout and operation flow.

RESOLUTION AND PROXIMITY CORRECTION

The resolution of electron beam lithography should not be a problem, as long as the beam size can be focused to an extremely small spot. However, in an MEMS-fabricated electro-static optics, according to the fundamental electromagnetic principles, the beam current trades off the spot size in 4th power. We then cannot avoid using a relatively larger spot size to write the features to maintain the throughput. Using a SEM-converted writer, Elionix ESM9000, we have proven the resolution down to 30 nm HP at 5 keV and 25-nm spot size by using 30-nm thick Hydrogen Silisesquioxane (HSQ), which is a negative-tone non-chemical amplified resist, as shown in Figure 3.[6][7] The high energy of the electron beam enables a wide

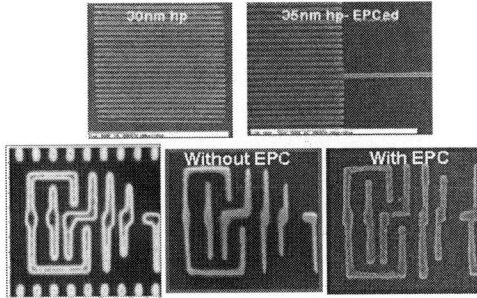

Figure 3. Imaging result of 5-keV EBDW by using HSQ.

range of selection of resist materials. This HSQ has been proven for ultra high resolution and it may provide a benefit of high etch selectivity in some specific etching process. The resist thickness for the low keV EBL has to be so small because the electron scattering effect produces a blur to the image. Nonetheless ultra-thin resists with aspect ratio below 3 is required to avoid collapsing with other lithography systems unaffected by scattering blurs. The edge-based proximity effect correction (PEC), like the conventional model-based OPC, has also been proven feasible in eliminating the electron scattering effect. The 5-keV MEBML2 is possible to go further down to 22-nm HP, with 20-nm beam size and 20-nm resist thickness, from simulation.

WAFER HEATING EFFECT

The wafer heating effect induced by the deposited high-energy electrons is a big concern at such high throughput. The estimated heating power for absorption in the resist and the remaining power through the resist into the substrate were benchmarked among 5 keV, 50 keV, ArF and EUV, as shown in Figure 4. The estimation was

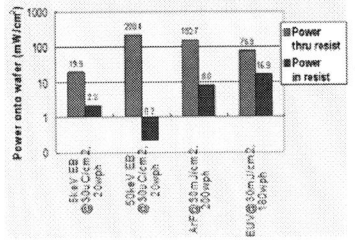

Figure 4. Radiation induced heating power in resist and through resist onto the wafer.

done for the worst radiation loading density for each case, 25% for e-beam, 50% for EUV, and 75% for ArF as there is not yet good negative resists for ArF and EUV. The freeware CASINO v2.42 was used to simulate the absorption energy inside the resist for both electron beam cases, while for ArF and EUV, the absorption of resist is calculated by assuming the absorbance at $\alpha \sim 1$ and 4 μm^{-1} respectively. Although electrons contain much higher energy than photons, the average heating power for a 5-keV, 20-WPH MEBML2 tool is not higher than that of an ultra-high throughput 200-WPH optical scanner. While 50-keV e-beam will have the highest heating power if 20-WPH is reached but much lowest heating in the resist as it is almost transparent to the 50-keV electrons. The higher energy

absorption inside resist at 5 keV also implies higher exposure sensitivity. Experiments have proven the benefits of around 10X reduction of dosage on the identical resist from 50 to 5 keV. These explain why the low energy e-beam at 5 keV is preferred than 50 keV for pushing the MEBML2 throughput.

DATA RATE

Data rate is another big challenge. To achieve high resolution, the pixel sizes have to be small in the data format. For 32-nm HP node at pixel size of 2.25 nm, a black-and-white bitmap file for a field will be around 21 Tera-Bytes (TB), including the redundant area. Pushing the throughput to >10 WPH, a data rate of >10 TB per second is required. Provide each beamlet an optical fiber, the data transmission rate per channel will be at 1~10 GHz. The data speed of the optical fiber and transmitter is well within the present-day communication technology capability. But the challenge is to integrate the 13,000 channels at such high-speed and synchronized, at a reasonable cost. Since MEMS-made electron optical columns shall be far cheaper than DUV optical lenses, which typically consumes more than one third of the scanner cost, the big cost component shifts from the optical lenses to the data path consisting of a huge memory and computing power such as FPGA or GPU. The computing power is required for decompressing the layout to bitmap and adding in process and equipment corrections in real time. So the wonder for using MEBML2 technology is that its cost drops as fast as that of the wafer price by following the same Moore's law, while the tool price for mask-based optical solutions will continue to go up.

CONCLUSIONS

Based on the maturing MEMS capabilities and electronics technologies, the cost effective high-throughput MEBML2, at >100 WPH and footprint similar to an optical scanner, can be realized. Resolution, proximity correction, wafer heating and data rate shall not be problems for 5 keV at such high throughput. Another big advantage of focusing on MEBML2 as the lithography solution for 32-nm HP node and beyond is that it only needs investments on developing this tool. Unlike EUV and double patterning, which need enormous investments on the mask infrastructure and process development, besides just the cost of the lithography tool. However, the success of the MEBML2 technology still requires enormous industrial support and investments, which may happen only when it is commonly viewed as one of the mainstream technologies for high-volume manufacturing. To catch up manufacturing of the 32-nm HP node, the clustered platform has to be ready by 2012, which needs big platform suppliers' involvement very soon.

REFERENCES

[1] Burn J. Lin, Proc. SPIE 6520-02 (2007).
[2] M.J. Wieland, G. de Boer, P. Kruit, G.F. ten Berge, A.M.C. Houkes, R.J.A. Jager, T. van de Peut, J.J.M. Peijster, E. Slot, S.W.H.K. Steenbrink, T.F. Teepen, A. van Veen, B.J. Kampherbeek, Proc. SPIE 6921-92 (2008).
[3] Paul Petric, 52nd International Conference on EIPBN (2008)
[4] C. Klein, E. Platzgummer, H. Loeschner, G. Gross, Proc. SPIE 6921-93 (2008).
[5] N. William Parker, Sematech 2008 Litho Forum.
[6] S.M. Chang, S.J. Lin, C.A. Lin, J.H. Chen, T.S. Gau, Burn J. Lin, P. Veltman, R. Hanfoug, E. Slot, M. J. Wieland, B.J. Kampherbeek, Proc. of SPIE 6921-19 (2008).
[7] S.J. Lin, S.M. Chang, T.Y. Fang, B.R. Luo, Faruk Krecinic, J.H. Chen, Proc. of SPIE 7271-54 (2009).

Status and Challenges of Extreme-UV Lithography

Kurt Ronse, Eric Hendrickx, Mieke Goethals, Rik Jonckheere, Geert Vandenberghe
IMEC
Kapeldreef 75
B3001 Leuven
Belgium

INTRODUCTION

Extreme UV lithography (EUVL) is considered today to be the leading candidate for printing the critical layers of the 22nm half pitch technology node. The most convincing progress in EUVL has recently been demonstrated through lithography results generated by the first EUV full field exposure tools, developed by ASML and Nikon. One of the ASML EUV alpha demo tools (ADT) is installed at IMEC and this paper will discuss the most important results obtained on this full field litho-cell[1,2].

Today, the following items are generally considered to be the focus areas requiring special effort and attention to bring EUVL to a manufacturable technology : *(i)* long term source operation with 100W at intermediate focus, *(ii)* defect free masks through mask lifecycle and availability of inspection/repair infrastructure, *(iii)* resist resolution, sensitivity and line edge roughness (LER), *(iv)* reticle protection during storage handling and use, and *(v)* protection and lifetime of illuminator optics and masks. Except for item *(i)*, which is beyond the scope of the ADT, all focus points are being addressed. In the next 3 sections, the imaging results obtained on the EUV ADT, the progress in resist performance, and the progress in reticle performance, will be discussed.

EUV ALPHA DEMO TOOL IMAGING

The characterization of the EUV ADT imaging performance started in 2008 using the 80nm thick Rohm&Haas XP6305-A baseline resist. The lithographic analysis focused on 40nm lines and spaces, exposed across the full field. Uniform wafer and focus-exposure matrices were used. The process windows for 40nm horizontal and vertical dense lines were measured and the overlapping process window determined. The individual process windows at one slit position typically exceed 20% exposure latitude and 300nm depth of focus. Due to slight focus and dose variations across the slit, the common window still has 19% exposure latitude and 300nm depth of focus.

For the uniformly exposed wafers, the printed linewith of 40, 45 and 50nm lines and spaces was measured over 52 positions within the die and for all dies on the wafer. The 3 sigma values from the resulting measurements, representing the CD uniformity (CDU), on the fields exposed at nominal focus are shown in Figure 1a. Note that the data in Figure 1a were obtained for the pooled H and V CD values after applying a CD correction for the shadowing effect to overlap the H and V CD distributions. When CD was measured in the nominal focus fields only, the 3 sigma values for 40nm horizontal and vertical LS combined are 3nm for both wafers. The 3 sigma values further decrease for the more relaxed 45nm and 50nm lines and spaces. All these CDU values are well within the +- 10% CD spec. From the CDU measurements, the evolution of the average CD along the slit can be determined by averaging over the dies exposed in nominal focus and over the different scan positions. This signature is shown in Figure 1b, along with the intensity at wafer level as measured by the slit uniformity test. From Figure 1b, it is clear that

the intrafield CD variation has a total range of 2nm. The normalized slit intensity has a range of 3%, and by comparing the slit signatures of CD and intensity in Figure 1b, the correlation between slit intensity and wafer CD is clear. An important part of the CD variation within die can thus be attributed to the dose variation along the slit.

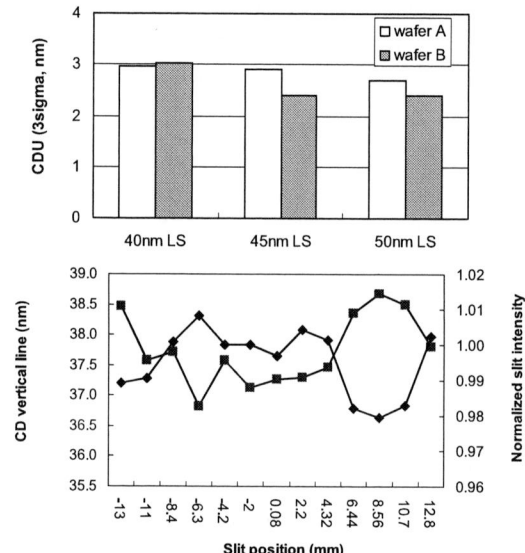

FIGURE 1. A (TOP) CDU IN NOMINAL FOCUS; B (BOTTOM) AVERAGE CD ALONG THE SLIT FOR NOMINAL FOCUS (SQUARES), ALONG WITH SLIT INTENSITY PROFILE (DIAMONDS).

A second contribution to the CD fingerprint is related to the variation along the slit of best focus position.

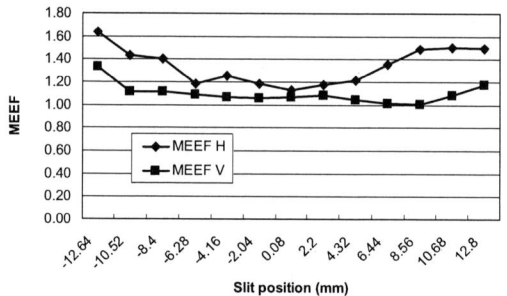

FIGURE 2. MEEF FOR 40NM HORIZONTAL AND VERTICAL LINES AND SPACES ACROSS THE EXPOSURE SLIT.

978-1-4244-2784-0/09 $25.00 © 2009 IEEE

A third factor contributing to the CD fingerprint can be the reticle fingerprint. Usually this contribution is very large in 193nm lithography, due to the large Mask Error Enhancement Factor (MEEF). The smaller MEEF in EUVL should lead to better CDU given similar reticle CD uniformity as available in 193nm lithography. Figure 2 shows the measured MEEF in EUVL for 40nm horizontal and vertical lines and spaces across the exposure slit in nominal focus. This MEEF is determined as the relation between the CD errors from the target as printed on the wafer, in relation to the CD errors from the target as measured on the reticle (and normalized to 1x). The MEEF numbers are indeed very low compared to 193nm lithography.

EUV RESIST PROGRESS

Another important task of the EUV ADT is to test the readiness of EUVL resists using full field exposures. So far most EUV resists have been tested primarily on micro-exposure tools or interference set-ups. The main challenge in resist development is to simultaneously achieve the resolution, sensitivity and LER requirements. Based on pre-screening of resists using interference lithography at the Paul Scherrer Institute[3,4], the progress in resist performance regarding these issues was followed-up and promising resist candidates were selected for testing on the ASML ADT.

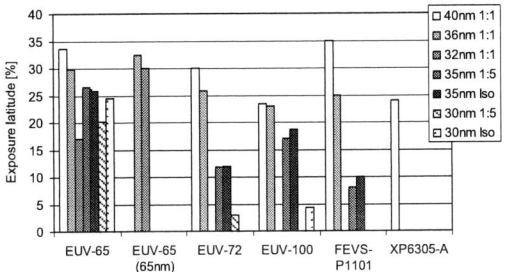

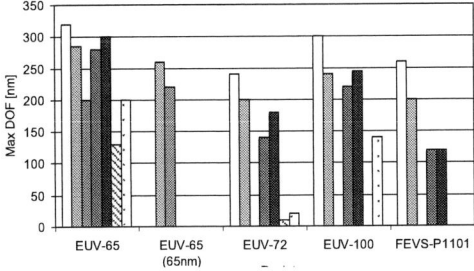

FIGURE 3. A (TOP) MAXIMUM EXPOSURE LATITUDES; B (BOTTOM) MAXIMUM DEPTH OF FOCUS; FOR VARIOUS DENSE AND (SEMI-) ISOLATED FEATURE SIZES AND VARIOUS RESISTS.

When screening resist on the EUV ADT, the performance of the new resists is compared to our current internal performance targets on the ADT (photospeed <15 mJ/cm^2, resolution 32nm LS, 30nm iso line and 40nm contacts, LER<4nm) and to the reference/baseline resists for the ADT, which are Rohm&Haas XP6305-A and Fuji FEVS-P1101. Processing conditions were used as recommended by the resist supplier. Resist thickness was 80nm for most of the evaluations up to recently. However, as pattern collapse was found to be an issue at sub-32nm dimensions, resist thickness reduction to 60-65nm is being considered.

A number of performance items from the benchmarking procedure are illustrated for the baseline resists (XP6305-A, FEVS-P1101) and for some new resists samples (EUV-65, EUV-72, EUV-100) in Figure 3. Process windows have been measured for different feature sizes; Exposure latitude and Depth-of-Focus have been calculated. Improved exposure latitudes are obtained with the new materials compared to the baseline for sub-40nm features (Figure 3a). Good focus latitudes are seen for dense features down to 32nm dense which was to be expected as EUV is imaging at low NA=0.25 (Figure 3b). Good exposure latitudes are seen for 32nm dense lines, as well as for 30nm isolated lines in case of the EUV-65 resist. Reduction of resist thickness to 65nm resulted in even wider latitudes on 32nm dense (EL=30% and DOF 220nm). Almost all resists have a linear resolution down to 32nm lines and spaces, only a few better resists can print 30nm isolated lines and a 28nm half pitch flash pattern.

EUV RETICLE PROGRESS

As discussed in previous publications[5,6], EUV reticles can have four types of defects: absorber defects, particles, local reflectivity loss and multilayer (ML) type defects. The printability of the latter is mainly determined by their size, the height of the ML disturbance at the top of the ML, and its cross-sectional profile [7]. Due to the lack of pellicle, the risk of capturing printable particles is much higher for EUVL masks compared to traditional 193nm photomasks.

CONCLUSIONS

In this paper, the experiences on full field EUVL lithography are reviewed. Besides the imaging performance of the EUV ADT at IMEC, also the progress in resists and reticles are discussed and compared to the production requirements for EUV lithography.

ACKNOWLEDGEMENTS

We acknowledge the ASML EUV alpha demo tool team We thank the IMEC Partners in the Advanced Lithography Program.

REFERENCES

[1] H. Meiling et al, "Performance of the Full Field EUV Systems", Proc. SPIE 6921 (2008)

[2] G.F. Lorusso et al, "Imaging performance of the EUV alpha demo tool at IMEC", Proc. SPIE 6921 (2008)

[3] Roel Gronheid, Harun H. Solak, Yasin Ekinci, Amandine Jouve, and Frieda Van Roey, *Microelectronic Engineering* **83** (2006), 1103-1106

[4] A.M. Goethals et al, "Implementing full field EUV lithography using the ASML ADT", Presented at 7th International Symposium on Extreme Ultraviolet Lithography, September 29-October 1, Lake Tahoe, USA (2008)

[5] R. Jonckheere et al, "Mask defect printability in Full Field EUV Lithography", Presented at the 2007 International Symposium on Extreme Ultraviolet Lithography, Sapporo, Japan (2007)

[6] R. Jonckheere et at, "Investigation of mask defectivity in full field EUV lithography", Proc. SPIE 6730 (2007)

[7] R. Jonckheere et al, "Mask defect printability in Full Field EUV Lithography – Part 2", Presented at the 7th international 2008 International Symposium on Extreme Ultraviolet Lithography, September 29-October 1, Lake Tahoe, USA (2008)

Recent Developments in NAND Flash Scaling

Krishna Parat
Intel Corporation
2200 Mission College Blvd., Santa Clara, CA, U.S.A. 95054
email: krishna.parat@intel.com

ABSTRACT

NAND Flash cell has scaled by >1000X in area since its inception over 2 decades ago. There are, however, several scaling challenges that need to be overcome to continue scaling below the 3X node. Many evolutionary and revolutionary approaches, such as high-K inter-poly-dielectric (IPD), engineered tunnel barriers, trap based charge storage devices, as well as 3-D structures are being pursued to overcome these scaling challenges. The paper will discuss some of these challenges and related developments.

INTRODUCTION

Flash memory has played an important role in the growth of the consumer electronics industry over the past two decades. The versatility of the Flash Memory in terms of non-volatility of the stored data, large memory density, high read/write endurance, fast read/write performance, inherent ruggedness, and continued price reduction through technology scaling has made it the memory of choice in many of these electronics devices.

Among the different types of non-volatile memories, the most extensively used memory type today is the NAND Flash memory based on the floating gate technology. NAND Flash memory [1] has a simple memory cell structure and an array architecture which is very amenable to scaling. The relatively simple structure and layout of the NAND cell (Fig 1) allows for the formation of a very small cell with bitline and wordline pitches of 2λ, where λ refers to the minimum feature on the technology, leading to a raw cell size of $4\lambda^2$ which is smallest for a memory cell.

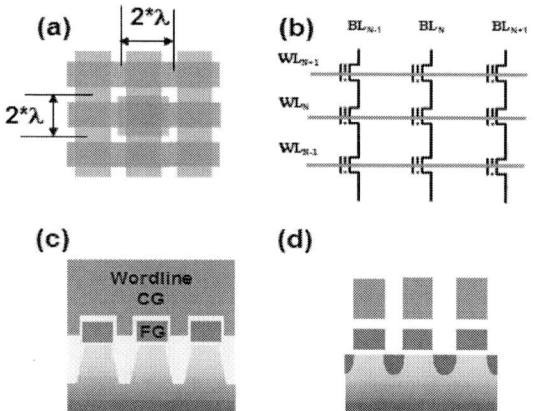

FIGURE 1. NAND Flash cell: a) layout, b) schematic, c & d) cross section along the Wordline and the Bitline directions

The historic scaling trend for the NAND cell is shown Fig. 2. The graph, which shows the effective cell size as a function of the Lithography node, shows the NAND cell size scaling down to the 4X and 3X nm nodes without showing any tendency for slowing down. Currently a significant part of the NAND Flash memory production

has already reached the 3X and 4Xnm Technology node with memory densities in the range of 16-32Gb [2-4].

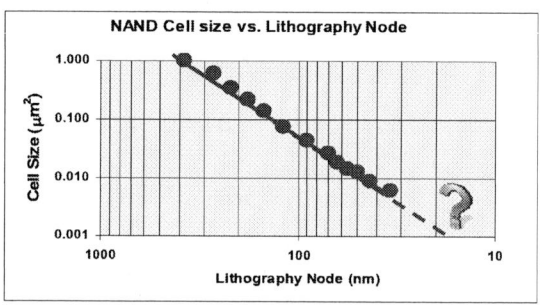

FIGURE 2. NAND Flash cell area scaling trend

FUTURE SCALING

While the NAND Flash has scaled very well over the past 2 decades per Moore's law, the future scaling is looking to be more difficult owing to the increasing number of scaling challenges. Some of these challenges are physical and some are electrical.

Physical Scaling

On the physical front, NAND Flash, owing to its simple and regular line and space layout, has enjoyed the use of many optical enhancement techniques and has led the industry in pitch scaling. This has also made NAND cell arrive at the end of the lithographic technology sooner than others. Going below the 4Xnm node requires new techniques such as double patterning [5] or use of spacers to achieve pitch reduction [6]. Once again the regular line space layout of NAND cell is conducive to this approach and as a result lithography is not expected to be a limiting factor for NAND scaling.

Another limiter on the physical scaling front is the cell structure of the floating gate NAND cell itself. Thus far the cell pitch had allowed for the control gate to wrap around the floating gate, thus providing good capacitive coupling. Going below the 3Xnm node, however, the cell pitch will be too small to allow this. As a result, the NAND cell will have to go to a planar structure that will need to employ a high-K IPD in order to maintain good coupling (Fig 3b).

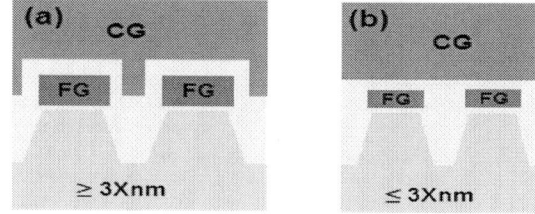

FIGURE 3. a) Conventional Floating Gate Flash Cell, b) Planar Floating Gate Flash Cell using high-K IPD layer.

978-1-4244-2784-0/09 $25.00 © 2009 IEEE

One of the key issues with the high-K IPD is the higher level of leakage through most of these high-K dielectrics owing to their lower barrier heights, which results in a lower cell Vt window. Another issue is related to the high trap densities in these films causing retention problems due to trap assisted tunneling [7].

Electrical Scaling

There are several other electrical issues as well which will have to be overcome in order to extend the NAND cell scaling into the sub 3Xnm regime. One of these is the cell to cell interference due to the increased proximity of the cells resulting from scaling. Due to the capacitive coupling between neighboring cells, the Vt of a given cell becomes dependent on the Vts of the surrounding cells (Fig 4).

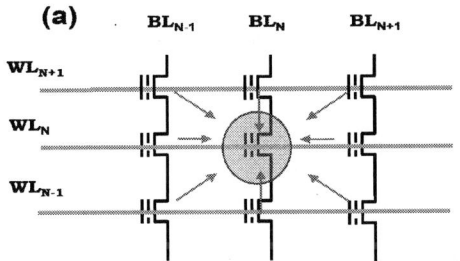

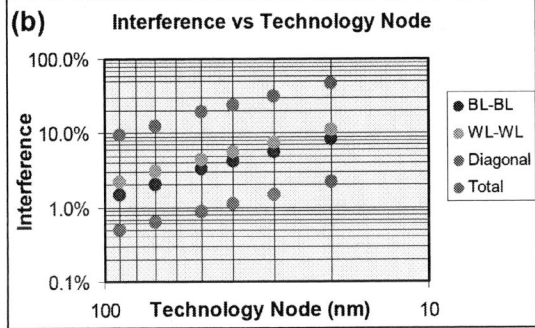

FIGURE 4. (a) Circuit schematic showing the interference effect, b) Interference as a percentage of the Vt shift of the various neighboring cells.

As shown in Fig 4, a simple scaling projection shows cell to cell interference becoming as much as 40% by the 20nm node, which is quite unacceptable. The interference can be reduced by going away from a floating gate type of NAND cell to a trap based NAND cell where the charge is stored in a trap layer such as a nitride layer or a discrete nano-crystal layer (Fig 5). While a trap based cell is not fully immune to cell to cell interference, it can significantly cut down the extent of interference to allow NAND cell scaling to be extended by another node or two [8].

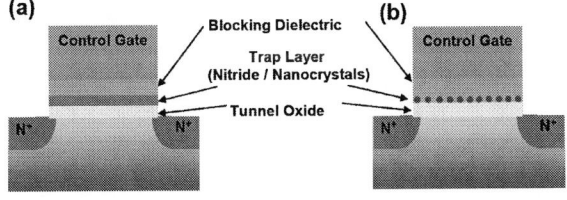

FIGURE 5. Trap Based Flash Cell. a) Nitride based Flash Cell, b) Nano Crystal based Flash Cell.

Some of the key issues with the Nitride based Flash Cell are the poor programming characteristics [9] and the high erase voltage required for hole erase which also leads to poor erase Vt. The erase voltage as well as the erase Vt issue can be adequately addressed through the use of a barrier engineered tunnel dielectric consisting of an Oxide-Nitride-Oxide multi-layer [10]. The key concern with the nanocrystal based cells is the density control of the nano crystals. With semiconductor nanocrystals there is the concern of quantum effects as well that can lead to poor program Vt window and retentions, which is not deemed to be an issue with metal nanodots [11]. However, metal nanodots introduce an additional concern of contamination for the process line.

Some of the other electrical scaling issues stem from the fact that cell scaling leads to decreasing cell capacitance and as a result, the number of electrons stored in the cell for a given Vt shift decreases. This makes the cell Vt more sensitive to single electron events such as trapping and de-trapping which can manifest in the form of 1/f noise [12] and an increased susceptibility of the cell to charge loss as well as other number fluctuation effects [13].

Three Dimensional Approaches

Given some of these issues with the scaling of the cell in the lateral dimension, there are on-going efforts towards achieving cost reduction via three dimensional approaches. Two diverse approaches have been considered. One approach is to stack two layers of NAND cells that share some of the elements of the NAND string such as the bitline and contacts as well as have a common periphery circuit [14]. Another approach is to make the NAND string into a vertical string with wordlines in the string stacked vertically instead of laterally as in the conventional technology [15]. While the former offers an evolutionary approach to scaling the latter is more extendable.

SUMMARY

Flash Memory has played an important role in the consumer electronics area and looking forward is expected to play an important role in the computing environment as well. This has happened as a result of the continued technology scaling over the past 2 decades. Future of scaling looks more challenging than before. There are, however, a large number of potential technical paths out there that can help overcome these challenges and extend the scaling further.

REFERENCES

[1] W. D. Brown, and J. E. Brewer, *Nonvolatile Semiconductor Memory Technology.* IEEE Press, 1997.
[2] M. Noguchi, et al., IEDM Technical Digest, pp. 445-448, 2007
[3] T. Kamigaichi, et al., IEDM Technical Digest, pp. 827-830, 2008
[4] R. W. Zeng, et al., *To be presented at ISSCC 2009*
[5] D. Kwak, et al., VLSI Tech Symp Digest, pp. 12-13, 2007
[6] M. F. Beug, et al., NVM Tech Symp Digest, 2008
[7] B. Govoreanu, et al. IEDM Technical Digest, 2006
[8] K. Kim, et al., IEDM Technical Digest, pp. 27-30, 2007
[9] A. Furnemont, et al., IEEE NVSM Workshop, pp. 96-97, 2007
[10] S-C. Lai, et al., IEEE NVSM Workshop, pp. 88-89, 2007
[11] P. K. Singh, et al., IEEE Elec. Dev. Lett, 29, pp. 1389-1391, 2008
[12] K.Fukuda, et al., IEDM Technical Digest, pp. 169-172, 2007
[13] C. Compagnoni, et al., IEDM Tech Digest, pp. 165-168, 2007
[14] S-M. Jung, et al. IEDM Technical Digest, pp. 37-40, 2006
[15] H.Tanaka, et al. VLSI Tech Symp Digest, pp. 14-15, 2007

Modeling and Scaling Evaluation of Junction-Free Charge-Trapping NAND Flash Devices

Yi-Hsuan Hsiao, Hang-Ting Lue, Kuang-Yeu Hsieh, Rich Liu and Chih-Yuan Lu

Emerging Central Lab, Macronix International Co. Ltd, 16, Li-Hsin Road, Hsinchu Science Park, Hsinchu, Taiwan, R.O.C.

E-mail: htlue@mxic.com.tw, or falaHsiao@mxic.com.tw

Abstract

The "junction-free" charge-trapping NAND Flash [1,2] is studied extensively. Simulation results show that the junction-free NAND Flash is scalable beyond 15 nm node (half pitch) with reasonable DC characteristics, while the conventional "with-junction" NAND device shows much worse short-channel effect. Simulation results show that lower p-well doping and smaller space (S) between the WL's are two key factors to enable the higher performance of junction-free NAND device. For the first time, we point out that the parameters of the region under the space (S) such as interface traps (Dit), parasitic trapped charge, and local p-well doping have strong impact on cell characteristics. Experimental results on junction-free BE-SONOS device showed some discrepancy with the simulation that may be due to non-ideal factors under the space. Finally, the feasibility of junction-free device on SOI for the future 3D NAND Flash is also examined.

I. Introduction

NAND Flash scaling will quickly go into the sub-30 nm node in the next few years. One important challenge is that the EOT of gate stack is not scalable, while channel length continues to decrease. In order to offer better short-channel effect, we proposed a novel junction-free BE-SONOS NAND structure [1]. Junction-free is only suitable for charge-trapping devices, since the very small space (S) between WL's for the conventional FG will cause severe coupling interference. It was also found that the junction-free device has much lower junction leakage, thus providing better self-boosting efficiency [2].

In this work, we have extensively carried out simulation to understand the operation principles and scaling capability of junction-free NAND Flash devices. Experimental results in 75 nm node BE-SONOS devices are compared to validate the simulation.

II. Device Structure and Sample Description

The junction-free device structure is shown in Fig. 1 (a). There is no junction inside the NAND array, while the diffusion junction is only at the outside region. Figure 1(b) shows the TEM picture of 75 nm junction-free BE-SONOS NAND device. A small space (<30nm) is fabricated to enable better performance. The ONONO under the space is cut-through to avoid any parasitic trapped charge in the space.

III. Simulation Results and Discussion:

(a) Operation Principle:

Figure 2 shows the simulated inversion electron density. When selected gate is turned-on, the inversion layers are connected together between each WL's. Therefore channel current can pass through. On the other hand, when the selected gate is turned-off, the p-well in the selected gate is in accumulation and the inversion layer is disconnected. Thus suitable ON/OFF switching is obtained without any diffusion junction. Figure 3(a) shows that smaller space (S) allows larger current and better subthresold slope. On the other hand, the conventional with-junction device shows a much worse short channel effect for a lightly doped p-well. This proves that the junction-free device is indeed superior to the with-junction device in terms of short-channel effect.

Figure 3(b) shows that a lighter doped p-well is better for the junction-free structure. Simulation in Fig. 4 shows that the inversion density significantly decreases in the space region with higher p-well doping. Thus lightly doped p-well is very important for the junction-free NAND because it is easier to punch-through the space.

(b) Scaling simulation:

In Fig. 5, we have carried out scaling simulation of the junction-free NAND Flash toward 15 nm node. At moderate p-well doping=5E17 cm^{-3}, the junction-free NAND preserves reasonable DC performance with S.S. <400 mV/decade, and DIBL ~600mV/V at 15 nm node. For a better Vt and S.S. for 15 nm node, a higher P-well doping or smaller EOT of gate stack can be used to adjust the performance.

(c) Non-ideal effects caused by the space region:

Since there is no n$^+$ diffusion junction under the space, the electrical properties at this region certainly have impact on the device characteristics. Figure 6 illustrates several factors in the space region. The first important factor is the possible interface trap (Dit) at the space region, which is possibly caused by the etching induced damages at the silicon surface. The second factor is the parasitic remnant charge above the space. This happens when the ONONO (or trap layer) is not completely removed during WL etching. During P/E cycling, there may be some remnant charge at the space region. This will causes issues in current punch through. The third effect is that the local well doping at the space region is different from the channel region.

Figure 7 illustrates the simulation results of these three factors. Dit significantly degrades S.S. and read current. Remnant charge with electrons causes decreased read current. This suggests that the ONO is better to be removed during WL etching. Furthermore, lower p-well doping in the space region is better for the read current and S.S.

It should be mentioned that the conventional with-junction NAND device is not sensitive to the space region, since the conducting n$^+$ S/D shorts out any such effect. Thus the space sensitivity is the major characteristic of the junction-free devices.

(d) Poly Gate thickness Effect:

The poly gate thickness effect is shown in Fig. 8. When the poly gate is sufficiently thick (>10nm), there is no thickness dependence. This suggests that the fringing field is induced without thick poly, and punch through in the narrow space is easily achieved.

(e) Fitting with the experimental result:

The experimental initial IV curve of junction-free BE-SONOS NAND at 75 nm [1] (Fig. 1(b), with S=30 nm) is compared with several simulation results. A retrograde p-well is used in the process simulator to include more accurate process condition. We find that the subthreshold slope is quite difficult to fit without considering the Dit of the space. Although the optimized simulation could approach the experimental result, there is still a certain mismatch of Vt. This suggests the junction-free devices are complicated and require more parameters for accurate modeling.

(f) Programming memory window with respect to scaling:

The simulated programming memory windows versus device dimension are shown in Fig. 10. The programming window gradually decreases when technology scales, even with a constant injection density. Other simulations (not shown here) show that significant fringing field exist in very small devices (because EOT is too large compared with dimension), leading to degraded memory window.

(g) Junction-free NAND Flash in SOI structure:

Junction-free structure on SOI has potential for future 3D NAND Flash device. Figure 11 shows the feasibility of this structure. Although the I-V curves are well behaved, however, the Vt decreases sharply with thinner silicon. This is because for the fully-depleted SOI, maximum depletion is limited and Vt is therefore dependent on the silicon thickness.

V. Summary

Junction-free charge-trapping device is a promising candidate for deeply scaled and 3D future NAND Flash.

Reference:

[1] H. T. Lue, et, al, VLSI Symposia, 2008, pp.140-141.
[2] C. H. Lee, et, al, VLSI Symposia, 2008, pp.118-119.

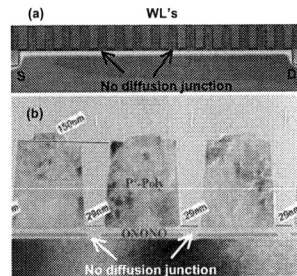

Fig. 1 **(a)** Simulation structure of the 16-WL junction-free BE-SONOS NAND Flash. Only the outer region has N⁺ diffusion junction, while the array region does not have diffusion junction. **(b)** The experimental TEM picture of 75 nm node (WL pitch=150nm) junction-free BE-SONOS NAND Flash. Very small space (<30 nm) is fabricated [1].

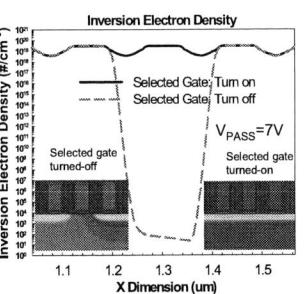

Fig. 2 The TCAD simulated inversion electron density for the junction-free NAND. When selected gate is turned-on, inversion layer is connected together, thus electrons can punch through the space (S). When selected gate is turned-off, inversion layer is suppressed.

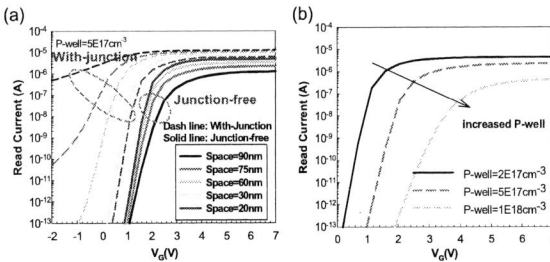

Fig. 3 **(a)** Comparison of devices with/without junction, and the effect of space (S) effect, for a 16-WL NAND array in 150 nm WL pitch (channel length Lg = 150nm –S). Smaller space (S) gives better read current and subthreshold slope (S.S.). Conventional "with-junction" NAND shows much worse short-channel effect, but also improves with smaller space. **(b)** Substrate doping effect. Higher P-well doping causes larger Vt and lower read current. Substrate doping <5E17 cm⁻³ is preferred in the junction-free NAND.

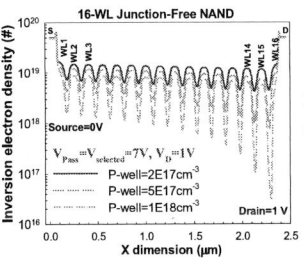

Fig. 4 Simulated electron density of a 16-WL junction-free NAND with various substrate doping. At the space region (S), the inversion density is smaller than the channel inversion. Higher substrate doping greatly decreases the inversion density under the space.

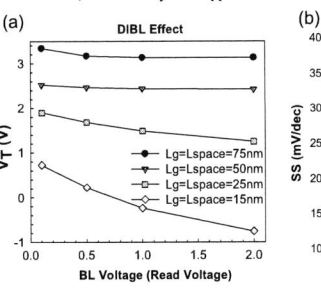
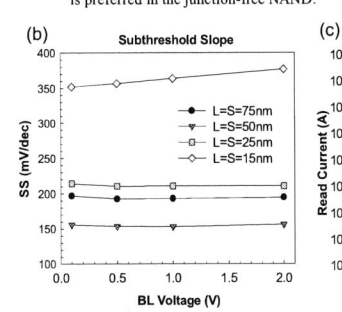
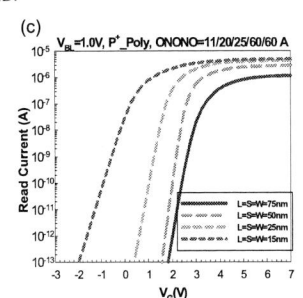

Fig. 5 Scaling simulation of junction-free BE-SONOS NAND Flash toward 15 nm node (half pitch =15 nm). WL poly line (L) = space (S) =channel width (W)=F. Substrate doping = 5E17 cm⁻³, ONONO=13/20/25/60/60 Å with P⁺-poly gate. NAND string has 16 WL's. **(a)** The DIBL effect. At 15nm node, DIBL effect is ~600mV/V. **(b)** Subthreshold slope (S.S.). S.S. is <400 mV/decade for 15 nm node. **(c)** The simulated I-V curve of selected gate (WL8) in a 16-WL NAND for various dimensions. At 15nm node, junction-free devices still show reasonable I-V curves. For very small devices, the low Vt can be adjusted by increasing the substrate doping.

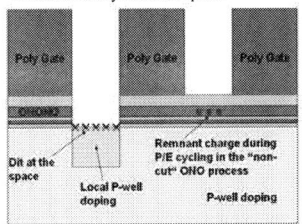

Fig. 6 Several non-ideal factors in the space region (S). **(1)** Interface trap density under the space (Dit). **(2)** remnant charge above the space. For example, if ONO is not cut-through at the space, there is possibly some remnant charge during P/E cycling. **(3)** The local p-well doping variation below the space.

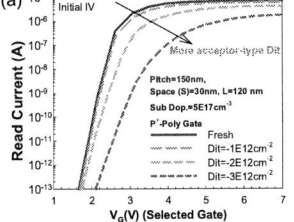
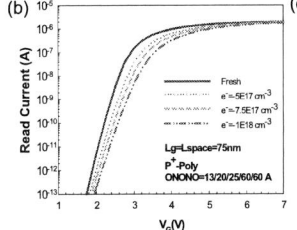
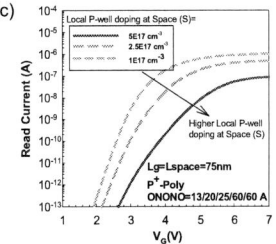

Fig. 7 Simulated I-V curves for the impacts of non-ideal factors in the space region (S). A 16-WL junction-free NAND with substrate doping =5E17 cm⁻³ is simulated. The selected gate is WL8, while other WL's are turned on by V_{PASS}= 6V. **(a)** Dit effect. Acceptor-type Dit causes degraded S.S. and read current. **(b)** Remnant electron traps at S causes degraded read current and S.S.. This suggests that the ONO should be cut through in the space region. **(c)** Local p-well doping effect under S. Lower local p-well doping is better for junction-free NAND Flash.

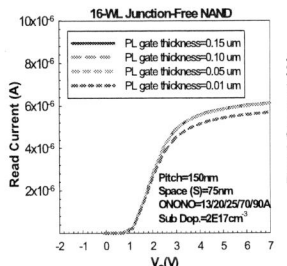

Fig. 8 Simulated I-V curves (in linear scale) for various poly gate thickness. Vt is independent of poly gate thickness. When poly gate thickness is sufficiently high (>0.01 um), poly gate thickness does not have any impact on the IV curve. This suggests that poly gate thickness is not important.

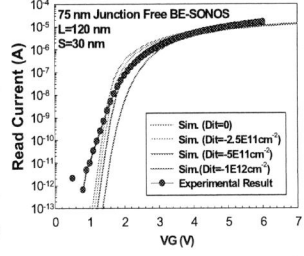

Fig. 9 Comparison of the experimental [1] and simulated initial I-V curve (fresh state) of 75-nm junction-free BE-SONOS NAND Flash device. A retrograde p-well is used in the simulation. Additional Dit is placed at the space (S) region. With Dit, the subthreshold slope is more consistent with the experimental result.

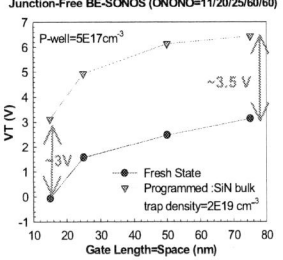

Fig. 10 Simulated Vt roll-off characteristics with various technology node (F). Space (S) = Lg is used. Junction-free NAND Flash has reasonable Vt roll-off down to 15 nm. At very small F, the programming window is slightly degraded even there is the same injected charge density (2E19 cm⁻³ in SiN).

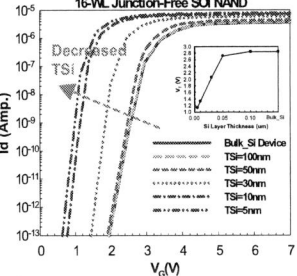

Fig. 11 Simulated I-V curves for the 75nm node SOI junction-free NAND. Inset shows the corresponding Vt vrsus TSi (silicon thickness). Thinner silicon thickness causes lower Vt. This is because the Si in SOI is fully depleted.

978-1-4244-2784-0/09 $25.00 © 2009 IEEE

Gate Last MOSFET with Air Spacer and Self-Aligned Contacts for Dense Memories

Jemin Park and Chenming Hu

Department of Electrical Engineering and Computer Sciences,
University of California, Berkeley, CA 94720, USA
Tel: (510) 643-2637, Fax: (510) 643-2636, e-mail: jemins@eecs.berkeley.edu

ABSTRACT

Gate-last metal-gate/high-k technology will allow MOSFET scaling to unprecedented levels. When the gate length is small, the dominant capacitance in the MOSFET is the gate to contact-plug capacitance. This is especially so with SAC (self-aligned contact) technology popular with high density memories. This papers proposes a compact SAC gate-last air-spacer structure that yield small size, high speed, and low switching energy. The improvement over the conventional SAC device increases dramatically with scaling.

INTRODUCTION

Metal gate and high dielectric constant (high-k) gate insulator will allow MOSFET gate length scaling to very small sizes. However, metal-gate/ high-k stack is easily degraded by high temperature processes such as the source/drain activation annealing. The gate-last process is attractive in this regards and has been put into production[1] [2]. Even without using metal-gate and/ot hig-k dielectric, the gate-last technology provides scaling benefits. In dense memory technologies, SAC is widely used for cell size reduction. However, SAC technology places the contact plug closer to the gate and the spacer material is nitride having a large dielectric constant. The gate to contact-plug capacitance can be the largest contributor to the bit-line and word-line capacitances in scaled technologies with serious consequences to speed and power. It will be critical to reduce the gate to SAC capacitance. Using the gate-first process technology, a nitride-spacer SAC MOSFET has 78% longer delay and 151% higher switching energy than an air- spacer SAC MOSFET [3].

We propose a novel air-spacer gate-last transistor that does not sacrifice the SAC density and reduces the gate capacitance, power, and delay to levels much lower than with the conventional SAC transistor with nitride-spacers. With this combination of density and performance, air-spacer SAC transistor could be attractive to not only DRAM (dynamic random access memory), but also SRAM (static random access memory), embedded SRAM, and perhaps even other applications.

CONCEPT AND PROCESS SIMULATION

The proposed air-spacer gate last process flow is as follows. Figure 1(a) shows that the sacrificial gate is patterned and source, drain and nitride spacer are formed. A very thin oxide liner (not shown) is deposited underneath the nitride spacer to protect the gate dielectric. ILD (inter-layer dielectric) is deposited and oxide CMP (chemical mechanical polishing) carried out. Figure 1(b) shows that the sacrificial gate is removed, gate dielectric is deposited, the gate material is deposited and etched back. Figure 1 (c) shows that another nitride spacer is formed on top of the gate to increase the air spacer size. Figure 1(d) shows that ILD is deposited, SAC is formed by high-selectivity contact hole etch, and contact plug filling. A novel steps of CMP to expose the top of the nitride spacer is performed. Figure 1(e) shows the selective etch of the nitride spacer without damaging the gate dielectric to create an air gap. Figure 1(f) shows that non-conformal ILD2 deposition has sealed the top openings and sealed the air spacers.

3D structures with SAC are constructed using the Sentaurus structure editor [4]. We compare two devices: SAC

devices with air spacers and nitride spacers. In order to study the effect of scaling on the benefits of the air spacer, we also compare 65, 45, 32 and 20nm gate-length structures. In each generation, two transistors having identical design parameters such as S/D (source/drain) and channel doping, equivalent oxide thickness, and L_g. Retrograde body doping is created with a 500 Å , $2e18/cm^3$ doped region and a 210 Å, $1e16/cm^3$ doped epitaxial layer to suppress the short channel effect at 20nm gate length. The thickness of the gate is 600 Å. Some other parameters of each generation are shown in Table 1. The PMOSFETs have same structures as the NMOSFETs except for the dopant.

DEVICE SIMULATION RESULTS

The characteristics of transistors and inverters are simulated with Sentaurus 3D device simulator [4]. Figure 3 shows that the I_{DS}-V_{GS} characteristics of the two transistors are little changed by the spacer/contact designs at each generation. Figure 4 shows a schematic of the inverter chains that were simulated with 3D mixed–mode simulation. The delay time of air spacer structure is decreased about 25% compared to nitride spacer structure. Figure 5 shows that delay time is increased according to reducing the gate length. The effect of air spacer over nitride spacer is more and more significant as the gate length is decreased.

Figure 6 shows that switching charge per area is increased as the gate length is decreased. The switching charge of air spacer at 20nm is smaller than that of nitride spacer at 65nm. The improvement of switching charge using air spacer is from 35% to 57% as the gate length is from 65nm to 20nm. Figure 7 shows that switching energy per area is also increased when the gate length is reduced. In nitride spacer technology, the switching energy is changed from 32fJ/um2 to 58fJ/um2 as the gate length is changed from 65nm to 20nm. In air spacer technology, the chage of switching energy is only 4fJ from 65nm to 20nm gate length. This air spacer technology is more and more important to not only delay time but also power consumption as the technology node is getting smaller and smaller. The summary of all data is shown in Table II. The characteristics of air spacer is excellent over all generation, especially at smller size.

CONCLUSION

Reducing the device capacitance will be an increasingly important way to improve the device speed and switching energy/power at smaller gate length. High density memories employ the SAC technology that requires the use of nitride spacer which significantly raises the delay and switching power. A novel SAC gate-last air-spacer structure that yield small size, high speed and low switching energy is proposed. Compared to an air spacer technology, a conventional nitride spacer transistor would have 41% longer delay and 129% larger switching charge and 155% larger switching energy at 20nm gate length. These benefits are more and more significant for smaller dense memories.

REFERENCES

[1] K. Mistry et al., IEDM Tech. Dig., pp.247-250, 2007.
[2] S.Mayuzumi et al., IEDM Tech. Dig., pp.293-296, 2007
[3] Jemin Park et al., SISPAD 2008
[4] SENTAURUS User Manual, Synopsys, Mountain View, CA.

978-1-4244-2784-0/09 $25.00 © 2009 IEEE

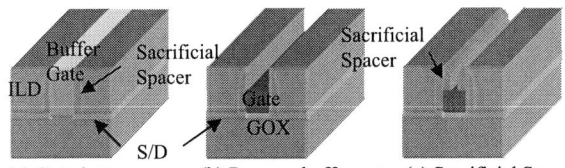 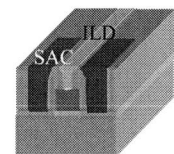  

(a) Gate-last process with sacrificial spacer   (b) Remove buffer gate and form gate material   (c) Sacrificial Spacer formation   (d) ILD deposition and making SAC   (e) Remove all Sacrificial Layers   (f) Deposition ILD2

Figure 1: Proposed process flow of the novel gate last with Air Spacer technology.

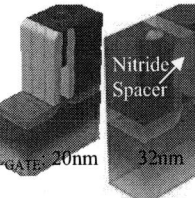

Figure 2: 3D transistor structures with Self-Aligned Contact and Air Spacer at four generations. 32nm and 65nm devices are shown with nitride spacers for comparison.

Table I: The several key specifications of each generation

Gate Length	65nm	45nm	32nm	20nm
T_{ILD} and T_{Gate}	ILD 60nm / GATE 60nm			
T_{Spacer}	35nm	27nm	18nm	12nm
Contact Size	70nm	50nm	40nm	30nm
G_{OX}	1.3nm	1.2nm	1.1nm	1nm
V_{DD}	1.2V	1.1V	1V	1V

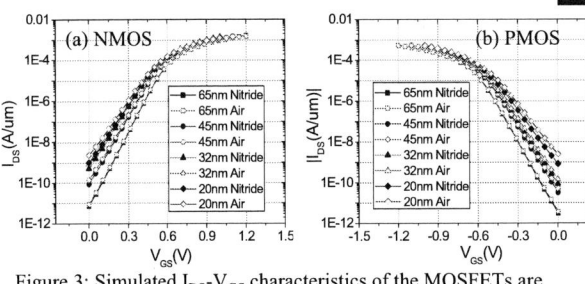

Figure 3: Simulated I_{DS}-V_{GS} characteristics of the MOSFETs are basically the same

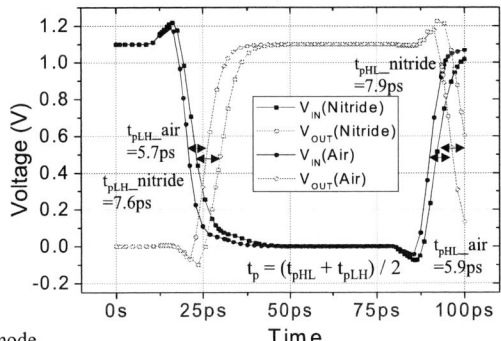

Figure 4: The mixed mode simulation of inverter delay. The delay of air spacer is decreased by 25% compared with nitride spacer.

Figure4 (a) The schematic of inverter chain

Figure4 (b) The mixed-mode simulation at 45nm gate length.

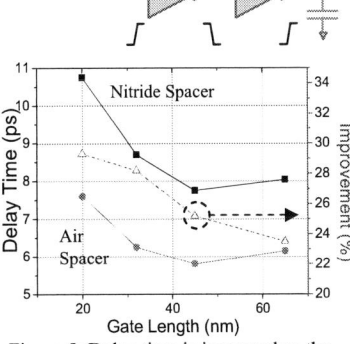

Figure 5: Delay time is increased as the gate length is decreased.

Table II: Summary of all data

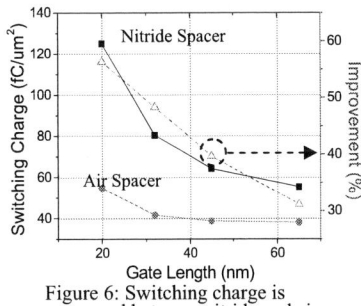

Figure 6: Switching charge is compared between nitride and air spacer structures.

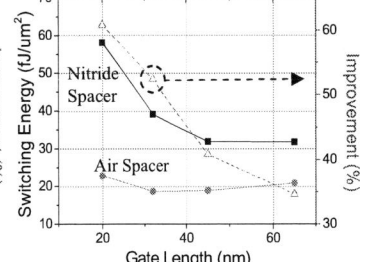

Figure 7: The benefit of switching energy using air spacer is much increased below 30nm of gate length.

Gate Length	65nm		45nm		32nm		20nm	
Spacer Type	Nitride	Air	Nitride	Air	Nitride	Air	Nitride	Air
NMOS I_{ON} (A/um)	1.61m	1.56m	1.46m	1.40m	1.39m	1.32m	1.35m	1.26m
NMOS I_{OFF} (A/um)	7.44p	8.76p	90.1p	0.14n	0.53n	0.87n	1.01n	2.28n
PMOS I_{ON} (A/um)	0.56m	0.55m	0.54m	0.53m	0.48m	0.47m	0.53m	0.514m
PMOS I_{OFF} (A/um)	3.07p	3.64p	32.6p	51.3p	87.5p	0.15n	0.83n	2.49n
Delay (ps)	8.04	6.15	7.75	5.8	8.7	6.25	10.75	7.6
Switching Charge (C/um²)	55.3f	38.1f	64.2f	38.7f	80.5f	41.6f	125.1f	54.7f
Switching Energy (J/um²)	31.8f	20.8f	31.9f	18.9f	39.1f	18.6f	58.1f	22.8f

978-1-4244-2784-0/09 $25.00 © 2009 IEEE   97

A Physics-Based Compact Model of Quantum-Mechanical Effects for Thin Cylindrical Si-Nanowire MOSFETs

Bastien COUSIN[1]*, Olivier ROZEAU[1], Marie-Anne JAUD[1], Jalal JOMAAH[2]

[1] CEA, LETI, MINATEC, F38054 Grenoble, France
[2] IMEP, MINATEC, INPG, 3 Parvis Louis Néel, BP257, 38016 Grenoble, France
*Phone: +33-4 38 78 17 82, Fax: +33-4 38 78 90 73, E-mail: bastien.cousin@cea.fr

Abstract – **Since we know that quantum-mechanical effects are predominant in surrounding-gate MOSFETs, a model should be developed. For the first time, this paper presents an analytic model of quantization for thin cylindrical Si-Nanowire MOSFETs by using a variational approach. The model is implemented into a surface potential like model. It is shown that results agree with the numerical simulations.**

INTRODUCTION

As transistor dimensions continue to shrink [1], bulk MOSFETs scaling is approaching the limit mainly imposed by short-channel effects. These unwanted effects increase leakage currents so new devices such as multiple gates, which represent a good alternative to get better gate control ability [2], are becoming intense subjects of research. Besides, integration of undoped silicon increases the carrier mobility in the channel [3]. Among these new devices, surrounding-gate MOSFETs, like cylindrical Si-Nanowire MOSFET, seem to be one of the most relevant and ambitious structure to replace CMOS classical devices for the 22nm node and beyond [4].

In the case of cylindrical Si-Nanowire MOSFET (Fig. 1), numerical simulations [5] have been performed in order to evaluate the influence of quantum-mechanical (QM) effects on the surface potential compared to the conventional (CONV) surface potential. In Fig. 2, we observe an increase of the surface potential when QM effects are included (+23% for a gate supply voltage of 1.2V in a 5nm radius nanowire MOSFET). Thus, QM effects are found to be dramatically significant and taking account of these effects is becoming a great relevance to accurately investigate thin nanowires.

Recently, two continuous analytic drain current models have been developed by D. Jiménez *et al.* [6] and B. Yu *et al.* [7, 8]. Both of them present continuous analytic current-voltage models for cylindrical undoped surrounding gate MOSFETs. Contrary to D. Jiménez, B. Yu has developed a surface-potential-based model (SPBM) that is more accurate particularly in the moderate inversion region. The aim of this paper is to present a simple QM model which can easily be implemented into a surface-potential-based like model.

QUANTUM-MECHANICAL EFFECTS MODELING

A model which takes into account quantization and volume inversion for thin Si-film Double-Gate (DG) MOSFETs has been developed by L. Ge *et al.* [9]. This model is particularly well-suited for our Si-Nanowire transistors because in these devices, carriers are subjected both to the radius R dependence (structural confinement) and to the transverse electric field dependence (electrical confinement) as illustrated in Fig. 3(a)-(b). Moreover, even if it has been developed for thin Si-film Double-Gate (DG) MOSFETs, it has been shown that for small nanowire radiuses (R<10nm), quantization in the device follows the same behavior as observed in DG MOSFETs [10].

Modeling QM effects with a very high accuracy involves solving the coupled Poisson and Schrödinger equations self-consistently. Because it requires numerical simulations and thus, very long computing times, this method can not be implemented in a compact model. Therefore, to model QM effects, the method used in this paper refers to the variational approach which involves an analytical approximation of the ground-state energy (Fig. 3(c)) based itself initially on the choice of a trial eigenfunction. Moreover, in order to simplify the different equations, we can focus exclusively on the

lowest-energy subband eigenfunction [11].Consequently, we propose an eigenfunction $\Psi_0(x)$ as

$$\psi_0(x) = a_0 \sqrt{\frac{1}{R}} \sin\left[\frac{\pi(x+R)}{2R}\right] \exp\left[\frac{-b_0(-x+R)}{2R}\right] \quad (1)$$

where a_0 and b_0 are undetermined parameters.

Normalization of (1) meaning $\int_0^R \psi_0^2(x)dx = 1$ gives an expression of a_0 according to b_0. The unknown parameter b_0 will be determined later.

Then, we follow the variational approach to solve the coupled Poisson and Schrödinger's (P-S) equations. Accordingly to the cylindrical structure represented in Fig. 1, we should solve these equations in two dimensions in order to compute the quantization according to the x direction and the θ angle and so, all over the circle's area. But, solving the 2D P-S equations by using the variational approach is very complex. A good approximation of the b_0 parameter is then really difficult to get and thus, the computation of the lowest subband total energy is not sufficiently accurate. To improve the model, we propose another method which consists to solve the P-S equations, in only one dimension (according to x).

$$\frac{\partial^2 \phi(x)}{\partial x^2} + \frac{1}{x}\frac{\partial \phi(x)}{\partial x} = \frac{qN_{inv}(x)}{\varepsilon_{si}}|\psi_0(x)|^2 \quad (2)$$

where $\phi(x)$ is the electric potential in the Si-nanowire, $N_{inv}(x)$ is the inversion-electron area density, q is the electron charge and ε_{si} is the Si permittivity. In (2), 1D Poisson's equation is expressed, for a cylindrical Si-Nanowire MOSFET, according to the eigenfunction obtained in (1). In this way, consider the electric potential obtained from the model of B. Yu *et al.*, $N_{inv}(x)$ can be computed. Solving analytically (2), we obtain an expression of the potential according to b_0.

To calculate b_0, the 1D Schrödinger equation is used and written as

$$-\frac{\hbar^2}{2m_x}\frac{d^2\psi_0(x)}{dx^2} + (-q)\phi(x)\psi_0(x) = E_0\psi_0(x) \quad (3)$$

where $\hbar$ refers to the Planck's constant, we assume $m_x = 0.916m_0$ is the effective mass of electrons for the lower energy level in the <100> Si direction [12] and E_0 is the lowest energy subband. Values of the kinetic and potential energies are respectively written as

$$\langle E_{0(kin)} \rangle = -\frac{\hbar^2}{2m_x}\int_0^R \psi_0(x)\frac{d^2\psi_0(x)}{dx^2}dx \quad (4)$$

$$\langle E_{0(pot)} \rangle = -q\int_0^R \psi_0^2(x)\phi(x)dx \quad (5)$$

From (4) and (5), the lowest energy subband's value is written as

$$E_0 = A\langle E_{0(cin)} \rangle + B\langle E_{0(pot)} \rangle \quad (6)$$

where A and B are smooth-functions valuable for all R values deducted from numerical simulations. A and B provide then an accurate correction both on the kinetic and potential energies computed from the 1D P-S equations. The aim of this method is to simulate the 2D P-S computation as previously mentioned. The b_0 parameter can be now approximated by minimizing E_0, setting $dE_0/db_0 = 0$ [13].

$$b_0 \cong R\left(\frac{5}{6}\frac{q^2 m_x \pi^2 N_{inv}}{C\varepsilon_{si}\hbar^2}\right)^{1/3} \quad (7)$$

where C is a smooth-function according R as well as A and B. Another smooth-function D is finally applied to reproduce the shift on the threshold voltage and is directly add to the gate voltage value as [14].

As a result, the QM model can be easily computed and implemented in the original SPBM by making a modification of the intrinsic carrier concentration n_i^{CONV} [15]. From (6), the new carrier concentration is then written as

$$n_i^{QM} = n_i^{CONV} \exp(E_0/2kT) \qquad (8)$$

RESULTS

To calculate the surface potential with QM effects for a compact model, the procedure is:

1. Compute the conventional surface potential.
2. Compute the inversion charge using boundary conditions.
3. Compute b_0 using (7).
4. Compute kinetic and potential energies using analytical expressions (4) and (5).
5. Compute lowest energy subband's value E_0 using (6).
6. Compute intrinsic carrier concentration n_i^{QM} using (8).
7. Compute new surface potential using conventional surface potential resolution with new n_i value.

As observed, this new QM model is well-built compared with the numerical simulations on the lowest energy subband (Fig.4) and surface potential (Fig.5). In addition, this QM model was validated for a wide range of R and gate supply voltage which equals to a gate oxide thickness (t_{ox}) variation.

CONCLUSION

We have developed, in this paper, a QM model for cylindrical Si-Nanowire MOSFETs. It has been shown that results obtained by our method used to model QM effects are very accurate compared to numerical simulation ones. By improving both the equations resolution and the computing time, it makes suitable this simple and predictive model for its integration in a surface-potential-based compact model. Furthermore, it can be easy to implement it in circuit design tools.

ACKNOWLEDGEMENT

This work was supported by the French National Research Agency (ANR) through a Carnot funding. The authors would like to thank T. Ernst, T. Poiroux and P. Martin for helpful discussions.

REFERENCES

[1] International Technology Roadmap for Semiconductors, _www.itrs.net_.
[2] J.-P. Colinge, _SSE_, vol. 48, n° 6, pp. 897-905, 2004.
[3] O. Gunawan _et al._, _Nano Lett._, vol. 8, n° 6, pp. 1566-1571, 2008.
[4] E. Dornel _et al._, _Appl. Phys. Lett. 91_, 2007.
[5] ATLAS user's manual – _Dev. Simu. Soft._, SILVACO International In.
[6] D. Jiménez _et al._, _IEEE EDL_, vol. 25, n° 8, pp. 571-573, 2004.
[7] B. Yu _et al._, _IEEE TED_, vol. 54, n° 3, pp. 492-496, 2007.
[8] B. Yu _et al._, _IEEE TED_, vol. 54, n° 10, pp. 2715-2722, 2007.
[9] L. Ge _et al._, _IEEE TED_, vol. 49, n° 2, pp. 287-293, 2002.
[10] V. Trivedi _et al._, _IEEE EDL_, vol. 26, n° 8, pp. 579-582, 2005.
[11] F. Stern, _Phys.Rev. B_, vol. 5, pp. 4891-4899, 1972.
[12] J. Wang _et al._, _J. Appl. Phys._, vol. 96, n°4, pp. 2192-2203, 2004.
[13] R.G. Winter, _Quantum Physics_. Belmont, CA: Wadsworth, 1979.
[14] M.J. Van Dort _et al._, _IEEE TED_, vol. 39, n° 4, pp. 932-938, 1992.
[15] M.J. Van Dort _et al._, _SSE_, vol. 37, n° 3, pp. 411-414, 1994.

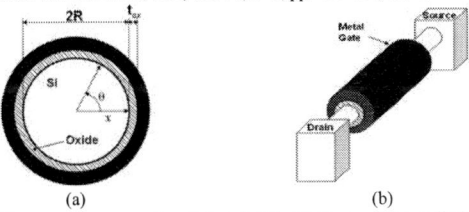

Fig. 1. Transverse cross-section (a) and 3D structure of the cylindrical Si-nanowire MOSFET (b).

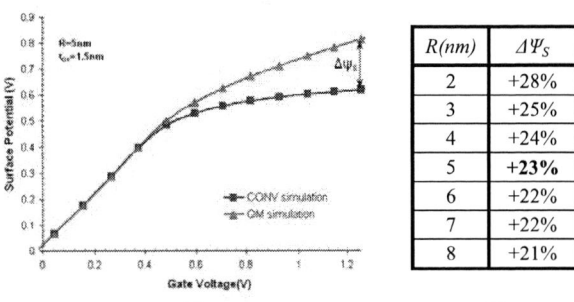

R(nm)	$\Delta\Psi_S$
2	+28%
3	+25%
4	+24%
5	**+23%**
6	+22%
7	+22%
8	+21%

(a) (b)

Fig. 2. A comparison between CONV and QM surface potentials versus gate voltage (a). Evolution of $\Delta\Psi_S$ (gap between conventional and QM surface potentials for a gate supply voltage of 1.2V) for different radiuses R (b).

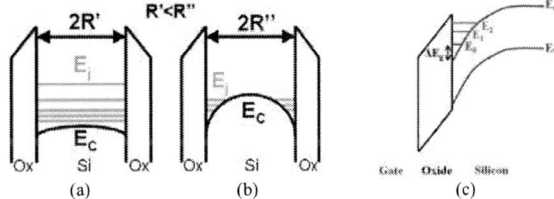

(a) (b) (c)

Fig. 3. Energy-band diagram of a cylindrical nanowire MOSFET showing a structural carrier confinement (a) and an electrical confinement (b). Energy-band diagram showing the energy difference ΔE_g between the ground-state energy level E_0 and the bottom of the conduction band (c).

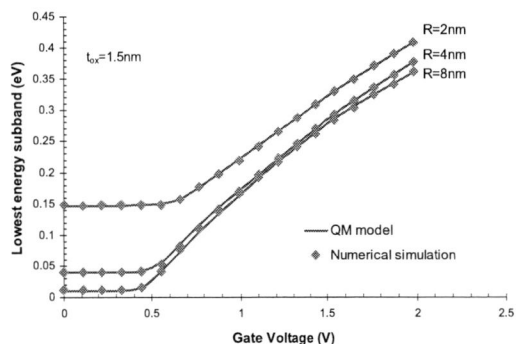

Fig. 4. Lowest energy subbands obtained from our QM model (solid curves) compared with those from the numerical simulation (dots).

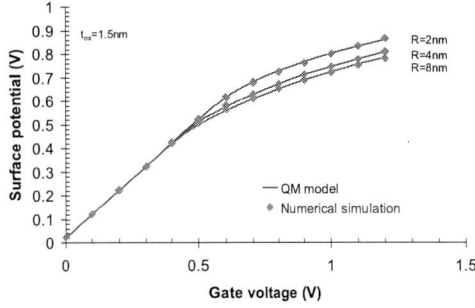

Fig. 5. Surface potentials obtained from our QM model (solid curves) compared with those from the numerical simulation (dots).

A New Technique to Extract the Gate Bias Dependent S/D Series Resistance of Sub-100nm MOSFETs

Dominique Fleury[1,2], Antoine Cros[1], Grégory Bidal[1,2], Hugues Brut[1], Emmanuel Josse[1] and Gérard Ghibaudo[2]

[1] STMicroelectronics, 850 rue Jean Monnet 38926 Crolles cedex, France

[2] IMEP-LAHC, 3 rue Parvis Louis Néel BP 257 38016 Grenoble cedex 1 France

Tel. : +33-4-3892-3314, Fax : +33-4-3892-2953, email : dominique.fleury@st.com

ABSTRACT

In this study, a new technique to extract the S/D series resistance (R_{sd}) from the total resistance versus transconductance gain plot $R_{tot}(1/\beta)$ is proposed. The technique only requires the measurement of $I_d(V_{gs})|_{Vgt}$ and β, allowing fast and statistical analysis in an industrial context. Unlike the usual $R_{tot}(L)$-based techniques, it has the advantage of being insensitive to the channel length and mobility variations and finally enables to extract very accurate values for $R_{sd}(V_{gs})$ and the effective mobility reduction factor $\mu_{eff}(V_{gt})/\mu_{eff}(0)$.

INTRODUCTION

The S/D resistance (R_{sd}) is a major concern for the MOSFET scaling as it plays a key role in device performance and power consumption [1]. Since the channel length is scaled down, the R_{sd}/R_{tot} ratio becomes higher and R_{sd} requires improved accuracy in extraction techniques to be assessed within a reasonable error. As described on Fig.1, a transistor can be modeled in linear regime by a channel resistance R_{ch} connected to the S/D series resistance $R_{sd} = R_s + R_d$ through which the drain current I_d flows ($R_{tot} = R_{sd} + R_{ch}$). Due to pockets implants, strain booster and neutral defects, the effective mobility (μ_{eff}) changes as a function of channel length (L_{eff}) [2]-[4] (Fig.2). As a consequence, R_{ch} is no more strictly proportional to the geometrical dimensions of the channel and all $R_{tot}(L)$-based techniques [5-7] fail when $\mu_{eff}(L)$ variations are not properly compensated for [8] (cf. Fig.3). To solve this issue, a new extraction technique based on the relationship between R_{tot} and the transconductance gain β of the transistor in linear regime is proposed. The technique is insensitive to the $\mu_{eff}(L)$, $L_{eff}(L)$ variations (which generally make the other techniques inaccurate) and provides a straightforward way to extract R_{sd} statistically.

THE $R_{TOT}(1/\beta)$ TECHNIQUE

The $R_{tot}(1/\beta)$ technique relies on the BSIM3v3 model (1) which reproduces the drain current behavior in linear regime. In (1), $V_{gt} = (V_{gs} - V_{th})$ is the gate overdrive, $\beta=\mu_{eff}(0).C_{ox}.W_{eff}/L_{eff}$ is the transconductance gain (where $\mu_{eff}(0)$ is the effective mobility extrapolated to $V_{gt} = 0V$) and (Θ_1, Θ_2) are the first and second order mobility attenuation factors, respectively.

$$I_d = \mu_{eff} C_{ox} \frac{W}{L} V_{gt} (V_{ds} - R_{sd} I_d) = \frac{\beta V_{ds} V_{gt}}{1 + \Theta_1 V_{gt} + \Theta_2 V_{gt}^2} \quad (1)$$

The channel resistance is defined as $R_{ch} = V_{d,0}/I_{d,0}$ (where the "$_0$" subscript refers to the intrinsic value of the parameter, for $R_{sd}=0$ $\Omega.\mu m$). From (1), R_{tot} can be expressed as (2).

$$R_{tot} = \frac{1}{\beta} \cdot \left(\frac{1 + \Theta_{1,0} V_{gt} + \Theta_{2,0} V_{gt}^2}{V_{gt}} \right) + R_{sd}(V_{gs}) \quad (2)$$

When V_{gt} is fixed once for a full set of devices with several channel lengths, the $R_{tot} = f(1/\beta)$ plot shows a linear behavior which returns the mobility reduction from the slope (3) and the $R_{sd}|V_{gt}$ from the y-axis intercept (2). By repeating the same extraction for several gate overdrives, $R_{sd}(V_{gs})$ and $\mu_{eff}(V_{gt})/\mu_{eff}(0)$ can be extracted

$$V_{gt} \cdot \frac{\partial R_{tot}}{\partial(1/\beta)}\bigg|_{V_{gt}} = 1 + \Theta_{1,0} V_{gt} + \Theta_{2,0} V_{gt}^2 = \frac{\mu_{eff}(0)}{\mu_{eff}(V_{gt})} \quad (3)$$

RESULTS

The following results were obtained by measurements on our 45nm node technology platform on the low stand-by power devices, featuring

1.7nm-EOT SiON gate dielectric with polysilicon gate and tensile contact etch stop layer for nMOS mobility optimization [9] (Fig.1). Extraction also been performed on FDSOI devices featuring metal gate (WN) with 2.5nm EOT $HfSi_xO_yN_z$ dielectric, 12nm thinned Si film and elevated S/D [10]. Statistical $I_d(V_{gs})$ measurements (72 dices) have been performed for lengths ranging from 35nm to 240nm and W=1μm. Strong pockets implants have been used in the process to increase the channel doping and limit the short channel effect in the smallest devices. V_{th} and β can be extracted from the McLarty's function [11] (4) or from the ξ-function [12] which have both the advantage of being insensitive to (Θ_1, Θ_2) when R_{sd} has a linear variation with V_{gs}.

$$\left(\frac{\partial^2 R_{tot}}{\partial V_{gs}^2} \right)^{-1/3} = \left(\frac{\beta}{2} \right)^{1/3} V_{gt} \quad (4)$$

Note that, as displayed in the inset of Fig.4, V_{th} deduced from McLarty's functions and ξ-function corresponds to the charge threshold voltage at strong inversion i.e. where $Q_{inv} = C_{ox}.V_{gt}$. $V_{th}(L_{eff})$ and $\beta(L_{eff})$ behavior are displayed on Fig.4 and Fig.5, where L_{eff} has been extracted from C-V measurements [13]. R_{tot} has been measured for each device at several gate overdrive ranging from 0.1 to 1.1V (the nominal voltage for this technology is $V_{gs} = 1.1$ V, i.e. $V_{gt} \approx 0.4V$). $R_{sd}(V_{gt})$ has been extracted from the $R_{tot}=f(1/\beta)$ plot, as described previously (2). The linear regression is displayed on Fig.5, where data has been filtered with a recursive normal filter within a ±3σ-tolerance (99% confidence). The points show a very good alignment which results in a very small error on the final result: $R^2>0.99$, $R_{sd} = (110 \pm 3)$ $\Omega.\mu m$. Fig.7 shows $R_{sd}(V_{gs})$, where V_{gs} has been approximated to $V_{gs} \approx V_{gt} + <V_{th}(L)>$, $<V_{th}(L)>$ being the average V_{th} for the set of devices: $V_{gs} \approx V_{gt} + 0.69 \pm 0.05$ V (cf. Fig.4). The behavior of $R_{sd}(V_g)$ is consistent with previous studies [14]. Results extracted for small gate overdrive ($V_{gt} \leq 0.2V$) show a slight deviation, which might be due to the limited accuracy in the V_{th}-extraction technique and/or non validity of strong inversion approximation close to V_{th}. Intrinsic mobility reduction factors have been extracted from (3) to be compared with the $\Theta(\beta)$ technique [15],[16]. As shown on Fig.8 and Fig.9, both techniques provide very close $\Theta_{1,0}$ values but R_{sd} extracted from $\Theta(\beta)$ shows a larger dispersion mainly induced by uncertainties on the Θ_1 parameter extraction. Finally, error resulting from the $<V_{th}(L)>$ approximation has also been quantified (Fig10) and R_{sd} has been estimated for the two extraction techniques. Results for bulk and FDSOI MOSFETs are summarized in Tab.1. As expected, FDSOI devices benefit from a lowered R_{sd} thanks to the elevated epitaxial S/D and an improved accuracy is confirmed for the $R_{tot}(1/\beta)$ technique compared to the $\Theta(\beta)$ one.

CONCLUSION AND PERSPECTIVES

This study demonstrates the ability of a new $R_{tot}(1/\beta)$ technique to provide $R_{sd}(V_g)$ and $\mu_{eff}(V_{gt})/\mu_{eff}(0)$ values with an improved accuracy thanks to statistical results. Unlike the $R_{tot}(L)$-based technique, the use of $1/\beta$ for the x-axis allows to correct any μ_{eff} or L_{eff} variations. The technique only requires to measure $I_d(V_{gs})|_{Vgt}$ and β on several channel lengths. The results match with the $\Theta(\beta)$ technique which suffers from a larger dispersion and requires full $I_d(V_{gs})$-curves measurements to extract R_{sd}. this technique is fully compatible with fast measurement techniques, offering new perspectives towards R_{sd} monitoring and large scale analysis in industrial environment.

ACKNOWLEDGMENTS

The authors would like to thank the Advanced Modules and Process Integration teams for providing the devices used in this work.

978-1-4244-2784-0/09 $25.00 © 2009 IEEE

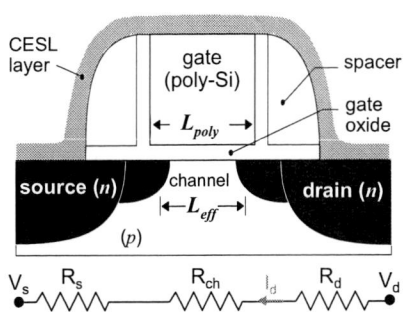

Fig.1 – Typical bulk nMOSFET with tensile contact etch stop layer (CESL).

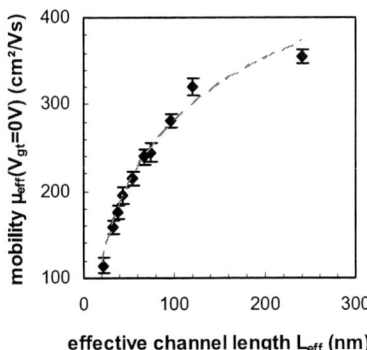

Fig.2 – decrease of the low field mobility for short channel length nMOSFETs

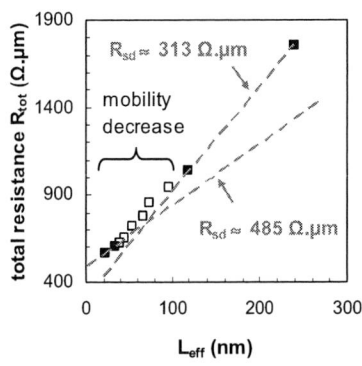

Fig.3 – uncertainty on the $R_{tot}(L)$ technique due to $\mu(L)$ degradation on short channels

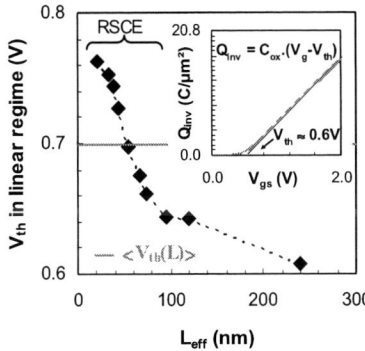

Fig.4 – $V_{th}(L_{eff})$ plot for nMOSFETs in linear regime, in inset: definition of V_{th}.

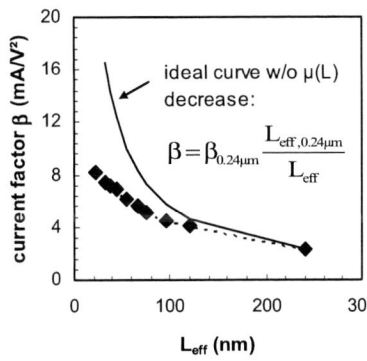

Fig.5 – $\beta(L_{eff})$ measurements. Continuous line: ideal behaviour w/o mobility reduction

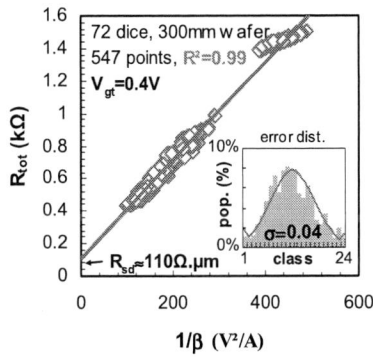

Fig.6 – $R_{tot}(1/\beta)$ plot for nMOSFETs. R_{sd} is extracted from the intercept with the y-axis.

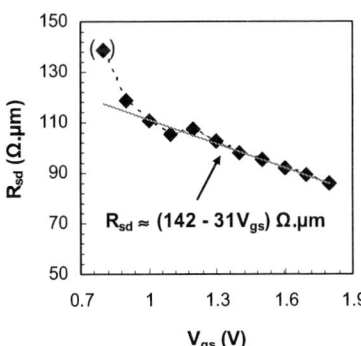

Fig.7 – $R_{sd}(V_g)$ behaviour extracted from the $R_{tot}(1/\beta)$ technique.

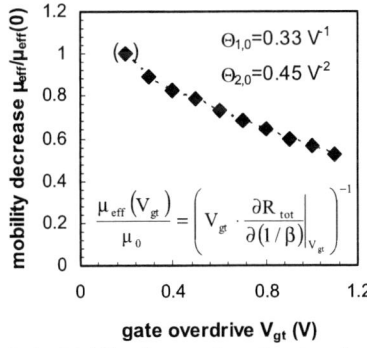

Fig.8 – Mobility decrease (as a function of V_{gs}) from the $R_{tot}(1/\beta)$ technique.

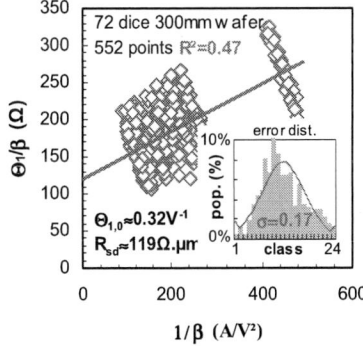

Fig.9 – R_{sd} and $\Theta_{1,0}$ extracted from the $\Theta_1(\beta)$ technique. In inset: error distribution

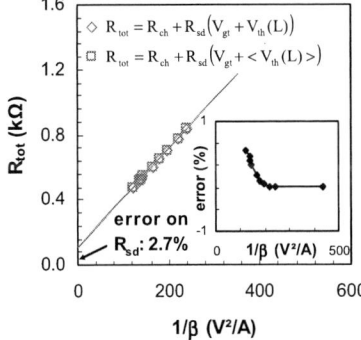

Fig.10 – Comparison between exact model ($\diamond$) and approximation ($\square$) using $<V_{th}(L)>$

R_{sd} (Ω.μm)	$R_{tot}(1/\beta)$	$\Theta(\beta)$
nMOS bulk	110 ± 3	119 ± 10
pMOS bulk	170 ± 5	155 ± 15
nMOS FDSOI	97 ± 5	126 ± 34
pMOS FDSOI	156 ± 5	208 ± 50

Tab.1 – $R_{sd}|V_{gs}=1.1V$ values extracted for bulk and FDSOI MOSFETs and compared to results obtained from the $\Theta(\beta)$ technique. As expected, the $R_{tot}(1/\beta)$ method gives more accurate results which remain in line with $\Theta(\beta)$.

References

[1] The ITRS, Semiconductor Ind. Assoc., 2006.
[2] K.M. Cao et al., IEDM.1999 pp. 171-174
[3] F. Andrieu et al., IEEE EDL, Oct 2005
[4] A. Cros et al., IEDM.2006 pp. 663-666
[5] Y. Taur et al., IEEE EDL, May 1992
[6] G.J. Hu et al., IEEE TED, Dec 1997
[7] Y.H. Chang et al., EDSSC 2007, pp.87-90
[8] J. Kim et al., IEEE TED, Oct 2008
[9] E. Josse et al., IEDM 2006, pp. 693
[10] D. Aimé et al., ESSDERC 2007, pp.255-258
[11] P.K. McLarty et al., SSE, Jun 1995
[12] D. Fleury et al., ICMTS 2008, pp.160-165
[13] D. Fleury et al., IEEE T-SM, Nov 2008
[14] S.D. Kim et al, IEEE TED, Mar2002
[15] G. Ghibaudo, Electronics Letters, Apr 1988
[16] C.Mourrain et al, ICMTS 2000, pp.181-186

ANALYTICAL MODELING OF ACCUMULATION-MODE SUSPENDED-GATE MOSFET AND PROCESS CHALLENGES FOR VERY LOW OPERATING POWER DEVICES

M. Collonge[1], M. Vinet[1], M. Ribeiro[1], J.-M. Pedini[1], B. Previtali[1], T. Ernst[1], S. Bécu[1] and G. Ghibaudo[2]

[1] CEA, LETI, MINATEC, 17 rue des Martyrs F-38054 Grenoble, France
[2] IMEP, INPG-INP Grenoble Minatec, 3 parvis Louis Néel, F-38016 Grenoble, France

ABSTRACT

For the first time, an analytical model of an Accumulation-Mode Suspended-Gate MOSFET is proposed. For very low power operation, adhesion energies of gate and gate oxide as low as $130\mu J/m^2$ are required as well as sub-2.3N/m doubly clamped gate. Experimentally a 0.2N/m suspended silicon nanowire was processed, opening perspectives for device downscaling.

INTRODUCTION

Invented more than 40 years ago [1], the Suspended-Gate Metal Oxide Semiconductor Field Effect Transistor (SG-MOSFET) structure presents interesting potentialities for very low operating power devices combining MEMS behavior into solid-state CMOS devices [2]. Its mechanical abrupt switching can be advantageously used to reduce device operating power. For the first time, an analytical model of an Accumulation-Mode Suspended-Gate Metal Oxide Semiconductor Field Effect Transistor (AM SG-MOSFET) is proposed. It highlights the process challenges required to benefit from mechanical abrupt switching for nanometer scale devices. First experiments to reach these specifications are detailed.

AM SG-MOSFET LOW POWER OPERATION

Device structure is based on standard solid-state MOS transistor whose gate has been disunited from gate oxide surface and is maintained suspended over the channel by two anchors (Fig.1). Depending on gate and drain biases (source electrode is grounded), mobile gate is more or less strained. Because of accumulation-mode operation, in the OFF state, the gate is pulled-down under electrostatic and adhesion forces, and contacts the gate oxide: it is gate down-state (Fig.2); Silicon on insulator device channel is fully depleted; so very low OFF currents can be reached. In the ON state, the gate is disconnected from the gate oxide and pulled-up thanks to restoring elastic force; gate capacitance is sharply decreased by additional air gap; thus drain current can flow into the intrinsically conductive channel: it is gate up-state (Fig.2).

Mechanical amplification of the channel potential allows overcoming the solid-state MOS transistor limitation for subthreshold slope: S<60mV/dec can be achieved (lowest reported value: S=2.16mV/dec [3]). Because of their abrupt switching, supply power of AM SG-MOSFET can be advantageously decreased for low operating power applications.

ANALYTICAL MODELING

Gate displacement results from equilibrium between three forces: a 1D elastic force, an electrostatic force resulting from potential difference between gate and channel surface and, adhesion forces such as Van der Waals forces and capillarity. For pulling-down to occur at low supply voltage, the gate spring constant needs to be minimized [4] down to a limit value relative to adhesion forces [5]. Fundamental principle of mechanics leads to the expression of gate voltage with respect to gate displacement (Table 1). Corresponding to an unstable position of gate, pulling-down displacement is solution of the equation: $dV_G/dx=0$. Gate pulling-up is chosen to occur for $V_G=0V$ (when electrostatic force is null, for flatband voltage). Thus it determines gate displacement analytically, depending on adhesion energy, channel doping, gate surface roughness and device design.

FROM DEVICE DESIGNING TO PROCESS CHALLENGES

AM SG-MOSFET operation is highly dependent on device design as it results from the equilibrium of electrostatic and adhesion attractive forces and elastic restoring one.

First, to benefit from abrupt switching, the mechanical gate pulling-down is chosen to occur for a higher voltage (more positive) than the threshold voltage. To satisfy this condition, Fig.3 shows that the adhesion energy needs to be lower than an upper energy value: $E_{MAX} \sim 130\mu J/m^2$ for $N=10^{18} at.cm^{-3}$. Controlling adhesion energy and surface roughness is prime order to reduce operating supply voltage. $E_{Adhesion}$ has to be decreased as much as possible below its usual value of $56mJ/m^2$ for standard oxide-coated structures [6]. Srinivasan, Houston et al. experimentally demonstrated that surface treatments can reduce it by three order of magnitude [6][7]. Solving these process challenges can lead to sub-1V supply voltage devices.

In addition, Fig.4 shows that gate spring constant lower than 2.3N/m is required for W=1µm and L=200nm. k_{MIN} (minimal gate spring constant for gate pulling-up to occur [5]) is proportional to the surface of contact WxL so sub-2.3N/m suspended beams would be required for device downscaling. k_{MIN} also depends on $E_{Adhesion}$ but it can not be used to tune k_{MIN} because $V_{Pulling-down}$, V_T and thus E_{MAX} are independent of WxL (Table 2). Fig.5 shows that pulling-down voltage does not strongly depend on channel doping; so doping level can not be used to increase E_{MAX}. Thus device downscaling mainly relies on the ability to process very supple beams.

EXPERIMENTAL DEVICES

We have experimented two paths to engineer low spring constant beams with low adhesion energy: a sacrificial SiO_2 layer had been removed first by wet HF etching to liberate low spring constant beams and then by HF-vapor etching to profit from low adhesion processing.

The interest of HF etching to make SG-MOSFET relies on its high selectivity. Thus nanoscale gate and air gap can be liberated. Fig.6 shows a silicon nanowire of sub 20nm-diameter and 1.6µm-long suspended over an 80nm-thick air gap by wet HF etching. This very low 0.2N/m spring constant beam opens perspectives for AM SG-MOSFET downscaling. In addition, HF etching allows HfO_2 integration as gate oxide. Fig.7 shows that a standard solid-state MOSFET gate stack $HfO_2/TiN/Poly$-Si was integrated in an SG-MOSFET. The deposition of the HfO_2 gate oxide before removing the sacrificial SiO_2 layer reduces the probability for the gate to collapse during its processing. Even more, HF-vapor etching can be advantageously used to reduce adhesion phenomena. It was applied to liberate Poly-Si/TiN beams over air gaps in the range of 20-50nm thick.

CONCLUSION

An Accumulation-Mode Suspended-Gate MOSFET is proposed for very low operating power applications. First analytical model of the device is presented and used to design transistors. It highlights process challenges required to benefit from mechanical abrupt switching and for device downscaling: adhesion energies of gate and gate oxide as low as $130\mu J/m^2$ are needed as well as sub-2.3N/m doubly clamped gate. Wet HF etchings were experimented to satisfy these requirements. A 0.2N/m suspended silicon nanowire was processed, allowing device downscaling. For the first time, standard solid-state MOSFET gate stack $HfO_2/TiN/Poly$-Si was integrated in an SG-MOSFET.

ACKNOWLEDGEMENT

This work was supported by the French National Research Agency (ANR) through Carnot Institute funding.

REFERENCES

[1] W.E. Newell, in *ISSC Conf. Dig.*, vol. IX, pp. 62-63, 1966.
[2] H. Kam *et al.*, in *IEDM tech. Dig.*, pp. 477-480, 2005.
[3] N. Abelé *et al.*, in *IEDM Tech. Dig.*, pp. 479-481, 2005.
[4] A.M. Ionescu *et al.*, in *IEEE ISQED Tech. Dig.*, pp. 18-21, 2002.
[5] M. Collonge *et al.*, in *IEEE ULIS Proc.*, pp. 53-56, 2008.

[6] U. Srinivasan *et al.*, in *J. of Microelectromechanical Syst.*, vol. 7, n°2, pp. 252-260, 1998.
[7] M. Houston *et al.*, in *J. of Appl. Phys.*, vol. 81, n°8, pp. 3474-3483, 1997.
[8] J.-P. Colinge, *Kluwer Acad. Publ.*, 3rd Ed., 2004.
[9] T. Ernst *et al.*, at *IEDM Conf.*, session 31, 2008.

TABLE 1: ANALYTICAL MODEL

Minimal gate spring constant [5]:

$$k_{MIN} = \frac{F_{Adhesion}^{At\ contact}}{t_{gap} - h - R} \text{ with } F_{Adhesion}^{At\ contact} = \frac{2WLE_{Adhesion}}{h + R}$$

Fundamental principle of mechanics :

$$F_{Elastic}(k_{MIN}, x) + F_{Electrostatic}(V_G, x) = 0$$

$$F_{Elastic} = k_{MIN}(t_{gap} - x) \quad F_{Electrostatic} = \frac{Q_{Depletion}^2 WL}{2\varepsilon_0}$$

$F_{Adhesion}$: *short distance forces only considered for gate pulling-up through expression of k_{MIN}*

Gate voltage versus gate displacement:

$$V_G = V_{FB} - \frac{q\varepsilon_{Si}N}{2}\left[\left(\frac{1}{C_{Gate}^{Eq}} + A\sqrt{t_{gap} - x}\right)^2 - \left(\frac{1}{C_{Gate}^{Eq}}\right)^2\right]$$

$$\text{with } \frac{1}{C_{Gate}^{Eq}} = \frac{t_{OX}}{\varepsilon_{OX}} + \frac{x}{\varepsilon_0}, \quad A = \frac{\sqrt{2\varepsilon_0 k_{MIN}/(WL)}}{q\varepsilon_{Si}N}$$

TABLE 2: ANALYTICAL MODEL

Gate displacement at pulling-down:

$$x_{Pulling-down} = t_{gap} - \frac{1}{36}\left[\varepsilon_0 A + \sqrt{\varepsilon_0^2 A^2 + 12(t_{gap} + t_{OX}\varepsilon_0/\varepsilon_{OX})}\right]^2$$

Pulling-down voltage:

$$V_{Pulling-down} = V_G(x = x_{Pulling-down})$$

Threshold voltage for channel fully depletion:

$$V_T = V_{FB} - \frac{qNt_{Si}^2}{2\varepsilon_{Si}} - qNt_{Si}\left(\frac{x_{Pulling-down}}{\varepsilon_0} + \frac{t_{OX}}{\varepsilon_{OX}}\right) \quad [8]$$

Parameters definitions and values:

h	Lennard-Jones length (3Å)
R	Roughness parameter for interaction between surfaces mean planes
t_{gap}	Initial air gap thickness as fabricated (10nm)
x	Variable air gap thickness ($x_{MAX} = t_{gap}$)
L	Gate length (200nm)
W	Gate width (1μm)
N	Channel doping level ($10^{18}at.cm^{-3}$)
V_{FB}	Flatband voltage (fixed at 0V)

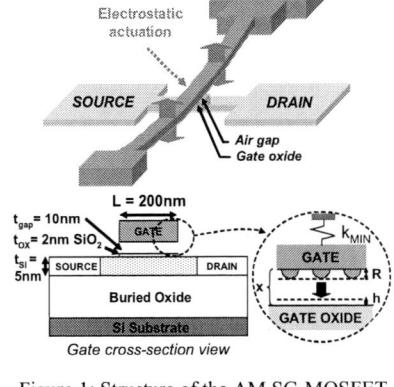

Figure 1: Structure of the AM SG-MOSFET in Silicon On Insulator technology.

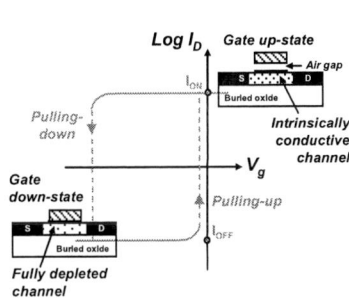

Figure 2: Hysteretic electro-mechanical operation of the AM SG-MOSFET.

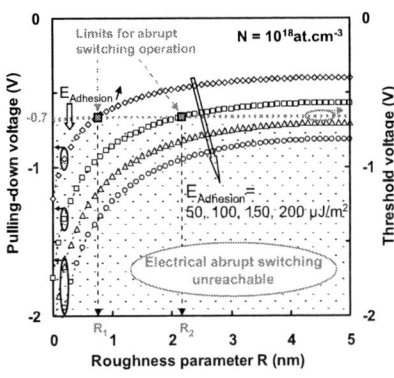

Figure 3: $V_{Pulling-down}$ and V_T sensibilities to adhesion energy as a function of the roughness parameter R.

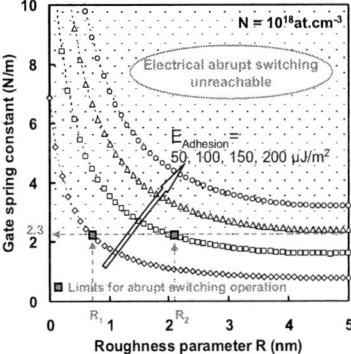

Figure 4: Required k_{MIN} values for gate pulling-up at V_G=0V as a function of the roughness parameter R for different $E_{Adhesion}$.

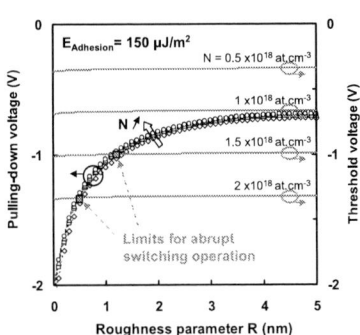

Figure 5: $V_{Pulling-down}$ and V_T sensibilities to channel doping N as a function of the roughness parameter R.

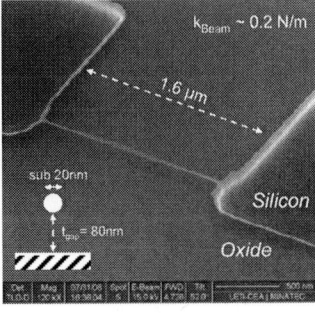

Figure 6: SEM picture of a silicon nanowire processed by wet HF etching of sacrificial SiO_2 [9]. Beam spring constant as low as 0.2N/m is made without gate collapsing.

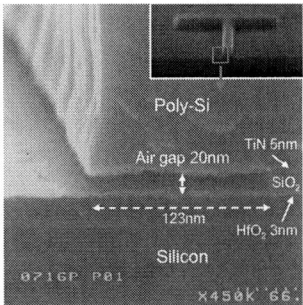

Figure 7: SEM pictures of a Poly-Si/TiN beam laterally liberated over a 20nm-thick air gap by HF-vapor etching of SiO_2. For the first time, HfO_2/TiN/Poly-Si gate stack is integrated in an AM SG-MOSFET.

978-1-4244-2784-0/09 $25.00 © 2009 IEEE

A New Robust Non-local Algorithm for Band-to-band Tunneling Simulation and its Application to Tunnel-FET

*†C. Shen, *L. T. Yang, *E.-H. Toh, *C.-H. Heng, *G. S. Samudra, *Y.-C. Yeo

* SNDL, ECE Dept., National University of Singapore, Singapore 119260. Email: yeo@ieee.org
† Cogenda Pte Ltd, Singapore. Email: shenchen@cogenda.com

Abstract—A new non-local algorithm for accurately calculating the band-to-band tunneling current suitable for TCAD semiconductor simulators is proposed in this abstract. The proposed algorithm captures the essential physics of multi-dimensional tunneling in a 2D structure, and is designed to be robust and to achieve independence on the mesh grid. The new algorithm enables accurate modeling of T-FET and investigation of its device physics. Application on T-FET is demonstrated. The physical origin of the saturation of I_d-V_d curve of T-FET is analyzed and clarified for the first time.

I. INTRODUCTION

Modeling of band-to-band tunneling (BTBT) current in TCAD semiconductor device simulators urgently needs an enhancement. Traditionally, it can be considered as a second order effect that contributes to the leakage current, as in the case of Gate Induced Drain Leakage (GIDL) in MOSFET. However, the recent interest in Tunnel Transistors (T-FET), which promises subthreshold swing below 60 mV/decade, demands more accurate calculation of strong BTBT current in 2D and 3D device strucutres, which is central to the operation of T-FET [1], [2].

Since a fully quantum mechanical treatment of the BTBT current is computationally too expensive, drift-diffusion TCAD simulators are usually used to study T-FET, and some semi-classical model is used to evaluate the tunnel current. Among these various models (summarized in Table I), the tunnel path based method is the most physical approach, but requires manual adjustment of meshing to align with the physical tunnel direction [3], which may not be known a priori, especially in advanced devices like T-FET.

Unfortunately, most commercial device simulators either use the inadequate local band-to-band tunneling model, or requires the user to specify the direction of tunneling current. These algorithms are not suitable for the 2D tunneling problem present in TFET, making the TCAD simulation of TFET to be widely known as "mesh-dependent".

In this abstract, we first propose a physical method that automatically identifies physical tunneling paths in 2D device structures. In the second part of the abstract, we present an analysis on the steady-state IV characteristics of T-FET. We identify the physical mechanism responsible for the high saturation voltage in the output characteristics. It is attributed to the presence of inversion charge at high gate to drain bias V_{gd}, which hinders the gate from effectively controlling the E-field at the tunnel junction.

II. PROPOSED ALGORITHM

It can be shown that the tunneling paths in an tunnel transistor vary significantly with different bias conditions. If the users are required to specify the special mesh for calculating tunnel current, as in many commercial simulators, Inaccurate tunnel paths specification would lead to erroneous underestimation of tunneling current. Therefore, a practically useful simulator must perform the path searching autonomously.

To obtain the most probable tunnel path starting from a point at one side of the barrier, the new algorithm uses the fact that tunneling current through the barrier is dominated by contribution in the vicinity of the most probable tunnel path originating from this point, and across the barrier. It can be derived under the WKB approximation, and the assumption that electrons strikes the tunnel barrier in the direction normal to the barrier, that the most probable tunnel path of the electron coincides with the Newtonian trajectory if one negates the sign of the potential, turning the barrier (forbidden region) to a classically allowed "valley" region [5], [6]. This may appears similar to the instanton method [5], but were derived as a multi-dimension extension of WKB method [5].

In this study, we implemented the above algorithm to search the tunnel paths, starting from various points at the conduction band front, end at the valence band front. The tunneling current along each path is evaluated using WKB approximation, as in [4] and integrated over all possible energies, while more complicated physical models can be introduced here to account for, for example, phonon-assisted tunneling. The obtained band-to-band tunneling current is directly coupled to the carrier continuity equation in a drift-diffusion simulator [7]. The new algorithm has demonstrated consistency on different meshes, and allows relatively sparse mesh to be used in the simulation.

III. APPLICATION ON TUNNELING-FET DEVICE PHYSICS

The tunneling FET structure shown in Fig.1 is simulated using the newly proposed algorithm. The I_d-V_g and I_d-V_d curves are shown in Fig.2. Electron and hole generation due to BTBT occurs very near the surface, as shown in Fig.1. Due to the low BTBT rate in silicon, the transistor current is limited by the tunnel junction, and does not depend on the channel length. Two notable features are observed in the I-V characteristics of Fig.2.

Firstly we notice that the output characteristics show saturation behavior at high V_d, although the saturation voltage is high compared to conventional MOSFETs. To understand the origin of the saturation behavior of T-FET, we consider the transition from bias condition A to B in Fig.2, and the corresponding band diagrams shown in Fig.3. At A, the gate to drain voltage is high enough to induce an inversion below the gate, which is populated with electrons supplied by the drain (see Fig.4). The conduction band in the channel can not be much lower than that in the drain due to the large amount of inversion electrons. In other words, the E_c in channel is effectively pinned by the drain voltage. An increase in V_d would allow the conduction band in the channel to be lowered further, leading to higher E-field at the tunnel junction and hence higher current. As V_d increases further (V_{gd} reduces), the density of inversion charge decreases, and at a point near B, the inversion layer starts to disappear, retreating first from the source end. The absence (depletion) of inversion charge implies that the drain voltage does not affect the electrostatic potential in the channel. It is obvious that a lower V_g leads to saturation at lower V_d. At bias C, the channel is shown to be completely depleted, and deeper into I_d saturation.

Secondly, we see that the drain current continues to increase exponentially at high V_g. While this can be easily understood for the case of high drain bias (e.g. from C to B), the case with lower drain bias is less obvious. At low V_d and high V_g, the channel is populated with inversion charge, density of which depends exponentially on the electrostatic potential. As discussed above, further increase of V_g does not pull down the E_c in the channel, but may bring more depleted region into inversion. This leads to the thinning of potential barrier, as illustrated in the band diagram at bias A. This barrier thinning in the presence of inversion charge is governed by a nonlinear Poisson's equation, and can not be solved with traditional quasi-2D analysis. We recently developed a technique to treat this problem analytically using variational method [8], which predicts that the potential barrier at the source junction would continue to thin down with increasing inversion charge, asymptotically according to $\sim N_{inv}^{-1}$.

IV. CONCLUSION

First, a novel algorithm of calculating BTBT current that automatically identifies the physical tunnel paths was proposed. Second, the algorithm was implemented in a device simulator to provide a versatile tool for simulating T-FET. Third, the device physics of T-FETs was studied. In particular, the physical origin of the saturation of the output characteristics of T-FET under high drain voltage was examined based on extensive simulations. It was revealed that the inversion carriers do not contribute to the drive current of T-FET, but instead screens off the gate from effectively controlling the tunnel junction. The delayed saturation in I_d-V_d characteristics is shown to be one of the consequences of the screening from inversion charge. Finally, the results of this work could enable further physical understanding and improvements of T-FET.

REFERENCES

[1] P. F. Wang, *et al.*, Solid State Electron., **48**, pp.2281-2286. (2004)
[2] K. K. Bhuwalka, *et al.*, T-ED, **51**, pp.279-282. (2004)
[3] M. Ieong, *et al.*, IEDM1998, pp. 733-736.
[4] W. Harrison, Phys. Rev. **123**, pp.85-89. (1961)
[5] Z. Huang, *et al.*, Phys. Rev. A **41**(1), pp. 32-41. (1990)
[6] M. Razavy, Quantum Theory of Tunneling, World Scientific. (2003)
[7] [Available online:] http://gss-tcad.sf.net/
[8] C. Shen, *et al.*, EDL, accepted for publication.

TABLE I. Summary of existing algorithms on BTBT calculation in TCAD simulators.

Model	Approach	Drawbacks
Local model	Generation of carriers according to local E-field	Numerically robust but unphysical
Nonlocal, node pair based	Search for nearest node pairs satisfying $E_{c,A} < E_{v,B}$. Generate electron and hole at A and B, respectively.	Highly sensitive to the position of mesh node, requires extremely dense meshes. Leads to unphysical results in non-uniform mesh.
Nonlocal, tunnel path based	Divide device into strips along the tunnel paths. Sum up contribution from all strips.	Requires the user to specify the tunnel paths explicitly, which are not obvious in advanced devices.
This work. Nonlocal, tunnel path based	Search for the most probable tunnel path autonomously, based on multi-dimensional WKB method.	

Fig. 1. a) Contour of electron and hole generation rate $(\mathrm{cm}^{-3}\mathrm{s}^{-1})$ in the T-FET device. b) the same plot at an expanded scale, focusing on the source junction. It is seen that electron generation peaked near the Si/SiO$_2$ interface, while the holes are generated at about 2 nm below the surface. The arrows indicate the direction of decreasing carrier generation rate.

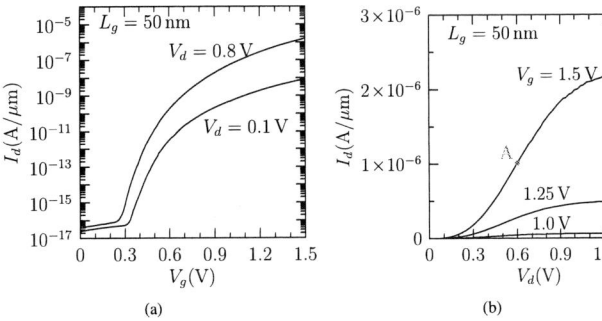

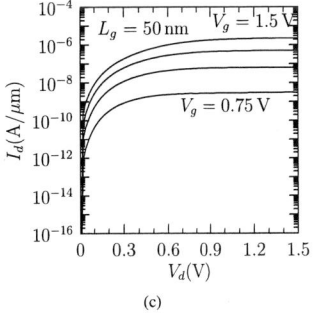

Fig. 2. $I_d - V_g$ and $I_d - V_d$ curves of the T-FET device with $L_g = 50\,\mathrm{nm}$ shown in Fig.1. It is seen from the $I_d - V_d$ curves that the drain current saturates at higher V_d compared to conventional MOSFET, as explained in the text. Also discussed is the continuing exponential increase of I_d with V_g at high V_g, though at progressively gentler slopes.

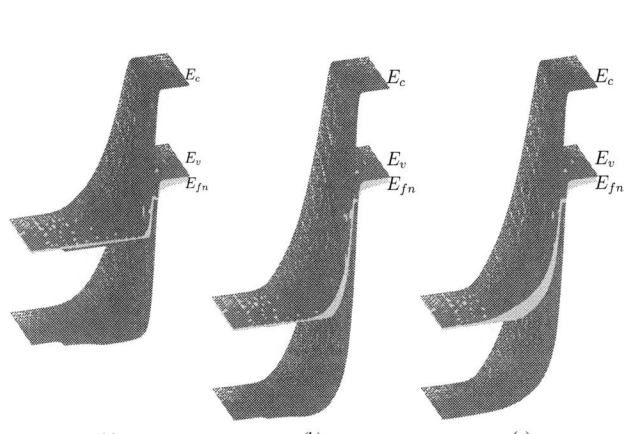

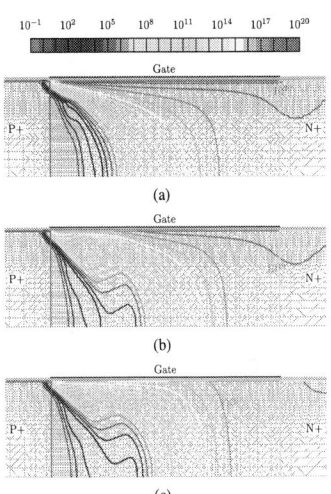

Fig. 3. Band diagram at the three bias conditions (A, B and C) shown in Fig.2 for the 50 nm T-FET. a) $V_g = 1.5\,\mathrm{V}, V_d = 0.6\,\mathrm{V}$, b) $V_g = V_d = 1.5\,\mathrm{V}$, and c) $V_g = 1.25\,\mathrm{V},\ V_d = 1.5\,\mathrm{V}$.

Fig. 4. Electron concentration (cm^{-3}) contour at the three bias conditions (A, B and C) shown in Fig.2. a) $V_g = 1.5\,\mathrm{V}$, $V_d = 0.6\,\mathrm{V}$, b) $V_g = V_d = 1.5\,\mathrm{V}$, and c) $V_g = 1.25\,\mathrm{V}$, $V_d = 1.5\,\mathrm{V}$. Electron concentration at the Si/SiO$_2$ surface decreases as the $I_d - V_d$ characteristics becomes deeper into saturation.

978-1-4244-2784-0/09 $25.00 © 2009 IEEE

Investigation of Static Noise Margin of Ultra-Thin-Body SOI SRAM Cells
in Subthreshold Region using Analytical Solution of Poisson's Equation

Vita Pi-Ho Hu, Yu-Sheng Wu, Ming-Long Fan, Pin Su[a] and Ching-Te Chuang

Department of Electronics Engineering, National Chiao Tung University, Hsinchu, Taiwan

[a]Tel:+886-3-5712121ext.54142, E-mail: pinsu@faculty.nctu.edu.tw

Abstract –This paper investigates the Static Noise Margin (SNM) of Ultra-Thin-Body (UTB) SOI SRAM cells operating in subthreshold region using analytical solution of Poisson's equation validated with TCAD simulations. An analytical SNM model for UTB SOI SRAM cells operating in subthreshold region is presented. Our results indicate that back-gate bias (V_{bg}) can mitigate the Read SNM (RSNM) variability of UTB SOI SRAM cells in the subthreshold region, and the improvement of SNM variability is more significant than superthreshold region. Increasing cell β-ratio shows limited improvement on RSNM and has no benefit on SNM variability for subthreshold operation. The UTB SOI 8T SRAM cell exhibits RSNM 2X larger than the 6T SRAM cell in subthreshold region.

I. Introduction

Subthreshold operation is an efficient technique to achieve ultra-low power consumption for circuits by lowering the power supply (Vdd) below the threshold voltage [1-2]. UTB SOI MOSFET with thin Buried Oxide (BOX) has emerged as a promising candidate to extend CMOS scaling [3-5]. Due to its better control of short-channel effects, lower subthreshold swing, and reduced leakage and Random Dopant Fluctuation (RDF) resulting from the use of un-doped (or lightly-doped) thin silison film, UTB SOI MOSFET is very attractive for subthreshold circuit applications. The SNM is a measure of SRAM cell's ability to retain its data state. The worst-case situation is usually under "Read-disturb" conditions. Several studies have discussed the methodologies for improving the Read Static Noise Margin (RSNM) in UTB SOI (FD/SOI) SRAM [6-8]. However, the stability of UTB SOI SRAM operating in subthreshold region has rarely been examined. In this work, we present an analytical framework that accurately describes the SNM for UTB SOI SRAM in subthreshold region. Based on our theoretical model, the methodology for improving the stability of UTB SOI SRAM in subthreshold region is examined.

II. Analytical Subthreshold Current Model

Our theoretical drain current for UTB SOI MOSFET is derived from analytical potential solution in the subthreshold region. Fig. 1 shows a schematic sketch of a UTB SOI MOSFET with thin BOX structure. The corresponding 2-D boundary value problem (Fig. 2) can be divided into two sub-problems, a 1-D Poisson's equation and a 2-D Laplace equation. Using the superposition principle, the channel potential solution can be expressed as

$$\phi_{ch}(x,y) = \phi_1(x) + \phi_2(x,y)$$

$$\phi_1(x) = (-qN_{ch}) \cdot x^2/(2\varepsilon_{ch}) + a \cdot x + b$$

$$\phi_2(x,y) = \sum_n \{[c_n \sinh(\gamma_n \cdot y) + d_n \sinh(\gamma_n \cdot (L_g - y))] \cdot \sin(\gamma_n \cdot x)$$

$$+ e_n \sinh(\lambda_n(T_{ch} + (\varepsilon_{ch}/\varepsilon_i)T_i - x)) \cdot \sin(\lambda_n \cdot y)\}$$

$$\gamma_n = (n\pi)/(T_{ch} + (\varepsilon_{ch}/\varepsilon_i)T_i)$$

$$\lambda_n = (n\pi)/L_g$$

where a, b, c_n, d_n and e_n are coefficients determined by material intrinsic parameters, device geometry, doping, and terminal biases.

Our analytical potential solutions have been verified with TCAD simulations [9]. Fig. 3 shows that our model is fairly accurate for various channel doping (N_{ch}). Based on the potential solution, the subthreshold current can be derived as

$$I_d = \frac{q\mu_n W(kT/q)(n_i^2/N_{ch})[1 - \exp(-V_d/(kT/q))]}{\int_0^{L_g} dy / \int_0^{T_{ch}} \exp[q\phi_{ch}(x,y)/(kT)]dx}$$

III. Methodology for Analytical Subthreshold SNM Model

With the subthreshold current model, the SNM can be derived by solving Kirchoff's current law at cell storage nodes VR and VL (Fig. 4), respectively: I(NR) = I(PR) + I(AR), I(NL) = I(PL) + I(AL). The SNM is obtained from the maximum possible square method (Fig. 5),

which is defined as the square with the longest diagonal, whose equation is given by VR = VL + c. Thus, the SNM is determined by the minimum value between SNM_L and SNM_R.

Fig. 6 demonstrates that our analytical subthreshold SNM model for UTB SOI SRAM provides accurate results across wide ranges of important device design parameters such as channel thickness (T_{ch}), buried oxide thickness (T_{BOX}), gate length (L_g) and Vdd. The subthreshold RSNM model for UTB SOI SRAM shows excellent agreement with TCAD simulations.

In the following section, we investigate the variation of RSNM caused by parameter variations in subthreshold and superthreshold regions for UTB SOI SRAM with N_{ch} = 1e16cm^{-3}, T_{BOX} = 10nm, L_g = 40nm, T_i = 1nm, T_{ch} = 10nm, at Vdd = 0.4V (subthreshold) and 1.0V (superthreshold) . Notice that a single mid-gap work function is used. The use of mid-gap gate material allows a single electrode for both NMOS and PMOS. It also provides high-V_T for both NMOS and PMOS to reduce the SRAM cell leakage, and enables subthreshold operation with a reasonable Vdd. Furthermore, it is consistent with the cost consideration for subthreshold SRAM for which the intended applications are cost-sensitive implantable devices, medical instruments, and wireless sensor network.

IV. RSNM for UTB SOI Subthreshold SRAM

Fig. 7 shows the impact of V_{bg} (back-gating) on RSNM variation (σ) for subthreshold (Vdd = 0.4V) and superthreshold (Vdd = 1V) UTB SOI SRAM. By using back-gating technique, the RSNM variation can be reduced by 7.3mV for UTB SOI subthreshold SRAM and 9.7mV for UTB SOI superthreshold SRAM, respectively. Fig. 8 shows the impact of V_{bg} on the % change in RSNM, which is defined as the ratio of σ to the nominal RSNM. Due to the smaller Vdd, the UTB SOI subthreshold SRAM has lower nominal RSNM than the superthreshold SRAM. As can be seen, the UTB SOI subthreshold SRAM shows 30% improvement due to back-gating, larger than 20% improvement for UTB SOI superthreshold SRAM. Thus, back-gating is very effective to suppress the RSNM variation in UTB SOI subthreshold SRAM. Fig. 9 illustrates the circuit schematic of the 6T SRAM with back-gating. Fig. 10 shows an 8T SRAM cell, which has been known to have better RSNM and low V_{MIN} due the decoupling of Read current path from the cell storage node. Fig. 11 compares the nominal RSNM, RSNM variation, and % change in RSNM of several cases for subthreshold SRAM cells listed in Table 1. The 8T cell (case e) shows 116.28% larger nominal RSNM as compared with the 6T cell (case a). The 6T UTB SOI SRAM cell with Vbgp = 1V and Vbgn = -1V (case c, Above-Vdd/Below-GND back-gate bias) mitigates the RSNM variation significantly by 42.7% as compared with (case a). Thus, (case c) and (case e) show comparable % change in RSNM. Cell β ratio is defined as the ratio of the driver transistors (NR, NL) width/length (W/L) to the access transistors (AR, AL) W/L. Increasing cell β ratio (case d) improves nominal RSNM by only 5.13% and shows no benefit on the RSNM variation suppression. Therefore, (case d) shows comparable % change in RSNM as (case a).

In summary, an analytical subthrehold SNM model for UTB SOI SRAM is presented to investigate the stability of several UTB SOI SRAM cells operating in subthreshold region. The study may provide insights for UTB SOI subthreshold SRAM design.

Acknowledgement

This work was supported in part by the National Science Council of Taiwan under Contract NSC 97-2221-E-009-162, and in part by the Ministry of Education in Taiwan under ATU Program.

References

[1] J. Kim et al., *IEEE TED.*, vol.51, p.1468, 2004.
[2] B. H. Calhoun et al., *IEEE JSSC.*, vol.41, p.1673, 2006.
[3] S.Monfray et al., *IEDM Tech. Dig.*, p.693, 2007.
[4] T. Ohtou et al., *IEEE EDL*, vol. 28, Aug., p.740, 2007.
[5] Y. Morita et al., *Symp. on VLSI Tech.*, p.166, 2008.
[6] M. Yamaoka et al., *IEEE JSSC*, vol. 41, p.2366, 2006.
[7] K. Samsudin et al., *SSE*, vol.50, p.660, 2006
[8] S.Mukhopadhyay et al., *IEEE TED*, vol 55, p.152, 2008.
[9] "ISE TCAD Rel. 10.0 Manual," DESSIS, 2004

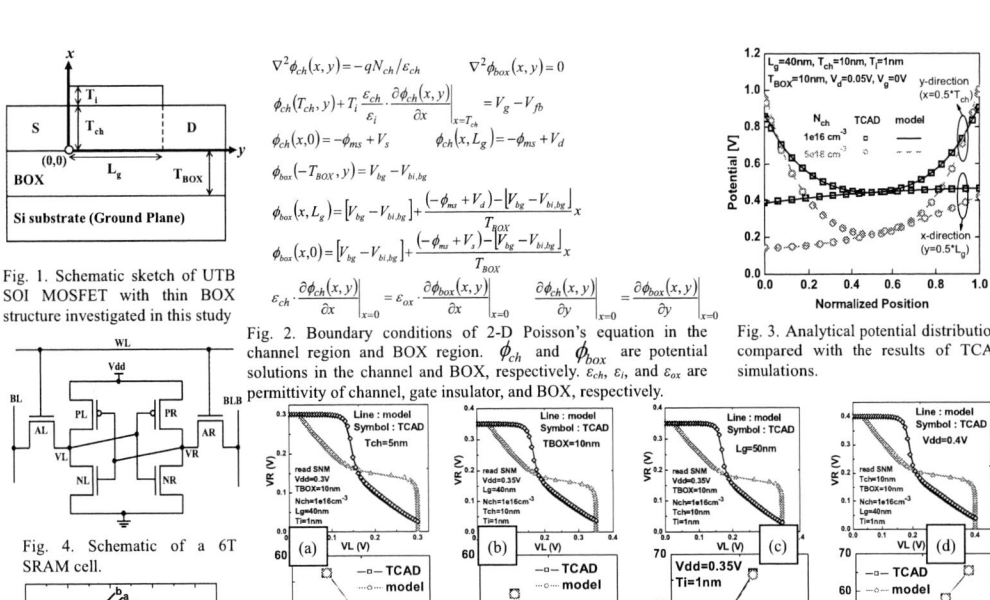

Fig. 1. Schematic sketch of UTB SOI MOSFET with thin BOX structure investigated in this study

Fig. 2. Boundary conditions of 2-D Poisson's equation in the channel region and BOX region. ϕ_{ch} and ϕ_{box} are potential solutions in the channel and BOX, respectively. ε_{ch}, ε_i, and ε_{ox} are permittivity of channel, gate insulator, and BOX, respectively.

Fig. 3. Analytical potential distributions compared with the results of TCAD simulations.

Fig. 4. Schematic of a 6T SRAM cell.

Fig. 5. The methodology for SNM calculation.

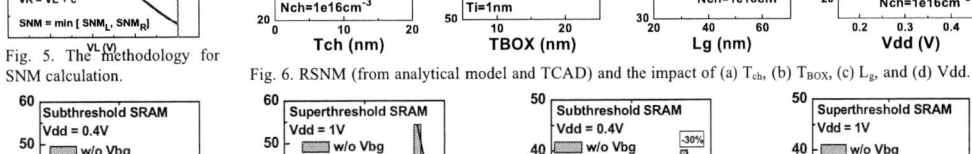

Fig. 6. RSNM (from analytical model and TCAD) and the impact of (a) T_{ch}, (b) T_{BOX}, (c) L_g, and (d) Vdd.

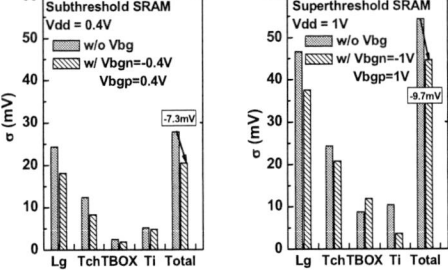

Fig.7. Impact of back-gate bias (V_{bg}) on RSNM variation (σ) for subthreshold and superthreshold UTB SOI SRAM. RSNM variation caused by parameter (P) variation $\sigma = $ [RSNM(P+20%) – RSNM(P-20%)]. (NMOS back-gates are biased below GND)

Fig.8. Impact of back-gate bias (V_{bg}) on % change in RSNM for subthreshold and superthreshold UTB SOI SRAM. % change in RSNM is determined by the ratio of RSNM variation (σ) to nominal RSNM = [σ/RSNM(nominal)]. (NMOS back-gates are biased below GND)

(a)	6T without Vbg
(b)	6T with Vbgp=0.4V(PMOS) Vbgn=-0.4V(NMOS)
(c)	6T with Vbgp=1V(PMOS) Vbgn=-1V(NMOS)
(d)	6T with Cell Beta Ratio = 2
(e)	8T without Vbg

Table 1. Subthreshold SRAM cells. (Vdd=0.4V)

Fig.9. Schematic of a UTB SOI 6T SRAM cell with V_{bg}.

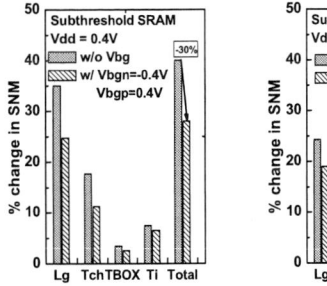

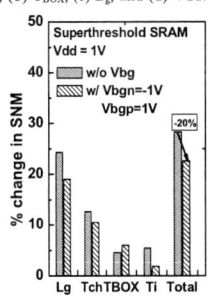

Fig.11. Comparison of different cell structure (listed in Table 1.) on RSNM and RSNM variation. The 8T SRAM (case e) shows 2X larger RSNM. The 6T SRAM with Vbgp =1V, Vbgn = -1V (case c, Above-Vdd/Below-GND back-gate bias) shows significant improvement in SNM variability. (case e) and (case c) shows comparable % change in SNM.

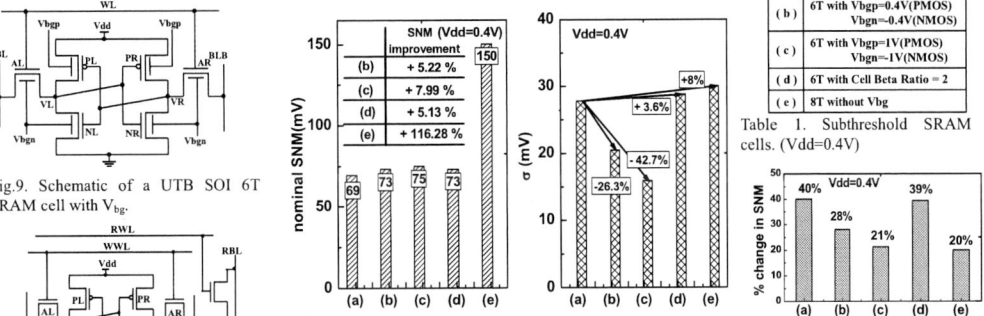

Fig.10. Schematic of an 8T SRAM cell.

978-1-4244-2784-0/09 $25.00 © 2009 IEEE

The Promise and Implementation of three Dimensional Integration (Invited)

Subramanian Iyer

IBM, USA

In many ways, three dimensional integration presents itself as a logical extension of planar monolithic integration – integration of additional function on the same die. Notwithstanding the remarkable advances in scaling we have witnessed over the last several decades, basic material limitations and lithography have slowed this trend down and the benefits of node to node migration need to be weighed against both technology development costs and complexity as well as the cost of design migration. Another consideration is die size which for high end applications such as high performance processors continues to increase well beyond the sweet spot dictated by yieldability, driven primarily by multiples cores and on-chip memory. Furthermore in such large die, long electrical paths cause significant delay and power draws. To address these limitations, three dimensional integration must be viewed beyond a simplistic packaging paradigm but rather as extension of silicon integration in the third dimension i.e., the introduction of low resistance, low inductance vertical interconnects between multiple active silicon strata that are co-designed in much the same way we design an SOC or ASIC today. This talk examines at the technology as it stands toady and the challenges going forward. These challenges include the development of fine pitch vertical interconnects and the degrees of integration they would permit. We will focus on the integration of three dimensional memory as the prototypical example of three dimensional integration and describe how these challenges are being met.

Optimization of the Channel Lateral Strain Profile for Improved Performance of Multi-Gate MOSFETs

L. De Michielis[1], K. E. Moselund[1], D. Bouvet[1], P. Dobrosz[2], S. Olsen[2], A. O'Neill[2], L. Lattanzio[1], M. Najmzadeh[1], L. Selmi[3] and A. M. Ionescu[1]

[1] Ecole Polytechnique Fédérale de Lausanne, Nanoelectronic Devices Laboratory (*Nanolab*), CH-1015 Lausanne, Switzerland, [2] Newcastle University, School of Electrical, Electronic and Computer Engineering, Newcastle, UK, [3] University of Udine, Dept. of Electrical, Managerial and Mechanical Engineering, 33100 – Italy, Tel: +41 21 693 3978, Fax: +41 21 693 3640, Email: Luca.DeMichielis@epfl.ch, Adrian.Ionescu@epfl.ch

Abstract - **We report for the first time the** *optimization of the channel lateral strain profile as a new technological booster* **for improved performance of multi-gate n-channel MOSFET. We find that quasi-uniform or flat-Gaussian-close-to-the-drain profiles are optimum for the I_{on} boosting of sub-50nm scaled MOSFETs, while the penalty on I_{off} and subthreshold slope is minimum. The reported predictions use realistic lateral uniaxial strain profiles, with peaks up to few GPa's and average values of hundreds of MPa's.**

I. INTRODUCTION

The enhancement of carrier mobility and thereby of device performance by strain is an integral part of front-end CMOS processes [1,2]. It is well known that uniaxial tensile strain increases electron mobility [3, 4], and recently we reported a 100% increase of low-field mobility, µ0, in *plastically deformed* [5] strained silicon nanowire (NW) gate-all-around (GAA) NMOSFETs, where µ0 ranged from 900cm^2/Vs for large (~100nm) GAA devices, to 500cm^2/Vs for 5nm wide GAA, and 400cm^2/Vs for larger non-strained tri-gate devices. So far research has attempted to correlate mobility enhancement and average amount of strain, and little work as been done on optimizing the strain profile for a fixed amount of strain, despite the fact that strain technologies used in fabrication rarely induce a uniform strain profile.

Here we demonstrate how high performance MOSFETs can improve their performance, working on optimization of lateral strain profile: and in particular using a lateral strain profile and its placement along the channel for constant average strain.

II. MODEL PARAMETERS

A SEM image of a typical strained (by *local bending*) nanowire device is shown in Fig. 1, while Fig. 2 is a schematic of the structure used for simulation. For simplicity, we simulate a 2D double-gate (DG) MOSFET rather than a 3D GAA structure using Sentaurus Device version A-2007.12 and we impose a variety of realistic lateral stress profiles in transistor channel. We simulate both a long device with parameters corresponding to fabricated devices to fit the model to experimental data, Fig. 4, and a short device scaled according to the 65nm node. We apply a uniaxial stress profile to the channel, and use both a model which accounts for bandgap shrinkage, mainly V_T shift, and a mobility tensor model to properly model I_{on}. The change in the band structure (i.e. conduction and valence band edge shifts) is taken into account according to the deformation potential theory [6]; moreover in order to describe the stress-dependent mobility, the simulations use the piezoresistance mobility model (enabled for electrons only) and the piezoresistivity coefficients of Silicon follow the doping and temperature dependence according to Kanda model [7]. The average value of simulated strain is realistically kept at 300MPa in order to stay within the limits of validity of the used models (enable peaks of Gaussian profiles of maximum 4GPa).

III. LATERAL STRAIN PROFILE

Measured strain curves along the channel are shown in Fig. 3. MOSFET performance is simulated for: a non-strained device,

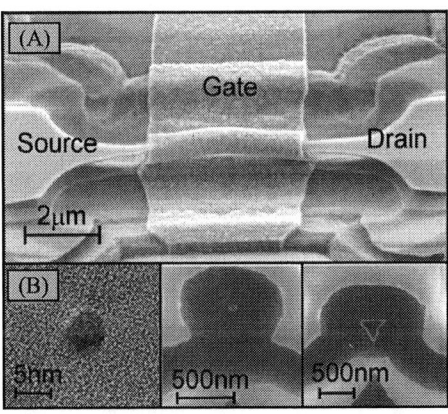

FIGURE 1 SEM IMAGES OF: (A) A GAA MOSFET WITH THE POLYSILICON GATE SHOWING VISIBLE BENDING OF THE NANOWIRE. (B) CROSS-SECTIONS OF INCREASINGLY LARGE WIRES GOING FROM A DIAMETER OF 5NM TO 250NM. FOR PROCESS DETAILS SEE [5].

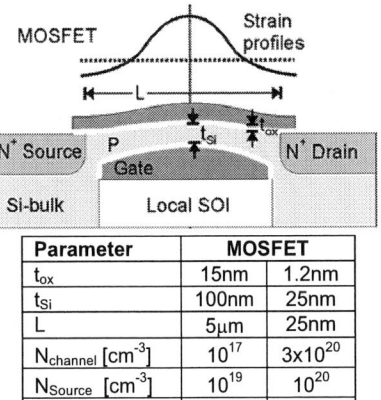

Parameter	MOSFET	
t_{ox}	15nm	1.2nm
t_{Si}	100nm	25nm
L	5µm	25nm
$N_{channel}$ [cm^{-3}]	10^{17}	3×10^{20}
N_{Source} [cm^{-3}]	10^{19}	10^{20}
N_{Drain} [cm^{-3}]	10^{19}	10^{20}

FIGURE 2: SCHEMATIC OF INVESTIGATED MOSFET GEOMETRIES ON BENDED NANOWIRES WITH LATERAL STRAIN PROFILE. SIMULATIONS ARE CARRIED OUT ON A 2D STRUCTURE WHOSE DIMENSIONS ARE SHOWN IN THE TABLE.

constant strain, sharp Gaussian (3σ=9% of the total channel length) and a flat Gaussian (3 σ =18%) profile, all with same average strain of 300MPa. For the Gaussian profiles the peak position is shifted along the channel. Fig. 4 validates the accuracy of the simulator and the used models; a constant tensile strain profile of 1.2GPa, similar to the one measured in bended nanowire transistors [5] provides a very good fit between experimental and simulated $I_{DS}(V_{GS})$ in GAA NW MOSFETs. However, for practical reasons motivated by the

978-1-4244-2784-0/09 $25.00 © 2009 IEEE

available strain levels by various stressor techniques in nano-CMOS and the maximum peaks experimentally observed in Si NWs, we choose to investigate various profiles with peaks up to 4GPa and average value of 300MPa. Fig. 5 shows the increase in low-field mobility; the placement does not influence the extracted value of μ0. In all cases, *flatter* the profile *greater* the enhancement. Overall, constant profile achieves more than twice the mobility gain of the Gaussian profiles for same average strain. This trend is much more pronounced in the scaled structures, where a sharp profile provides very little mobility gain, but substantially deteriorates the inverse subthreshold slope. The position of the strain profile along the channel has a strong influence on the off characteristics in the case of the 25nm long device, Fig. 6. A centered strain profile increases junction leakage, whereas a strain profile close to either source or drain shows better I_{off} performance than for constant strain. Fig. 7 shows the drain current enhancement as a function of gate overdrive for a short device, it is evident that the optimum profile is a constant strain profile, providing ~30% gain in I_{on} (for $V_{GS} > V_T$) in 25nm channel, compared with less than 10% gain in same device for Gaussian profiles. However, it appears that a Gauss profile close to the drain offers better performance at low gate overdrive.

IV CONCLUSION

Inspired by the constant and Gaussian-like strain profiles of bended silicon nanowire devices, we have investigated the influence of the lateral strain profile on double-gate MOSFET performance. A uniform strain profile is optimum for mobility and on-current characteristics, whereas either a constant profile or a flat Gaussian profile close to either of the junctions results in the best off-characteristics.

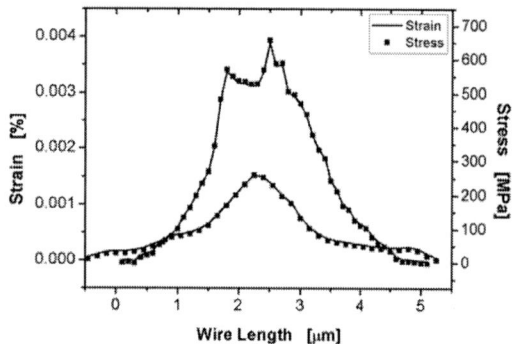

FIGURE 3: EXPERIMENTAL STRESS AND STRAIN LATERAL PROFILES FOR TWO SILICON NANOWIRES, MEASURED BY RAMAN SPECTROSCOPY. THE STRAIN LEVEL DEPENDS ON WIRE DIMENSION (W/L RATIO).

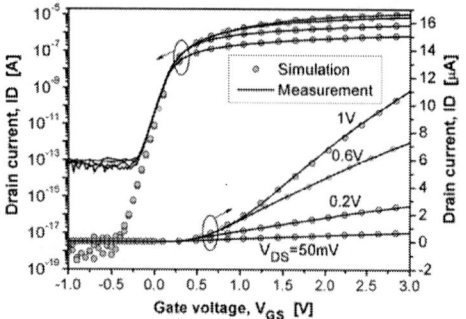

FIGURE 4 SIMULATED VERSUS EXPERIMENTAL DRAIN CURRENT, I_D, VERSUS GATE VOLTAGE, V_{GS}, CHARACTERISTICS; A 1.2GPA CONSTANT STRESS IN THE CHANNEL OF DG MOSFET PROVIDES A VERY GOOD FITTING OF SIMULATED DATA ON EXPERIMENTAL CURVES, FOR L=5μM.

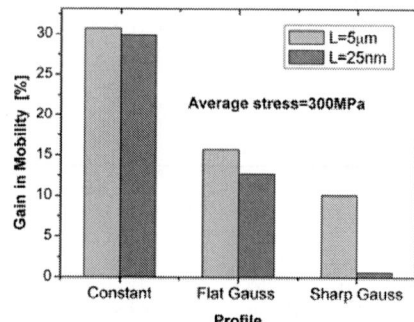

FIGURE 5: GAIN IN LOW-FIELD MOBILITY, μ0, OBTAINED IN STRAINED CHANNELS WITH DIFFERENT STRAIN PROFILES (CONSTANT, FLAT AND SHARP GAUSS), COMPARED TO THE NO STRAIN CASE. μ0 IS EXTRACTED FROM SIMULATED I_{DS}-V_{GS} AT V_{DS}=50mV AND THE PROFILE PLACEMENT IN A LONG CHANNEL HAS NO SIGNIFICANT INFLUENCE ON THE EXTRACTED VALUE.

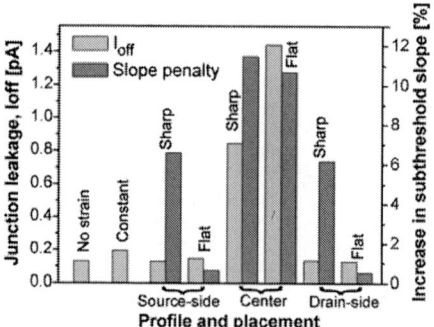

FIGURE 6 JUNCTION LEAKAGE CURRENT, I_{OFF} (V_{GS}=-1V, V_{DS}=100mV) AND PERCENTAGE OF DEGRADATION OF THE SUBTHRESHOLD SLOPE (V_{DS}=10mV) COMPARED TO THE NO-STRAIN CASE, BOTH SIMULATED IN A 25NM LONG DEVICE, WIDTH=1μM.

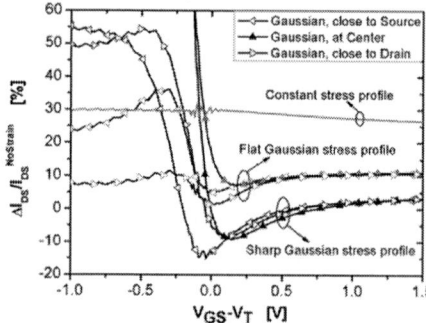

FIGURE 7 DRAIN CURRENT ENHANCEMENT COMPARED TO A NON-STRAINED DEVICE FOR L=25NM AND V_{DS}=100mV. THE CONSTANT STRESS PROFILE PROVIDES THE GREATEST ENHANCEMENT. IN SUBTHRESHOLD THE PLACEMENT OF THE PROFILE IS SEEN TO HAVE A GREATER INFLUENCE THAN THE PROFILE SHAPE. V_T ~0.15V.

REFERENCES

[1] T. Miyashita et al. IEDM '07, pp. 251-254, 2007
[2] S. Tyagi et al. IEDM '05, pp. 245- 247, 2005
[3] P. R. Chidambaram, Trans. Electron. Dev. 53 (5), pp. 944-964, 2006
[4] K. Uchida et al. IEDM '05, pp. 135–138, 2005
[5] K. E. Moselund et al., IEDM '07, pp. 191-194. 2007
[6] J. Bardeen et al., Physical Reviev, Vol. 80, n1, pp. 72-80, 1950
[7] Y. Kanda, Trans. Electron. Dev. 29 (1), pp. 64-70, 1982

A Novel Double-gated Nanowire TFT and Investigation of Its Size Dependency

Wei-Chen Chen[a], Chuan-Ding Lin[b], Horng-Chih Lin[a,b,*], and Tiao-Yuan Huang[a]

[a]Institute of Electronics, National Chiao Tung University, No. 1001, Ta Hsueh Rd., Hsinchu, Taiwan 300, R.O.C.
[b]National Nano Device Laboratories, No. 26, Prosperity Rd. I, Science-Based Industrial Park, Hsinchu, Taiwan 30078, R.O.C.
*Phone: +886-3-571-2121 ext. 54193, Fax: +886-3-572-4361, E-mail: hclin@faculty.nctu.edu.tw

ABSTRACT

A simple method for fabricating poly-Si nanowire (NW) TFT with multiple gates is proposed and characterized. In this structure, NW is formed mainly using both anisotropic and highly selective isotropic plasma etching. It is found that when the size of NW is scaled down, double-gated operation provides more improvement. Furthermore, by utilizing this unique independent double-gated configuration, the function of threshold voltage modulation is investigated.

INTRODUCTION

Inherent with ultra-thin body and high surface-to volume-ratio, Si nanowire (NW) poses as a promising candidate with great potential in device applications [1]. When combined with multiple-gated configuration, poly-Si NW TFT has demonstrated impressive electrical characteristics [2]. Previously, our group had proposed simple methods in fabricating poly-Si NW TFT without the necessity of advanced lithography tools [3][4]. Nevertheless, they all suffer from irregular shapes of NW, which when scaled down may lead to such problems as non-uniform carrier distribution and difficult theoretical modeling. This concern is relaxed by our new NW structure in this work, which resembles a rectangle.

DEVICE STRUCTURE AND FABRICATION

Figure 1 depicts schematic process steps of the proposed device. (i) First, gate stack consisting of nitride hard mask and n$^+$ poly-Si(1st gate) was grown on an oxidized wafer capped with a nitride layer. (ii) Following the gate stack patterning, highly selective plasma etching was used for lateral etching of 1st gate. (iii) Then 1st gate dielectric and amorphous Si were deposited, which subsequently underwent 600°C annealing in N$_2$ ambient for 24 hours. (iv) After S/D implant was conducted, (v) NWs and S/D were defined simultaneously by a reactive plasma etching step. Note that this etching completely removed poly-Si film outside the hard mask and the portion that resided underneath the hard mask would remain intact forming a rectangular NW channel. (vi) 2nd gate dielectric and gate were deposited and patterned. The device was completed after standard metallization and 3-hour NH$_3$ plasma treatment before characterization.

RESULTS AND DISCUSSION

To investigate electrical characteristics dependence on NW size, in this work, three devices (SA, SB, and SC) corresponding to different NW sizes accomplished by 12, 10, and 8sec lateral etching are fabricated and characterized. To demonstrate the feasibility of selective isotropic etching, Fig. 2(a) exhibits various lateral encroaching depths as a function of etching time and Fig. 2(b) shows one of the resultant image. By optimizing the choice of source gas, gas flow and applied RF bias, sub-100nm encroaching depths are easily obtained. Transfer characteristics of device SA, SB, and SC under various operational modes and their corresponding TEM images are shown in Fig. 3. From the TEM results, the NW widths are 52, 43, and 18 nm for device SA, SB, and SC,

respectively. The SG-1 mode denotes the scheme when 1st gate acts as the driving gate while 2nd gate is grounded, and similarly for SG-2 mode. On the other hand, in DG mode, two gates are connected together to serve as the driving gate. Ideally, SG-1 and SG-2 should display identical characteristics based on the nearly symmetric structure of NW. However, as shown in Fig. 4 which identifies the current conduction path for SG-1 mode, this route contains non-gated parts which drastically increase series resistance leading to lower on-current. Owing to the gate-coupling effect and larger current conduction width, the double-gated scheme dramatically enhances device performance in terms of smaller subthreshold swing (S.S) and threshold voltage. Nevertheless, this phenomenon is more evident when the size of NW is smaller, attributed to stronger coupling effect. Hence in SA, DG mode exhibits similar characteristics to SG-2 mode because the current conduction is now mainly dominated by 2nd gate while apparent performance improvement is observed in SC for which volume inversion is speculated to take place [5]. The extracted S.S is plotted in Fig. 5 and the smallest is 115mV/dec for DG mode in SC, which is quite impressive for poly-Si based devices. To further compare drain current improvement dependence on NW size, Fig. 6 shows the drain current ratio (defined as I$_{DG}$ divided by the sum of I$_{SG-1}$ and I$_{SG-2}$) versus V$_d$ for the three devices. For SA, the ratio is close to 1, which is consistent with previous Id-Vg curve in that V$_{th(SG-1)}$ is much larger than V$_{th(DG)}$ and thus I$_{DG}$ is almost equal to I$_{SG-2}$. In contrast, the ratio can be as large as 2 for SC, which further supports our volume inversion assumption. For the purpose of fluctuation inspection, SC devices with multiple channels are characterized as given in Fig. 7, indicating that our NW TFT indeed possesses excellent uniformity. Taking advantage of this independent double-gated configuration, V$_{th(SG-2)}$ modulated by 1st gate is next investigated. The result in Fig. 8 implies that the V$_{th(SG-2)}$ can be efficiently adjusted by 1st gate voltage due to the tiny volume of NW.

CONCLUSION

In all, a novel poly-Si NW TFT with simple fabrication process is proposed and characterized. Equipped with independent double-gated structure, impressive device performance is obtained and effective threshold voltage modulation is demonstrated, which profoundly increases the flexibility of device operation.

Acknowledgement -This work was supported in part by the National Science Council of the Republic of China (ROC) under contract NSC 96-2221-E-009-212-MY3.

REFERENCES

[1] F. L. Yang et al., *Symp. VLSI Tech. Dig.*, pp. 196, 2004.
[2] Maesoon Im et al., *IEEE Electron Device Lett.*, vol. 29, pp. 102-105, Jan. 2008.
[3] H. C. Lin et al., *IEEE Trans. Electron Devices.*, vol.53, pp. 2471-2477, Oct. 2006.
[4] H. H. Hsu et al., *VLSI-TSA*, pp.101-102, 2008.
[5] F. Balestra et al., *IEEE Electron Device Lett.*, vol. 8, pp. 410-412, Sep. 1987.

978-1-4244-2784-0/09 $25.00 © 2009 IEEE

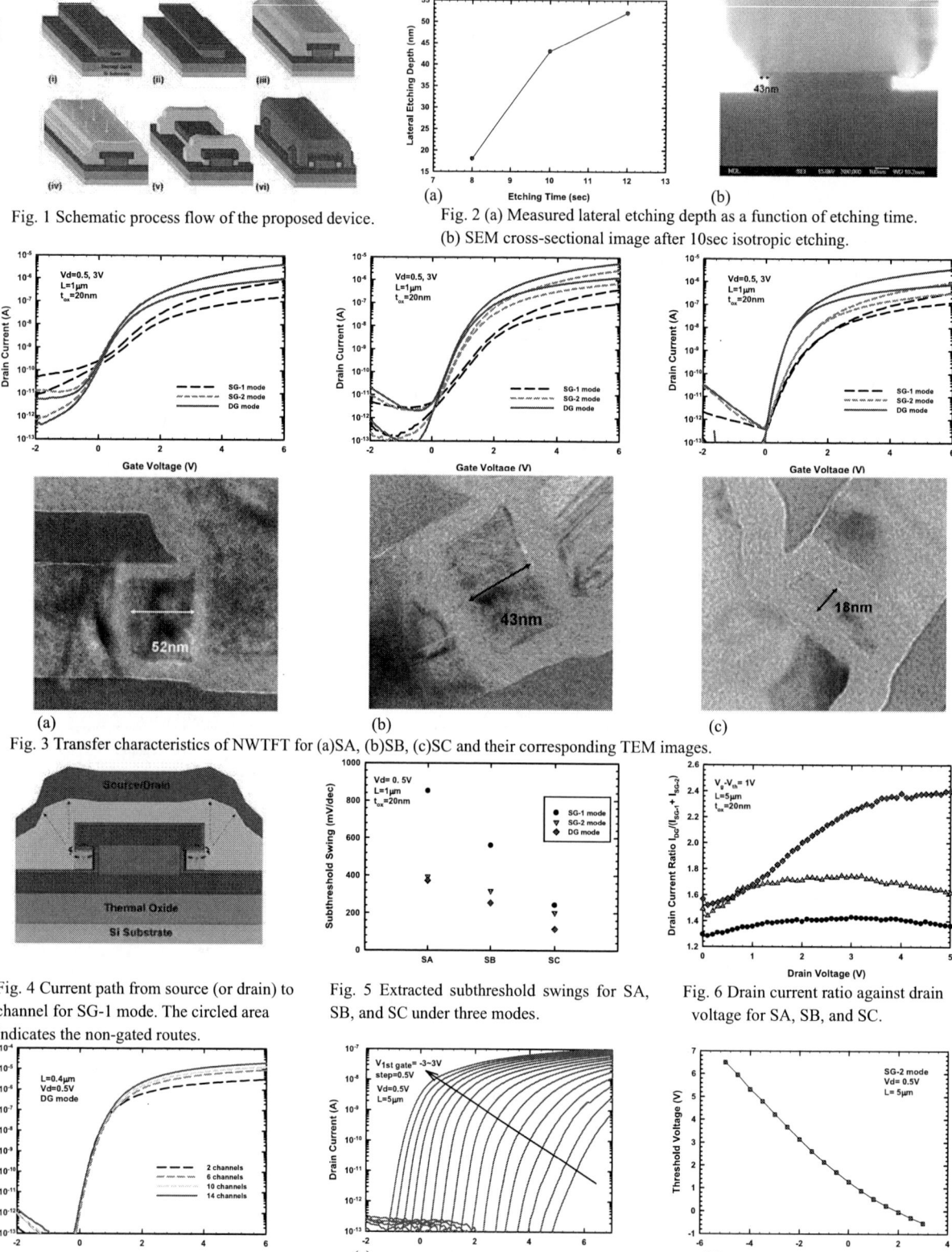

Fig. 1 Schematic process flow of the proposed device.

Fig. 2 (a) Measured lateral etching depth as a function of etching time.
(b) SEM cross-sectional image after 10sec isotropic etching.

Fig. 3 Transfer characteristics of NWTFT for (a)SA, (b)SB, (c)SC and their corresponding TEM images.

Fig. 4 Current path from source (or drain) to channel for SG-1 mode. The circled area indicates the non-gated routes.

Fig. 5 Extracted subthreshold swings for SA, SB, and SC under three modes.

Fig. 6 Drain current ratio against drain voltage for SA, SB, and SC.

Fig. 7 Transfer characteristics of SC device with multiple channels.

Fig. 8 (a) Transfer characteristics of SC device with varying 1st gate bias.
(b) Extracted V_{th} v.s. 1st gate bias.

978-1-4244-2784-0/09 $25.00 © 2009 IEEE

Fermi Level Depinning For the Design of III-V FET Source/Drain Contacts

Jenny Hu, Ximeng Guan, Donghun Choi, James S. Harris, Krishna Saraswat, and H. -S. Philip Wong

Department of Electrical Engineering, Stanford University, 420 Via Palou Mall, Stanford, CA, 94305, USA

Email: jennyhu@stanford.edu

INTRODUCTION

High mobility III-V compounds is a strong contender for extending high performance logic beyond the 22 nm technology node [1-3]. However, demonstrations of exceptional III-V performance required device footprints on the μm-scale despite nm-scale gate lengths, in order to avoid source/drain shorting during contact alloying. The scaling of III-V FETs is severely limited by the unacceptably large lateral diffusion of the multi-layer alloyed structures typically used for ohmic contacts [4]. In our recent work, we introduced a novel non-alloyed, highly scalable contact structure through the use of Al as a low workfunction metal on an unpinned Fermi level [5]. We use GaAs as a baseline III-V material, where the developed contact techniques can be extended to InGaAs and InSb, materials which are more technologically important [6]. **In this work, we explain in detail the unpinning mechanisms and the rationale for the material selection. We demonstrate the same method can be applied to a variety of metals, Y, Er, Al, Ti, W, and Pt, providing much flexibility in the design of an ideal source/drain contact for III-V HEMTs/MOSFETs and Schottky Barrier FETs.**

SCHOTTKY BARRIER MODULATION

In metal semiconductor junctions, metal-induced gap states (MIGS) pin the semiconductor Fermi level towards the charge neutrality level E_{CNL}[7-8], causing Schottky barrier heights to be roughly independent of the metal [9]. To unpin the Fermi level, we insert an ultrathin insulator to decay the penetration of the metal electron wavefunction into the semiconductor (Fig.1). This reduces the electron states available to charge the interface, so fewer interface dipoles are created to drive the band lineup towards E_{CNL}. This method for depinning was first proposed for Si by Connelly [10], followed by demonstration on Ge by Kobayashi [11], and our demonstration on GaAs [5].

After the Fermi level is unpinned, the metal workfunction (Φ_M) can be used to tune the effective barrier $\Phi_{B,eff}$ of the contact. Low Φ_M metals are used for a near zero $\Phi_{B,eff}$ contact to n-GaAs. This contact structure can be viewed as two resistances in series: a tunneling resistance through the dielectric (R_T) and a resistance associated with the barrier (R_{SB}). Since R_T and R_{SB} are inversely dependent upon t_{SiN}, there exists an optimal insulator thickness to minimize the contact resistance R_C (Fig.2). Initial insertion of the insulator unpins the Fermi level and reduces $\Phi_{B,eff}$, which exponentially increases current as more carriers can be thermionically emitted over the reduced barrier, and thus decreases R_C. Upon further increase in thickness, the R_T begins to dominate, increasing R_C. The optimal thickness is the critical balance between a lower $\Phi_{B,eff}$ that decreases R_{SB} and a thicker tunneling barrier that increases R_T.

The ideal insulator for this application has a conduction band offset ΔE_C=0 to minimize the tunneling resistance penalty in R_C, and a maximum pinning factor S=1 to minimize potential pinning of the dielectric. S accounts for dielectric screening, and is inversely correlated with the dielectric constant ε_∞ [12]. We investigated silicon nitride (SiN) since it: 1.) unpins the Si and Ge Fermi level, 2.) has a small ΔE_C =1.5eV, 3.) has a large S~0.6, and 4.) is dense enough to serve as a good oxidation barrier.

FABRICATION OF CONTACT STRUCTURE

To emphasize the thermionic barrier effect, we used lightly doped (2×10^{16}cm^{-3}) MBE grown n-GaAs. Samples underwent organics degrease, HCl native oxide removal, and $(NH_4)_2S$ sulfur passivation. SiN was sputtered, and diodes defined by metal evaporation through shadow masks (Fig.3).

MEASUREMENT RESULTS

The R_C vs. t_{SiN} plot for various metals (Fig.4) confirms the unpinning effect is not limited to Al. The decrease in R_C reflects a reduction in $\Phi_{B,eff}$, where without unpinning, R_C would only increase with t_{SiN} due to increased R_T. One key point is the change in the optimal t_{SiN} with different metals due to the dependence of the height of the tunneling barrier Φ_T on Φ_M. A lower Φ_M has a smaller Φ_T, so R_T is reduced compared to that of a higher Φ_M. On the R_C vs. t_{SiN} plot, this corresponds to a shift of the R_T branch towards the right (Fig.6), so for a lower Φ_M the minimum R_C occurs at a thicker SiN.

In studying the effects of doping, we found there is a wider t_{SiN} process window for higher doped substrates (Fig.5). With higher doping, the tunneling current increases significantly, which lessens the tunneling resistance penalty in R_C.

$\Phi_{B,eff}$ is extracted through temperature measurements (Fig.7). As expected, ideality factor n increases with t_{SiN} as conduction is due to tunneling, rather than purely thermionic emission (n=1). $\Phi_{B,eff}$ decreases with increasing t_{SiN}, confirming reduced Fermi level pinning. The difference between the minimum achieved $\Phi_{B,eff}$ and the ideal (Fig.8) suggests there still exists minimal pinning by either extrinsic defect states, or remaining MIGS.

DISCUSSION

The focus of our work is to demonstrate unpinning of the III-V Fermi level, and to systematically study how to minimize R_C. Therefore, we report R_C ratio values rather than absolute R_C due to the use of a lightly doped substrate to study the effect of t_{SiN} on $\Phi_{B,eff}$. The R_C of a lightly doped substrate is not meaningful since R_C can be easily improved through a higher electron concentration/doping and the thinner insulator needed to depin the Fermi level. We achieved a minimum specific contact resistivity ρ_C of 2.0 Ω-cm^2, but from NEGF simulations [13] we see ρ_C can be reduced by five orders of magnitude by increasing the doping to 1×10^{19}cm^{-3}, and reduced two more orders by a reduction of the optimal t_{SiN} to 1.1 nm (Fig.9).

CONCLUSION

In summary, we successfully demonstrate the unpinning of the n-GaAs Fermi level for a variety of metals, through the insertion of a thin insulator to reduce the penetration of MIGS from metal into the semiconductor. Modulation of the barrier heights was verified through reduction of $\Phi_{B,eff}$ and the dependence of R_C on t_{SiN}. There exists rich design flexibility in the choice of insulator and metals, where the optimal insulator thickness and lowest achievable R_C depends on both materials. We expect this method can be used to make scalable non-alloyed low resistance ohmic contacts for III-V MOSFET/HEMTs, and can tune the barrier height for Schottky Barrier FETs.

978-1-4244-2784-0/09 $25.00 © 2009 IEEE

ACKNOWLEDGEMENTS

This work is supported in part by the Focus Center Research Program (MSD), Intel Corporation, and NSF (ECS-0501096). J. Hu is additionally supported by the Stanford Graduate Fellowship (SGF) and the National Defense Science and Engineering Graduate (NDSEG) Fellowship.

REFERENCES

[1] M. Hudait et al, IEDM Tech. Dig., pp.625, 2007.
[2] Y. Xuan et al, IEDM Tech. Dig., pp. 637, 2007.
[3] N. Waldron et al, IEDM Tech. Dig., pp.633, 2007.
[4] A. Baca et al, Thin Solid Films, pp.599, 1997.
[5] J. Hu et al, DRC Conf. Dig., pp.89, 2008.
[6] R. Chau et al, IEEE Trans. On Nanotech., pp.153, 2005
[7] W. Mönch, PRL, pp. 1260, 1987.
[8] J. Tersoff, PRL, pp.465, 1984.
[9] J. Roberston et al, JAP, pp.14111, 2006.
[10] D. Connelly et al, IEEE Trans. on Nanotech., pp.98, 2004.
[11] M. Kobayashi et al, Symp.VLSI Tech., pp.54, 2008.
[12] Y.Yeo et al, JAP, pp.7266, 2002.
[13] S. Datta, "Electronic Transport in Mesoscopic Systems", Cambridge University Press, 1995.

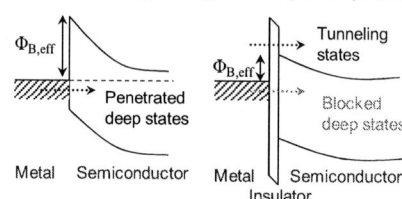

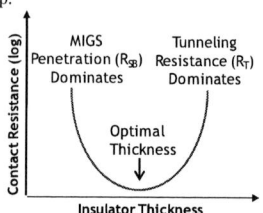

Fig. 1. (a) In direct contact, the electron wavefunction of the metal decays into the semiconductor, creating MIGS that pin the Fermi level. (b) With an insulator, the metal wavefunction is attenuated in the gap states.

Fig. 2. There exists an optimal insulator thickness for minimal contact resistance, where MIGS is partially blocked, but free carrier tunneling still exists.

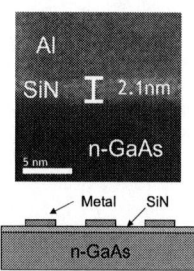

Fig. 3. (a) Cross sectional TEM image of the Al/SiN/n-GaAs contact. (b) Side view showing SiN passivation of the surface to reduce surface leakage.

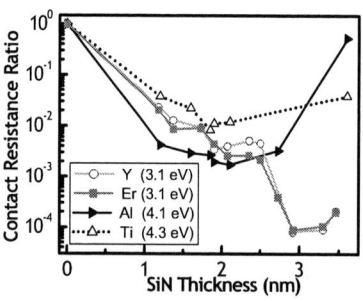

Fig. 4. R_C measurement using the contact end resistance method. The R_C ratios are relative to the Schottky case. Optimal SiN thickness is found to depend on the metal workfunction.

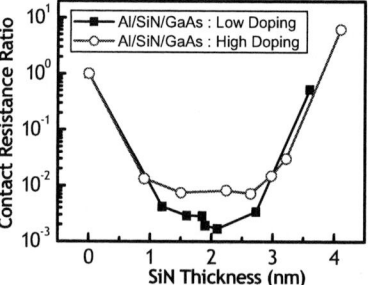

Fig. 5. Compared to lower doped substrates $(2 \times 10^{16}$ cm$^{-3})$, using higher doping $(2 \times 10^{17}$ cm$^{-3})$ shifts the optimal t_{SiN} lower and increases the range of suitable SiN thicknesses, without sacrificing Rc.

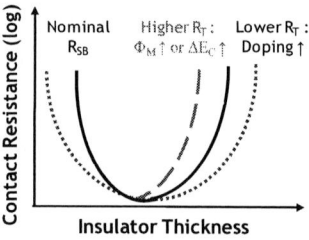

Fig. 6. Effect of different metals, dielectrics, and substrates on the tradeoff between R_C and insulator thickness, where R_T is the tunneling resistance through the insulator, and R_SB is the resistance barrier related resistance.

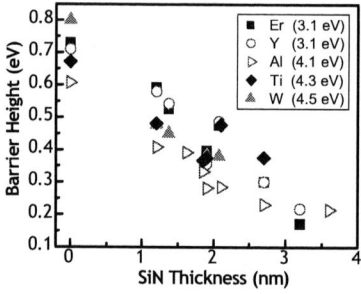

Fig. 7. The extracted barrier heights decrease with increasing SiN thickness. This confirms the ability for SiN to modulate the barrier for a variety of metals.

	Φ_M	Optimal t_{SiN}	Minimum Achieved $\Phi_{B,eff}$	Ideal Upinnined $\Phi_{B,eff}$	Pinned Schottky $\Phi_{B,eff}$
Al	4.1 eV	1.9 nm	0.21 eV	0.08 eV	0.61 eV
Ti	4.3 eV	2.1 nm	0.27 eV	0.15 eV	0.67 eV
Y	3.1 eV	2.9 nm	0.17 eV	0.08 eV	0.71 eV
Er	3.1 eV	2.9 nm	0.22 eV	0.08 eV	0.73 eV

Fig. 8. Summary of values illustrating the tradeoff between using a lower workfunction metal and the thicker SiN required to obtain the minimum Rc. The optimal thickness can easily be reduced by using a metal with a slightly higher Φ_M or using a higher doped substrate. The ideal and obtained $\Phi_{B,eff}$ are compared, where the minimum ideal $\Phi_{B,eff} = E_c - E_F = 0.08$ eV is due to the low doped substrate used.

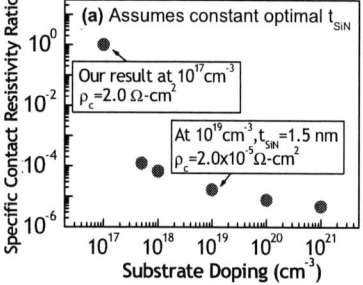

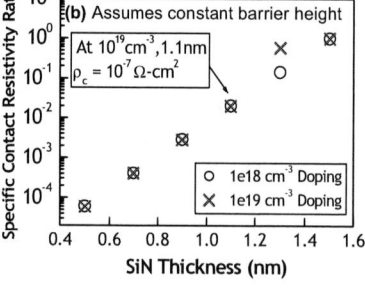

Fig. 9. Dependence of the tunneling limited contact resistance on substrate doping is simulated by a fully self-consistent NEGF simulation of Al/SiN/n-GaAs using the effective mass approximation. (a) Ratios are taken relative to 1e17 cm^{-3} doping. Specific contact resistivity is reduced by 5 orders of magnitude when the doping is increased from 1e17 to 1e19cm^{-3}. (b) Increasing the doping decreases the optimal insulator thickness, where the specific contact resistivity further decreases exponentially.

978-1-4244-2784-0/09 $25.00 © 2009 IEEE

Influence of Gate Misalignment on the Electrical Characteristics of MuGFETs

Chi-Woo Lee, Aryan Afzalian, Ran Yan, Nima Dehdashti, Weize Xiong*and Jean-Pierre Colinge
Tyndall National Institute,Lee Maltings, Prospect Row

Cork, Ireland
* Texas Instruments Inc., SiTD, 13121 TI Boulevard
Dallas, TX USA

ABSTRACT

This work studies the influence of gate misalignment on the electrical properties of MuGFETs using both measurements and 3D simulations. Electrical characteristics such as a DIBL, V_{th} and drain breakdown voltage are shown to be dependent on the gate misalignment due to the resulting change in fin width. Devices that have a Widening of the fin at the Drain side (DW) decreased due to weakening of gate control. Widening of the fin at the Source (SW), however, does not alter the device characteristics.

INTRODUCTION

Many semiconductor manufacturers are currently carrying out research in multi-gate SOI MOSFETs (MuGFETs) for deep submicron CMOS applications, as these devices hold the promise for pushing the limits of silicon integration beyond the limits of classical planar technologies [1]. In a MuGFET the gate electrode is wrapped around a silicon wire, called "finger" or "fin", forming a multi-gate structure with excellent control of the channel potential. However, a fin layout has inherent non-uniformity fin width due to lithography proximity effects [2]. In this paper, we study the influence of gate misalignment on the electrical characteristics of MuGFETs using both measurements results and 3D simulation.

DEVICE FABRICATION AND SIMULATION

Standard Unibond SOI wafers with a 60-nm top silicon layer and a 150-nm buried oxide were used for device fabrication. The SOI film is p-type with a boron doping concentration of approximately $2\times10^{15}cm^{-3}$. Fins with a width down to 11nm were patterned using 193nm lithography and Reactive Ion Etching (RIE). H_2 annealing was used to round the fin corners and to smooth the fin sidewalls. A 1.8-nm gate oxide was then grown by wet oxidation. The channels of some devices was doped by boron ion implantation to reach a concentration $N_a=2\times10^{18}/cm^{-3}$. To form the gate metal stack, a 6-nm thick TiSiN gate layer was deposited by low-pressure chemical vapor deposition (LPCVD) on the gate oxide and capped with a 100-nm polysilicon layer. The work function of the TiSiN gate is 4.65 eV, which makes it a midgap gate material. Gate electrodes were patterned using lithography and etched. A combination of dry and wet chemistry was used to etch the poly capping layer and metal gate. Some gates were intentionally misaligned from 0 to ±60nm from the centre of the fin to evaluate the impact of gate misalignment on electrical characteristics. Figure 1 shows the original layout and the actual fin shape after processing in MuGFETs with intentional gate misalignment (L_{GM}).[2] The physical fin width of the devices used in this letter ranges from 11nm to 45nm, and each device consists of 20 fins operating in parallel. The electrical characteristics of the devices were simulated using the ATLAS 3-D numerical device simulator.[3] The simulated structure is shown in figure 2. Abrupt source and drain junctions are used in the simulations. Physical models accounting for the electric field dependence of carrier mobility, SRH recombination /generation, band gap narrowing, and electric-field dependent impact ionization were included in the simulation.

RESULTS AND DISCUSSION

Reducing the fin width is the most important parameter for avoiding short-channel effects in MuGFETs, and this study shows that thinning the fin also improves drain breakdown voltage (BV_{DS}). Figure 3 shows the effect of gate misalignment on the transfer characteristics of MuGFETs with W_{si}=11nm and L_g=55nm. Devices with a wider fin near the drain (L_{GM} =+60nm) have a larger DIBL than if the fin is wider near the source (L_{GM} =-60nm) because of a larger penetration of drain field lines at high drain voltage when the channel region is larger. Figure 4 shows good agreement between measured and simulated DIBL values. The BV_{DS} was measured by applying V_{GS} =0V and V_{GS2} =0V and by applying a voltage ramp to the drain. We defined the BV_{DS} using similar technique to that used to extract threshold voltage [4]: BV_{DS} is taken at the maximum of $d(log(I_D))/dV_{DS}$ plotted as a function of drain voltage. Figure 5 shows measured $V_D$$I_D$ characteristic and $d(log(I_D))/dV_{DS}$ as a function of drain voltage with L_{GM}=0nm and ±30nm. One can see drain wider device is more sensitive on a high drain voltage. BV_{DS} is plotted as a function of misalignment in Figure 6; clearly, a gate shifted towards the source does not degrade BV_{DS}, but a shift towards the drain does. This is clearly related to the widening of the fin near the drain. This is confirmed by the measurement of BV_{DS} in devices with different fin widths and gate lengths: the reduction of BV_{DS} is observed in wider fins for all gate lengths (Figures 7 and 8). The decrease of breakdown voltage is caused by both an increase of impact ionization rate and a decrease of the potential barrier in the channel region. The latter increases the parasitic NPN BJT gain, which reduces the open-base collector breakdown voltage (Figures 9 and 10). Figure 11 shows the impact ionization rate underneath the gate in devices with three gate misalignment values. One can see that the gate shifted towards the drain presents a larger region with high impact ionization rate. Figure 12 shows simulated breakdown characteristics that outline the premature breakdown of devices with devices having the gate shifted towards the drain.

CONCLUSIONS

Misalignment of the gate electrode in MuGFETs leads to a variation of fin width that affects both drain breakdown voltage and DIBL. Structures where gate misalignment yields a widening of the fin on the drain side have more pronounced short-channel effects than structure where the fin is wider at the source side.

References

[1] J.P. Colinge Solid-State Electron, vol.48, p.897, 2004.
[2] T. Schulz *et al.* IEEE SOI Conference, p.154, 2005.
[3] http://www.silvaco.com/
[4]. H.S Wong *et al.* Solid-State Electron vol.30, p.953, 1987.

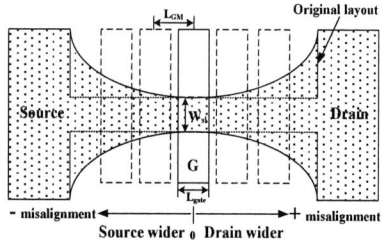

Figure 1 – Schematics of original layout and actual pattern after device processing with intentionally misaligned gates (L_{GM} is the gate offset).

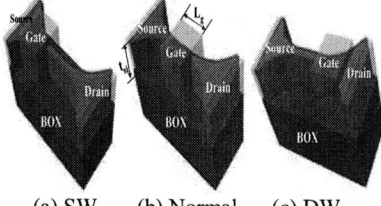

(a) SW   (b) Normal   (c) DW

Figure 2 – Simulated 3D device schematic view with wider source (a), L_{GM}=0 (b) and wider drain (c). t_{si}=60nm, W_{si}=11nm, L_g=55nm, t_{ox}=2nm.

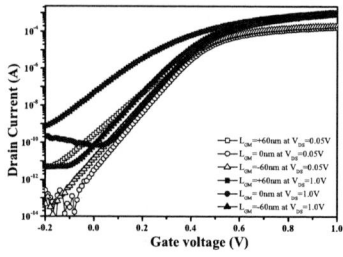

Figure 3 – Measured effect of gate misalignment on the $I_D(V_G)$ characteristics in log scale in MuGFETs with W_{si}=11nm and L_g=55nm.

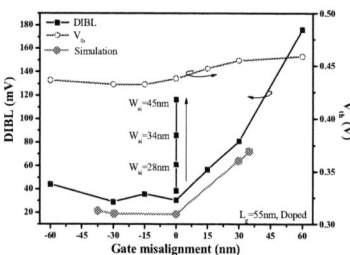

Figure 4 – Measured and simulated DIBL of 55nm gate length MuGFETs as a function of L_{GM}.

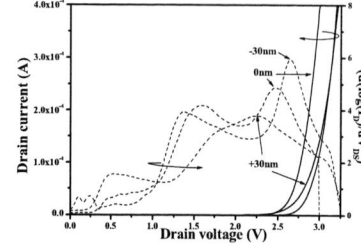

Figure 5 – Drain breakdown voltage extraction on 55nm gate length MuGFETs with L_{GM}=±30nm, L_{GM}=0nm.

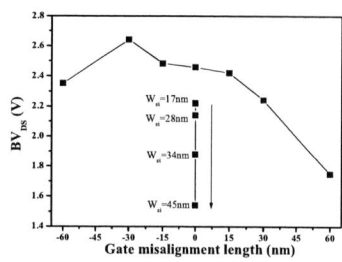

Figure 6 – Measured drain breakdown voltage as a function of gate misalignment length.

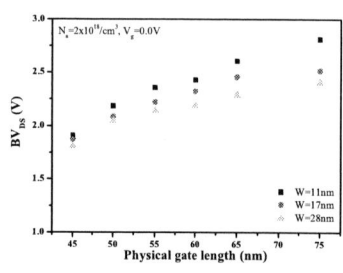

Figure 7 – Measured breakdown voltage as a function of gate length L_g for MuGFETs with W=11, 17 and 28nm and N_a=2x10^{18}/cm^3.

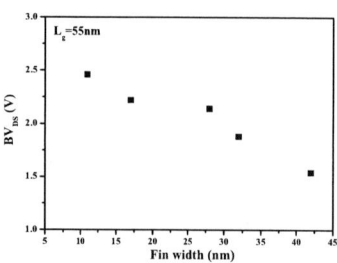

Figure 8 – Measured breakdown voltage as a function of fin width.

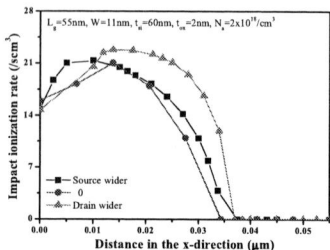

Figure 9 – Simulated impact ionization rate along a horizontal cut line at the center of the device.

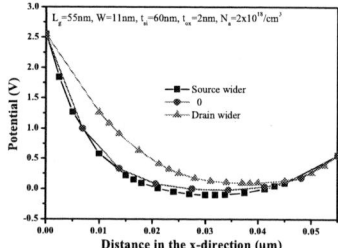

Figure 10 – Simulated potential along a horizontal cut line at the center of the device. V_{DS}=2.0V, V_{gs}=0V.

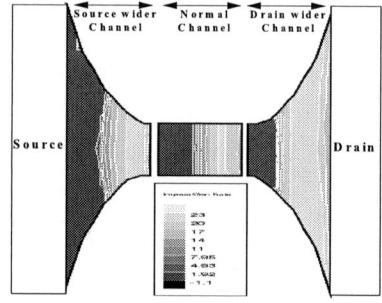

Figure 11 – Simulated impact ionization rate underneath the gate electrode along a horizontal cut. V_{DS}=2.0V, V_{gs}=0V.

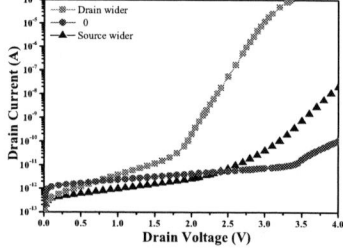

Figure 12 – Simulated drain current of 55nm gate length MuGFETs as a function of drain voltage with wider drain, wider source and regular devices.

978-1-4244-2784-0/09 $25.00 © 2009 IEEE

FinFET Resistance Mitigation through Design and Process Optimization

Cindy Wang, Josephine Chang, Chung-Hsun Lin, Arvind Kumar, Andreas Gehring[*], Jin Cho[*], Amlan Majumdar, Andreas Bryant, Zhibin Ren, Kevin Chan, Thomas Kanarsky, Xinlin Wang, Omer Dokumaci, Michael Guillorn, Marwan Khater, Qingyun Yang, Xi Li, Munir Naeem, Judson Holt, Yongsik Moon[*], John King, John Yates, Ying Zhang, Dae-gyu Park, Christine Ouyang, Wilfried Haensch

*AMD, Inc., IBM Research Yorktown Heights, NY

ABSTRACT

The intrinsic FinFET device structure can provide an estimated 10-20% reduction in delay relative to planar FETs at the 22nm technology node due to superior electrostatics. However, FinFETs are more prone to parasitic resistance and capacitance due to the thin body channel and 3-dimensional device architecture. Here we present strategies for minimizing FinFET parasitic resistance, and discuss overall device design optimization. Using FinFETs built at 45nm node dimensions, we have demonstrated FinFETs with an NFET/PFET external resistance of 230/350 Ω-um. (Keywords: FinFET, parasitic resistance).

INTRODUCTION

Layout constraints posed by gate pitch at the 22nm technology node demand gate length scaling. Both bulk and partially depleted SOI 22nm planar devices will likely require heavy channel doping to control off-state leakage, resulting in degraded device performance [1]. We predict from simulation that improved electrostatics in FinFETs can be exploited to reduce circuit delay by as much as 10 to 20%. However, this performance advantage will be offset by increased parasitics. Gate-to-source/drain capacitance (Cgs) in FinFETs is expected to present at least a 6% performance degradation [2]. Increased parasitic resistance will further degrade FinFET performance.

$R_{parasitic}$ in FinFETs can be roughly broken down into two main components: extension resistance (R_{ext}), and resistance in the source/drain region R_{SD} ($R_{epi} + R_{co}$). Contact resistance R_{co} is affected by resistance and geometry of the heavily doped silicon in the source/drain regions, as well as the silicide-to-silicon contact barrier height (Fig. 1). In this paper, we report on strategies for minimizing Rext through process optimization and device design, and study the predicted impact of Rext at dimensions relevant to the 22nm technology node.

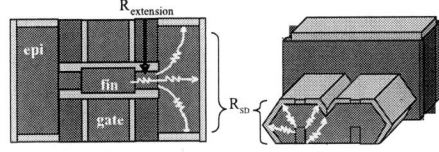

FIGURE 1. $R_{parasitic}$ is composed of R_{ext} and R_{SD} ($R_{epi} + R_{co}$).

RESULTS

FinFETs were fabricated on silicon-on-insulator (SOI) substrates, using sidewall image transfer (SIT) to pattern fins with sub 20nm fin width and 70nm fin pitch, and standard 45nm lithography to pattern all other levels (Fig. 3). In order to simplify the contact scheme and mitigate series resistance, S/D regions were merged with epitaxial silicon. Epitaxially fattened S/D regions are vital for Rparasitic reduction.

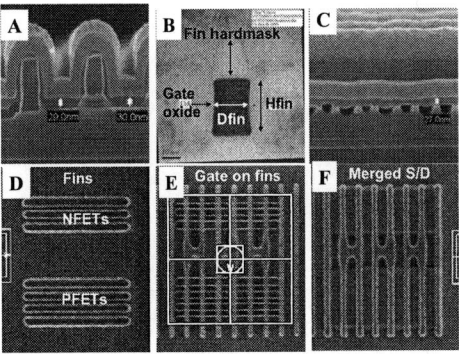

FIGURE 2. Device cross-sections (A) through the gate, (B) through a fin under the gate, and (C) through the merged S/D region. Top-down view of devices (D) after fin definition, (E) after gate definition, and (F) after merging of S/D.

FinFET device characteristics for undoped channel devices are given in Figure 3 and 4. Due to the negative threshold voltage of these devices, a shifted Vgs window is used. Since metal gates can be used to set V_T without the use of channel doping, we used undoped channel devices in this study. $R_{parasitic}$ of the fabricated devices is extracted from the dependence of Ron on L_{gate} (Fig 4) to be 230/350 Ω-um (NFET/PFET). To reduce $R_{extension}$, minimizing the length of the extension region while maximizing its cross-sectional area proved most effective. For PFETs, boron diffusion into the buried oxide resulted in a significant resistance penalty which was addressed by reducing incidental thermal cycles during device fabrication. R_{SD} is reduced by optimizing post-epi implants and anneals and by reducing the vertical epi height. As shown in Fig 5, R_{SD} can be optimized separately without affecting L_{eff}. $R_{parasitic}$ reduction through process optimization is shown in Fig 6. While some processes achieved lower $R_{parasitic}$ at the expense of worse electrostatics, DIBL (Drain-Induced Barrier Lowering) for an L_{gate} of 32nm remains well below 200mV for the most optimized case.

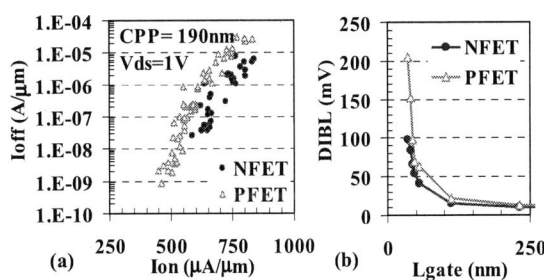

FIGURE 3. (a) FinFET Ioff vs. Ion using a shifted Vgs window (PFETs: -0.4V to 0.6V; NFETs: -0.3V to 0.7V) (b) DIBL.

978-1-4244-2784-0/09 $25.00 © 2009 IEEE   117

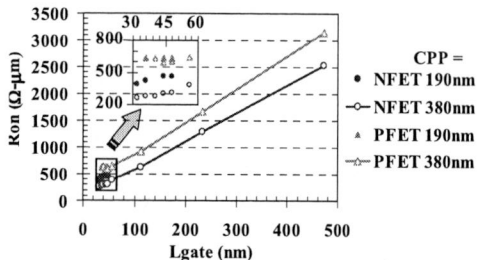

FIGURE 4. Rparasitic is extracted from experimental data using the y-intercept of Ron vs. Lgate, where Ron is the total resistance of the device at $Vgs-V_T=0.7V$, Vds=50mV. Lgate is determined from cross-section SEM.

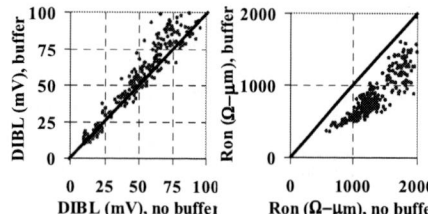

FIGURE 5. R_{SD} reduction through buffer implants on NFETs.

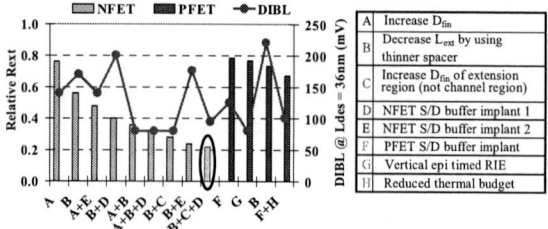

A	Increase D_{fin}
B	Decrease L_{ext} by using thinner spacer
C	Increase D_{fin} of extension region (not channel region)
D	NFET S/D buffer implant 1
E	NFET S/D buffer implant 2
F	PFET S/D buffer implant
G	Vertical epi timed RIE
H	Reduced thermal budget

FIGURE 6. Rparasitic (190nm CPP data) reduction through process optimization.

DISCUSSION

The dependence of $R_{parasitic}$ on contacted gate pitch (CPP) shown in Figure 4 underlies the importance of R_{co} in FinFETs, especially as dimensions are scaled to 22nm-like dimensions. Figure 7 shows the

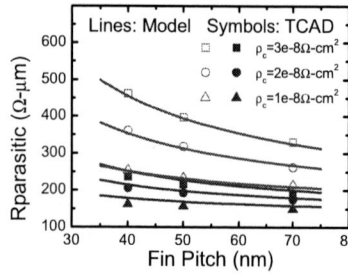

FIGURE 7. FinFET Rparasitic scaling from 45nm node to 22nm node dimensions. Closed symbols: CPP=190nm, Lg=45nm, Df=20nm; Open symbols: CPP=90nm, Lg=25nm, Df=12nm.

predicted increase in Rco as contact length is reduced with CPP, and current crowding increases with increased fin pitch (FP). Reduction of contact barrier height becomes increasingly important.

The tradeoff between Cparasitic and Rparasitic is given in Fig 8, revealing opposite dependencies on both fin height (H_f) and FP. Using a more optimal source/drain and silicide topology can improve both Cparasitic and Rparasitic simultaneously. For this optimized structure, the effect of Rparasitic and Cparasitic on the performance of a FinFET technology is given in Figure 9. We predict an optimal design point at FP=40nm, Hf = 30nm.

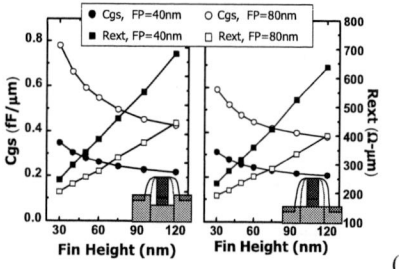

(a)   (b)

Fig 8. Tradeoff between Cparasitic and Rparasitic for FP and Hfin for (a) merged source/drain structure and planar top-surface silicidation (b) no vertical epi growth in the S/D.

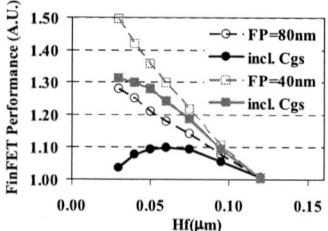

FIGURE 9. Device design optimization for Cparasitic and Rparasitic for a FinFET with merged source/drain and no vertical epi-growth (Fig 8b). Dashed lines: calculated effect of Rparasitic on FinFET performance; solid lines: FinFET performance when Cparasitic is considered as well. An optimal design point is seen at FP = 40nm, Hf = 30nm.

CONCLUSIONS

Reasonably low Rexternal of 230/350 Ω-um has been demonstrated for NFET/PFET at 45nm dimension. In order to balance Cparasitic and Rparasitic components of the device, we have developed an analytical model calibrated to both TCAD and hardware which allows tradeoffs to be calculated.

ACKNOWLEDGMENTS

This work was performed by the Research Alliance Teams at various IBM Research and Development Facilities.

REFERENCES

[1] W. Haensch, E. J. Nowak, R. H. Dennard, P. M. Solomon, A. Bryant, O. H. Dokumaci, A. Kumar, X. Wang, J. B. Johnson, and M. V. Fischetti, "Silicon CMOS devices beyond scaling," *IBM J. Research & Development*, vol. 50, pp. 339, 2006.

[2] M. Guillorn, et al, "FinFET Performance Advantage at 22nm: An AC perspective," *VLSI 2008*, 2.1.

978-1-4244-2784-0/09 $25.00 © 2009 IEEE

Sub-100μW Low Power Operation of Vibrating Body FETs

Daniel Grogg, Adrian Mihai Ionescu

Nanoelectronic Devices Laboratory (Nanolab), Ecole Polytechnique Fédérale de Lausanne (EPFL),
CH-1015 Lausanne, Switzerland,
Email: daniel.grogg@epfl.ch, adrian.ionescu@epfl.ch

ABSTRACT

This paper reports the low power operation of Vibrating Body Field Effect Transistors as active resonators for communication applications. For the first time we report active resonators operating at 2MHz and 20MHz with power consumption less than 100μW and Quality factors in the order of 3000. This performance opens new applications of devices for wireless sensor networks.

INTRODUCTION

Microelectromechanical resonators could replace quartz crystal resonators in electronic systems due to their potential for miniaturization and integration [1]. Further miniaturization faces the problem of decreasing signal levels, especially for capacitively coupled devices, as the motional resistance of such devices depends on the electrode area [2]. Among the different strategies proposed to lower the impedance of MEM resonators, piezoresistive [3] or FET [4-6] based current modulation seem best suited. In both methods, the resonator can be built in single crystal silicon without Q-factor reducing elements, such as solid gaps [7] or piezoelectric material [8]. However, they suffer from the drawback of static power consumption in the resonator structure, for instance in [3] a power consumption >1.9mW is reported. In this paper, we demonstrate sub-100μW power consumption for VB-FETs and provide detailed frequency characteristics for low power operation.

DEVICE STRUCTURE AND STATIC CHRACTERISTICS

The Vibrating Body Field Effect Transistor (VB-FET) shown in Fig.1 is fabricated in a 1.25μm thick SOI using the fabrication process previously described [9] to create 180nm lateral air gaps as shown in Fig.1b. A low-doped body region is placed in the center of the beam structure and designed to allow the formation of two independent lateral channels with a length of 1μm and a width of 1.25μm. As the body is n-type doped, the VB-FET behaves similar to a normally-on type transistor, which is represented by a resistor in parallel with two FETs (Fig.1c). Length and width of the beams determine the in-plane flexural vibration frequency and are used as naming scheme in this paper. The four beams characterized are 100μm or 30μm long and 3μm or 4μm wide.

Fig.2 shows the measured power consumption (I_D*V_D) of the devices calculated from the experimental I_D-V_D characteristics shown in the inset of Fig.2. At a V_D of 50mV, the total power dissipated in the beam is around 1μW from which it is increasing to attain 1mW at a $V_D > 1.8V$. The series resistance (R_S) for source and drain, given by the length, the cross section area and the beam doping is the reason for the differences observed in the I_D-V_D and power-V_D plots.

The transconductance (gm) (dI_D/dV_G) of the devices versus V_G is shown in Fig.3, which corresponds to a double gate operation.

All structures show a considerable modulation of gm with a maximum around VG=0V. For the 100μm long beams R_S seems to be limiting the maximum gm. For negative values of V_G this limitation disappears, as R_{chan} increases and the measured gm of all devices overlap.

Fig.4 reports the transconductance values for different drain voltages at V_G=0V (maximum value), which is rising with V_D as predicted by the theory of FETs. The slope starts to saturate at lower V_D for the beams with a high series resistance R_S.

FREQUENCY CHARACTERISTICS

The frequency spectra of all four VB-FET resonators are shown in Fig.5. The resonance frequencies vary between 1.5MHz (a) and 25.7MHz (d). The V_G values was kept just below the pull-in voltage for the two 100μm long beams (10V and 16V for a and b respectively) and set to 30V for the two 30μm beams (c and d). The noise in Fig.5 a and b is explained by the very low power handling capability of the two 100μm beams (around -80dBm). As the MOSFET detection is an amplitude dependent detection mechanism, the signal transmission is inversely proportional to the Q-factor and the resonance frequency. Therefore, the measured motional resistances are strongly dependent on the resonance frequency of the device, as with traditional capacitive detection. The improvement over capacitive detection has been detailed elsewhere [6, 10] and is limited by the maximum gain of the integrated FET.

The output impedance ($1/g_{DS}$) of the VB-FET was measured to be slightly above 2kΩ for low V_D. In the measurement of Fig.5 the device is connected directly to the 50Ω impedance of the Vector Network Analyzer using a bias-T. This results in a poor impedance matching and additional losses. To improve this we used an external buffer amplifier with unity gain (BUF-634p). Fig.6 gives the frequency spectra of the 100μm long and 3μm wide device without the buffer (a) and with the buffer (b) under equal biasing conditions. In both cases, the strong influence of V_D on gm is observed as reduced signal attenuation. The frequency changes slightly with the changing V_D due to electrostatic effects. Most important is the improvement of the signal attenuation at resonance by 15dB independent of the Q-factor reduction observed.

CONCLUSION

The unique characteristics of the VB-FET are detailed with respect to possible sub-100μW low power applications. On these first generation devices, a motional resistance of 1.8kΩ at a frequency of 2MHz is obtained with a power consumption of only 30μW. Further, the importance of impedance matching is shown using a simple voltage buffer.

REFERENCES

[1] Wan-Thai Hsu, Proc. of IEEE Freq. Control Symp. 2008, pp. 392-395.
[2] C. T. C. Nguyen, IEEE Trans. Ultrasonics, Ferroelectrics and Frequency Control, Vol. 54, 2007, pp. 251-270.
[3] J.T.M. van Beek et al, IEDM Techn. Dig. 2007, pp. 411–414.
[4] H. C. Nathanson, W. E. Newell, R. A. Wickstrom, and J. R. Davis, Jr., IEEE Trans. Electron Devices, vol. 14, 1967, pp. 117-133.
[5] N. Abele et al., IEDM Tech. Dig., 2005, pp. 479-481.
[6] D. Grogg et al, Digest of DRC 2008, June 2008, pp. 155-156.
[7] L. Yu-Wei et al.,Proc. Joint IEEE Int. Freq. Contr./Precision Time & Time Interval Symp., 2005, pp. 128-134.
[8] G.K. Ho et al., IEEE Journal of Microelectromechanical Systems, Vol. 17, April 2008, pp. 512–520.
[9] D. Grogg et al., Proc. of ESSDERC 2008.
[10] D. Grogg et al., to appear, IEDM 2008.

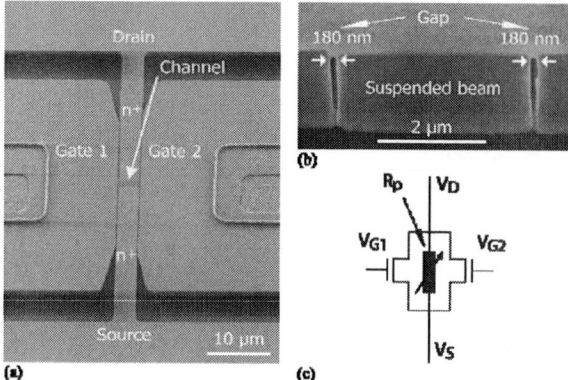

(a) **(c)**

Fig. 1: a) SEM image of a double gate Vibrating Body. The beam is 50μm long amd 4μm wide and the FET-body indicated with a small square is 1μm long. b) A FIB cross-section through a VB-FET beam showing 180nm air-gaps on both sides. C) The n-type doping of the FET-body acts as a resistance in parallel with the two channels formed on the lateral sides of the beam.

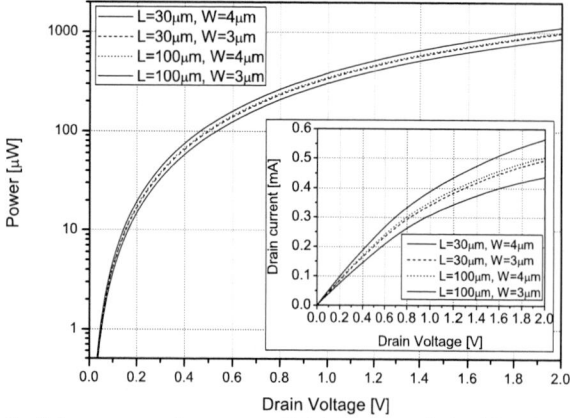

Fig. 2: Power consumption versus applied drain voltage calculated from the I_D-V_D characteristic (inset) of the four beams. Due to the normally-on nature of the current generation VB-FET operation at low V_D can reduce the power consumption by orders of magnitude.

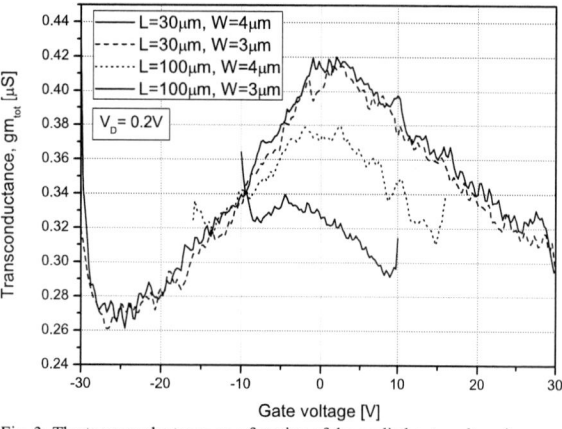

Fig. 3: The transconductance as a function of the applied gate voltage in a double gate configuration shows a maximum around $V_G=0V$ for all beams. The series resistance to source and drain caused by the beam seems to limit the gm for 100μm long beams.

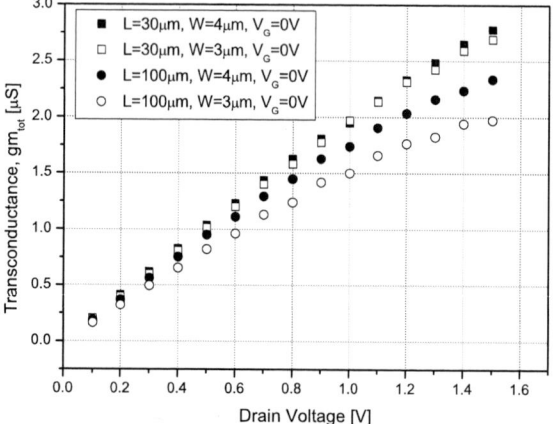

Fig. 4: Measured transconductance versus V_D shows a linear trend for all VB-FETs for low values of V_D. The transconductance value of the long beams starts to saturate at around 1V due to the series resistance of drain and source.

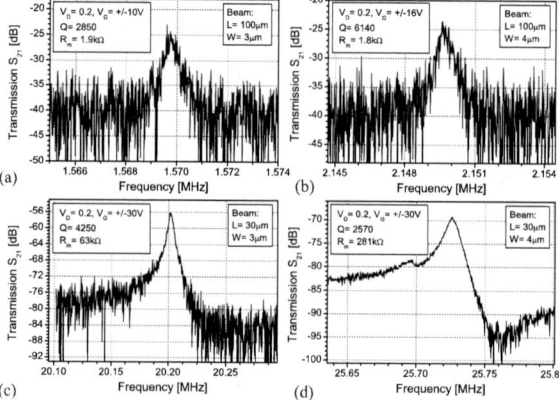

Fig. 5: Transmission scattering parameters measured at a low drain voltage of $V_D=0.2V$ for all four VB-FETs. A motional resistance of ~2 kΩ is obtained for low frequency structures with a power dissipation of 30μW in the beam.

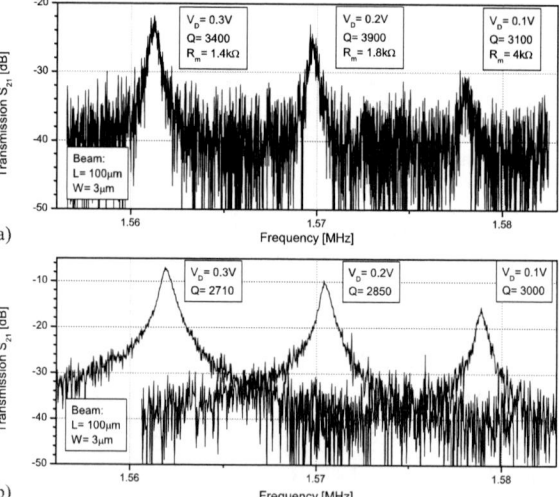

Fig. 6: The effect of impedance matching was investigated using a buffer amplifier. It is found that (i) a considerable improvement is obtained and (ii) Q-factor loading starts to be a limiting factor.

Inversion-channel GaN MOSFET using atomic-layer-deposited Al_2O_3 as gate dielectric

Y. C. Chang[a], W. H. Chang[a], H. C. Chiu[a], Y. H. Chang[a], L.T. Tung[a], C. H. Lee[a], M. Hong[a]*, J. Kwo[b]*, J. M. Hong[c], and C. C. Tsai[c]

[a] *Dept. Materials Sci. and Eng, and* [b] *Dept. of Physics, Natl Tsing Hua Univ., Hsinchu, Taiwan 30013,*

and [c] *HUGA Optotech Inc., Taichung, Taiwan 407*

*Authors to whom the correspondence is addressed: mhong@mx.nthu.edu.tw (M. Hong) and Raynien@phys.nthu.edu.tw (J. Kwo)

ABSTRACT

For the first time, inversion-channel GaN MOSFETs using atomic-layer-deposited (ALD) Al_2O_3 as a gate dielectric have been successfully fabricated, showing well-behaved drain $I-V$ and transfer characteristics. The drain current was scaled with gate length, showing a maximum drain current of 10 mA/mm in a device of 1 μm gate length, at a gate voltage (V_{gs}) of 8 V and a drain voltage (V_{ds}) of 10V. High I_{on}/I_{off} ratio of 2.5×10^5 was achieved with a very low off-state leakage of 4×10^{-13} A/μm. In addition, depletion-mode (D-mode) GaN MOSFETs have also been demonstrated, showing a very low on-resistance of 2.5 mΩ·cm², a high mobility of 350 cm²/Vs, and a high maximum drain current of 300 mA/mm in a device of 4 μm gate length.

INTRODUCTION

GaN, with a high saturation velocity at high electrical fields ($v_{sat} \sim 3 \times 10^7$ cm/s), a high critical electrical field (up to 3 MV/cm), good thermal conductivity, and epi-layers grown on Si, has been feverishly studied for applications in high-power and high-temperature devices. Owing to its wide energy band gap (3.4 eV), which alleviates the adverse affects like drain-induced barrier lowering (DIBL) and band-to-band tunneling (BTBT), GaN is now also being considered as a strong channel contender for the next generation CMOS devices beyond the 22-16 nm node technology. Furthermore, by taking into account of the short channel effect with the cutoff frequency (f_T) given by $f_T = v_{sat}/2\pi L$ (where L is the gate length), GaN MOSFET's may outperform its counterparts of Si and GaAs in further scaled-down devices, despite the fact that GaN offers no special advantage in electron mobility.

EXPERIMENT

The ALD-Al_2O_3 deposition includes a substrate temperature of 300°C and a chamber pressure of 1 Torr using alternating pulses of $Al(CH_3)_3$ (TMA) and H_2O as precursors. The oxide thickness and interfacial microstructure were studied using x-ray reflectivity (XRR), and cross-sectional high-resolution transmission electron microscopy (HR-TEM). Capacitance-voltage ($C-V$) and leakage-current-density versus gate-electrical-field ($J-E_g$) measurements of GaN MOS capacitor (MOSCAPs) have been carried out to assess the interfacial quality between Al_2O_3 and GaN. The process flow of inversion-channel GaN MOSFET is listed in Table I. Fig. 1(a) shows a cross-sectional schematic of the device structure, with its planar view (by SEM) shown in Fig. 1 (b). Fig. 9 shows a cross-sectional schematic of the D-mode GaN MOSFET.

RESULTS AND DISCUSSION

Well-behaved drain $I-V$ characteristics of a GaN MOSFET are clearly shown in Fig. 2(a) with a clean pinch-off and enhancement mode operation. For a device of a 4μm gate length and a 100 μm gate width, the maximum drain current ($I_{d,max}$) is 3.5 mA/mm at V_{gs} of 10V, and V_{ds} of 15V. The $I_{d,max}$ was improved to ~10 mA/mm in an 1 μm gate-length device, measured at V_{gs} of 8 V and V_{ds} of 10 V. The measured drain current is scaled with gate length, with the scaling dependence displayed in Fig. 2(b). At V_{ds} of 0.1V, a high I_{on}/I_{off} ratio, 2.5×10^5, is achieved at a very low off-state leakage of 4×10^{-13} A/μm shown in Fig. 3. The threshold voltage of 2.8V, mainly due to difference in work function between p-GaN and gate metal of Pt, was extracted along with the sub-threshold slope of 290 mV/dec.

The overall performances of these devices are markedly improved over the previously reported results of the inversion-channel GaN MOSFETs based on high κ dielectrics [1,2].

The temperature dependence of the operation for the inversion-channel GaN MOSFET was studied. The device worked very well with good pinch-off characteristics and very low off-state leakage (V_{ds}=0.5V) even at 550K (Fig. 4(a)). At elevated measurement temperatures, $I_{d,max}$ and threshold voltage (V_{th}) were improved (Fig. 4(b)). This is perhaps the highest operation temperature for the GaN MOSFET with high κ gate dielectrics. Both MOSFET and MOSCAP have shown very low leakage current densities of 10^{-8}-10^{-9} A/cm² even at a gate-electrical-field (E_g) of 4MV/cm, indicating the highly insulating characteristics of the ALD-Al_2O_3 dielectrics (Fig. 5). The schematic view of the Pt/Al_2O_3/p-GaN MOSCAP is shown in the inset of Fig. 5. $C-V$ curves with frequencies varying from 1 kHz to 100 kHz show clear accumulation, depletion, and deep depletion (Fig. 6). The deep depletion occurred due to the low intrinsic carrier concentration ($n_i \sim 10^{-10}$ cm⁻³). The D_{it} was calculated to be ~ 5×10^{11} cm⁻²eV⁻¹ near the midgap by the conductance method, shown in the inset of Fig. 6.

The Al_2O_3/GaN interface remains atomically smooth after high temperature 750°C annealing, with the oxide thickness and the roughness determined from XRR to be 12 nm and 0.5 nm. A cross-sectional HR-TEM image also showed an abrupt transition from amorphous Al_2O_3 to crystalline GaN even after high temperature annealing, in agreement with the XRR data (Fig. 7). The conduction-band offset (ΔE_C) for Al_2O_3/GaN was determined to be 1.9 eV by Fowler-Nordheim tunneling analysis. The valence-band offset (ΔE_V) of 1.3 eV was determined by the valence-band maximum (VBM) of Al_2O_3 on GaN using x-ray photoelectron spectroscopy analysis. The schematic representation of the energy-band parameters for Al_2O_3/GaN is shown in Fig. 8. Al_2O_3, with a considerably larger band gap than other high κ dielectrics, gives higher energy-band offsets with GaN, resulting in reduced gate leakage currents given the same thickness of high κ dielectrics.

Ring-gate D-mode GaN MOSFETs (a schematic shown in Fig. 9) with the same gate dielectric ALD-Al_2O_3, were fabricated with only two-step processing (gate metal formation and S/D contact metal). TiN/Al_2O_3/n-GaN MOSCAP exhibits excellent $C-V$ curves with different frequencies, as well as a dielectric constant of 9. (Fig. 10). For a device of a 4μm gate length and a 200 μm gate width, the $I_{d,max}$ is 300 mA/mm at V_{gs} of 8V, and V_{ds} of 20V (Fig. 11). The on-resistance (R_{on}) of 2.5 mΩ·cm² is the lowest value, compared to previous published results for GaN MOSFETs. The maximum G_m of the device is 27.5 mS/mm at V_{gs} of 3V, and V_{ds} of 20V. The mobility of 350 cm²/Vs for the D-mode MOSFET is given by $G_m = (W_g/L_g)\mu_n N_d T_{ch}$.

CONCLUSION

In summary, remarkable device performances of inversion-channel and D-mode GaN MOSFETs using ALD-Al_2O_3 dielectrics under various temperatures have been demonstrated. These are due to excellent material properties, as also been confirmed with the excellent $C-V$ and $J-E_g$ characteristics obtained in the MOSCAPs. These devices are promising for CMOS, high temperature, and high power applications.

REFERENCES

[1] Y. Irokawa, et al, Appl. Phys. Lett. **84**, 2919 (2004)
[2] Y. N. Saripalli, et al, Appl. Phys. Lett. **90**, 204106 (2007)

978-1-4244-2784-0/09 $25.00 © 2009 IEEE

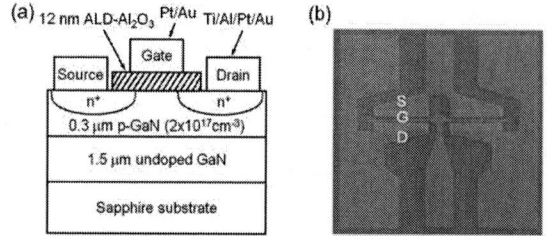
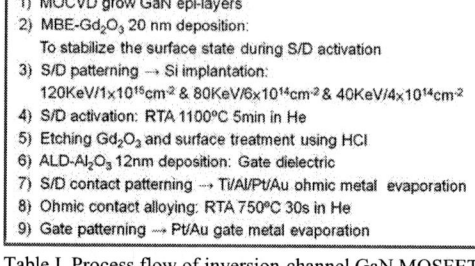

Fig. 1 (a) Schematic view and (b) planar view of the inversion-channel GaN MOSFET with ALD-Al$_2$O$_3$ as gate dielectric

Table I. Process flow of inversion-channel GaN MOSFET

1) MOCVD grow GaN epi-layers
2) MBE-Gd$_2$O$_3$ 20 nm deposition:
   To stabilize the surface state during S/D activation
3) S/D patterning → Si implantation:
   120KeV/1×10^{15}cm^{-2} & 80KeV/6×10^{14}cm^{-2} & 40KeV/4×10^{14}cm^{-2}
4) S/D activation: RTA 1100°C 5min in He
5) Etching Gd$_2$O$_3$ and surface treatment using HCl
6) ALD-Al$_2$O$_3$ 12nm deposition: Gate dielectric
7) S/D contact patterning → Ti/Al/Pt/Au ohmic metal evaporation
8) Ohmic contact alloying: RTA 750°C 30s in He
9) Gate patterning → Pt/Au gate metal evaporation

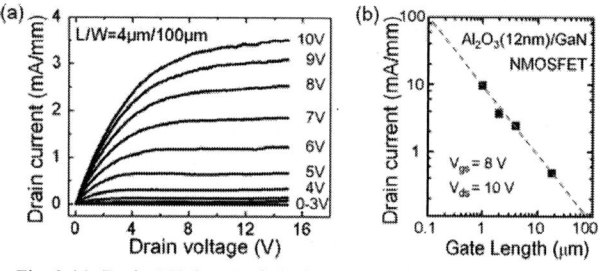
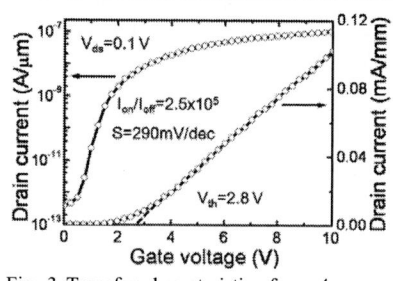

Fig. 2 (a) Drain *I-V* characteristic for a 4 μm gate length GaN MOSFET (b) The scaling characteristic of drain current vs gate length

Fig. 3 Transfer characteristics for a 4 μm gate length GaN MOSFET at V_{ds} =0.1 V.

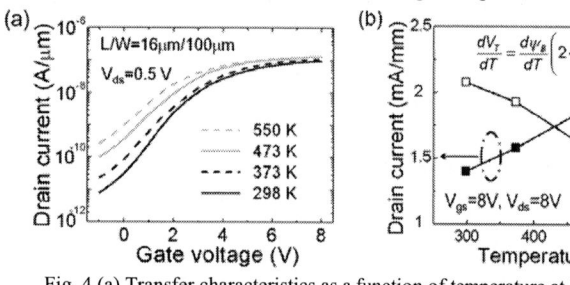
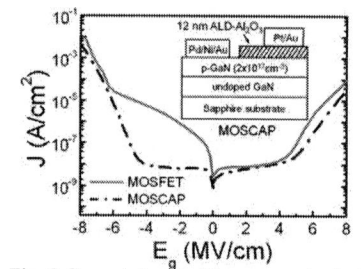

Fig. 4 (a) Transfer characteristics as a function of temperature at V_{ds}=0.5 V (b) Temperature dependence of I$_{d,sat}$ (V$_{gs}$=8V, V$_{ds}$=8V) and threshold voltage for a 16 μm gate length GaN MOSFET

Fig. 5 Current-density (*J*) versus gate-electrical-field (E_g) for both MOSCAP and inversion-channel MOSFET with 12 nm Al$_2$O$_3$

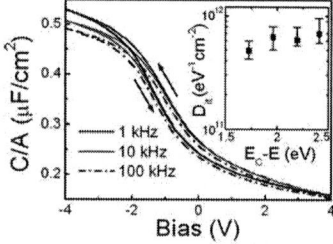
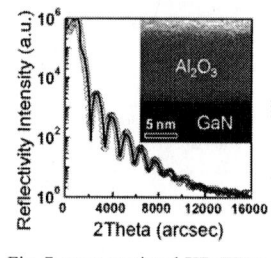
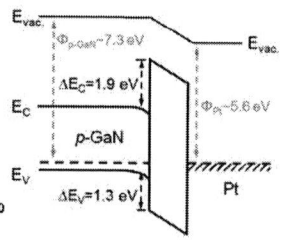
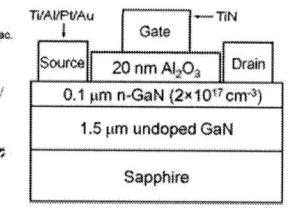

Fig. 6 *C-V* curves for *p*-GaN MOSCAP under various frequencies, with inset showing D_{it} value

Fig. 7 cross-sectional HR-TEM image and low angle XRR of Al$_2$O$_3$ on GaN

Fig. 8 Schematic representation of energy-band parameters for Al$_2$O$_3$ on GaN

Fig. 9 Schematic view of the D-mode GaN MOSFET with Al$_2$O$_3$ as gate dielectric

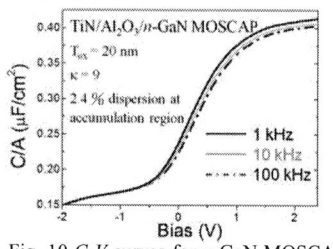
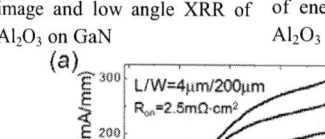
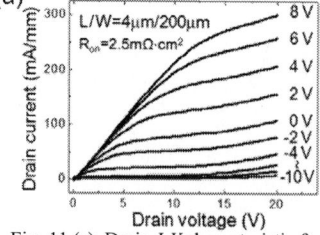
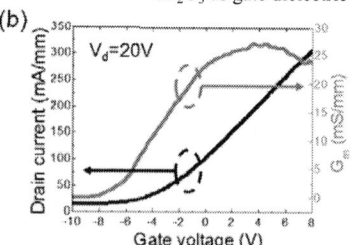

Fig. 10 *C-V* curves for *n*-GaN MOSCAP under various frequencies

Fig. 11 (a) Drain *I-V* characteristic for a 4 μm gate length D-mode GaN MOSFET (b) Transfer characteristics for a 4 μm gate length GaN MOSFET at V_{ds} =20 V

978-1-4244-2784-0/09 $25.00 © 2009 IEEE

A Comparison Study of Silicon Nanowire Transistor with Schottky-Barrier Source/Drain and Doped Source/Drain

Zhaoyi Kang, Liangliang Zhang, Runsheng Wang, and Ru Huang*
Department of Microelectronics, Peking University, Beijing 100871, China
Email: *ruhuang@pku.edu.cn

1. INTRODUCTION

Schottky Barrier (SB) S/D Nanowire Transistor (SB-NWT) is considered to be one of the candidates for future CMOS technology due to the combination of the advantages both from Silicon Nanowire (NW) Transistor (SNWT) [1][2] and metal S/D [3]-[6]. However, previous studies, both simulations [6][8] and experiments [9], show that the electrostatic properties (EPs) of SB-NWTs are not promising. In this paper, SB-NWTs is comprehensively studied in comparison with SNWTs. The SB impact is found to be responsible for the degraded EPs and contributes to unexpectedly large linear region resistance (R_{lin}). Several approaches to minimize the SB impact are studied and limitations of the approaches are investigated in comparison with SNWTs. Whether SB-NWTs can show better performance than SNWTs will be further discussed.

2. DEVICE STRUCTURE AND SIMULATION

P-type SB-NWTs and SNWTs (Fig.1) are focused. Gate-oxide thickness t_{ox} is 1nm, channel length L_{ch}=10~30nm and cross-section diameter D_m=4~8nm. Supply voltage V_{dd}=-0.6V and channel doping N_{chan}=1×10^{15}cm^{-3}. Gate barrier is 0.55eV. For SNWTs, SDE length L_{SDE}=10~20nm with doping N_{SDE}=1×10^{20}cm^{-3}. For SB-NWTs, S/D SB height φ_{Bh}=0.03~0.29eV (0.25eV is default) and effective holes tunneling mass m_{th}=0.01~0.171m_0 (0.171m_0 is default) and tunneling mass of electron is 0.19m_0, according to the calibration on a planar 27nm PtSi S/D SB-MOSFETs [4] (Fig.2). The device simulation is based on Synopsys Sentaurus TCAD tools [7] and tunneling current is calculated based on WKB approximation [8].

3. RESULTS AND DISCUSSIONS

3.1. Performance Limiting Factors of SB-NWTs

Short channel SB-NWTs with D_m=4nm, L_{ch}=10nm is focused and the principal mechanisms and formulas are shown in Fig.3, which J_{tunnel} and J_{therm} are the tunneling and thermionic emission current, respectively. The SB-NWTs show worse Sub-threshold Slope (SS) (>100mV/dec) and DIBL (>100mV/V). Principally, this can be attributed to the SB S/D, the NW structure or the channel mobility (μ_{chan}) degradation. In this work, the intrinsic limiting factor will be clarified by the comparisons between different structures. Firstly, SB-MOSFETs with NW and Double-Gate (DG) are compared. The NW is better in terms of On-Off-ratio (7.2×10^5 >2.75×10^5), SS (103<150mV/dec), which show that NW cannot be the cause of the degraded EPs. Secondly, SB-NWTs with different μ_{chan} are compared and I_{ON} is found to have little dependence on μ_{chan}, while μ_{chan} dependence in SNWTs is obvious, which means μ_{chan} degradation cannot be the cause of the degraded EPs in SB-NWTs as well. Thirdly, SB-NWTs are compared with SNWTs in term of SS and SS$_{SB-NWTs}$ is much worse than that of n-SNWTs in [2]. This suggests that EPs problems may come from SB impacts. Further, EPs can be enhanced by SB modifications such as barrier lowering (BL) and m_{th} reduction (MR). Meanwhile, SB is proved to the cause of other degradations in EPs of SB-NWTs including the convert region (Fig.4) due to the barrier height stiffness and the remarkably large R_{lin} (Fig.5) in which the equivalent resistances distributed at SB (R_{SB}) are significant component. Therefore, SB is confirmed to be the cause of the EPs degradation in SB-NWTs.

3.2. Limitation of the Improving Approaches

Here in this work, BL and MR as EPs enhancing methods are studied and investigations will be given on whether it is reasonable to replace SNWTs by SB-NWTs.

3.2.1. The Limits in SS improvement

Firstly, SS improvement under MR is focused and it is also found that the SS>75mV/dec even when m_{th}≈0 while the SS$_{SNWTs}$~63mV/dec with L_{SDE}=10nm, which means that MR cannot be promising method to improve EPs. Secondly, SS under BL can be improved to 65mV/dec, but impractically requiring that φ_{Bh} close to zero. Moreover, convert region, especially when it shifts to the ON-state as φ_{Bh} reduces, will critically cause SS degradation. Here, one figure of merits (FoM) to evaluate EPs is defined as FoM=(DIBL×SS)$^{-1}$ and its improvement, although acceptable, is suppressed by more than 5% under BL. These indicate that SS degradation is an inevitable problem for SB-NWTs.

3.2.2. DIBL&FoM Degradation

Unexpectedly, under MR, obvious degradation of DIBL (more than 40%) is found and the FoM enhancement is limited. This degradation is found to be more obvious under shorter L_{ch} (Fig.7) due to the enhanced V_d impact. Physically, DIBL$_{SB-NWTs}$ characterizes the impact of ΔV_d, which induces Fermi-gap change ($\Delta[\Delta f_{SD}(E)]|_{Vd}$) and $\Delta T_{TL}(E)|_{Vd}$, versus the gate controllability, which induces only $\Delta T_{TL}(E)|_{Vg}$ (Fig.7) on tunneling current ($\Delta J|_{Vd}$ versus $\Delta J|_{Vg}$). Under MR and shorter L_{ch}, $\Delta T_{TL}(E)|_{Vd}$ is more dominant due to the enhanced drain influence on SB, causing $\Delta J|_{Vd}$>$\Delta J|_{Vg}$ and worse DIBL effect. Moreover, DIBL$_{SB-NWTs}$ degradation is also more dominant under shorter D_m (Fig.8) due to that the extracting point of V_{th} comparatively shifts to the convert region as m_{th} reduces, which results in larger DIBL effect. These unexpected degradations will threaten device scalability and circuit application. Moreover, for BL it is found that DIBL degrades under large D_m due to the reduced gate controllability and enhanced $\Delta[T_{TL}(E)]|_{Vd}$. Thus, SB brings intrinsic limits in EPs improvement of SB-NWTs and neither BL nor MR can be promising solution to the drawbacks of the device.

3.2.3. The Limits in R_{lin} Reduction

An important motivation of the replacement of SNWTs by SB-NWTs is the lower contact resistance. However, unexpected limits in R_{lin} improvement are induced by SB that limit the application of SB-NWTs. Firstly, the output curve of SB-NWTs peculiarly shows two stages (Fig.9), in which Stage (I) (where R_{lin} is extracted) the transport is limited with less $T_{TL}(E)$ due to drain SB backscattering, which indicates that the extracted R_{lin} can be an intrinsic drawback causing extra RC-delay in digital circuits application. Then, since BL and MR are believed to improve the EPs of SB-NWTs, we calculated the R_{lin} under different φ_{Bh} and m_{th} (Fig.11). Although the available φ_{Bh} and m_{th} that make $R_{lin-SB-NWTs}$<$R_{lin-SNWTs}$ increases as L_{SDE} increases, $R_{lin-SB-NWTs}$ is always worse than $R_{lin-SNWTs}$ once L_{SDE} is practically short enough (e.g. L_{SDE}<10nm). Therefore, the replacement of SNWTs by SB-NWTs cannot be promising. In addition, since the equivalent advantage of Modified Schottky Barrier (MSB) or Dopant Segregated Schottky Barrier (DSSB) is BL. Therefore, neither MSB nor DSSB can make the replacement of SNWTs by SB-NWTs more acceptable.

4. CONCLUSION

In this work, SB-NWTs are comprehensively studied in comparison with SNWTs. The EPs of SB-NWTs are shown to be unpromising and $R_{\text{lin-SB-NWTs}}$ is found to be unexpectedly large, which can both be attributed to SB impact. Then, BL and MR are studied and the improvement of SB-NWTs is investigated. It is found that the SS of SB-NWTs cannot be superior to that of SNWTs and unexpected DIBL/FoM degradation may be induced. In addition, $R_{\text{lin-SB-NWTs}}$ is found to be always worse than $R_{\text{lin-SNWTs}}$ if L_{SDE} of SNWTs is designed to be adequately short (e.g. $L_{\text{SDE}}<10\text{nm}$). The results show that the replacement of SNWTs by SB-NWTs, even the MSB-NWTs and DSSB-NWTs cannot be promising.

REFERENCES

[1] Y. Cui *et al.*, *Nano Lett.*, vol. 3, no. 2, pp. 149-152, 2003
[2] Y. Tian *et al.*, *IEDM Tech. Dig.*, 2007, pp. 895–898
[3] M. Shin, *IEEE TED*, vol. 55, no. 3, pp. 737-742, 2008
[4] S. Xiong *et al.*, *IEEE TED*, vol. 52, no. 8, pp. 1859-1867, 2005
[5] R. Vega *et al.*, *IEEE TED*, vol. 55, no. 10, 2665-2677, 2008
[6] E. Tan *et al.*, *IEEE EDL*, vol. 29, no. 10, 2008
[7] ISE User's Manual, Synopsys, Mountain View, CA, 2005
[8] M. Ieong *et al.*, *IEDM Tech. Dig.*, 1998, pp. 733–736.

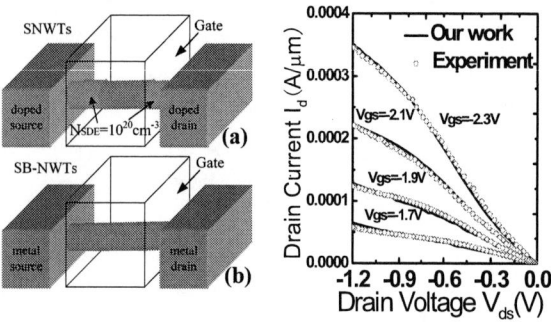

Fig.1 Schematic view of SNWTs and SB-NWTs in this work.

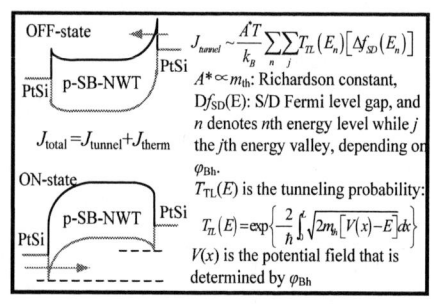

Fig.2 Calibration with previous experimental work with well fitting.

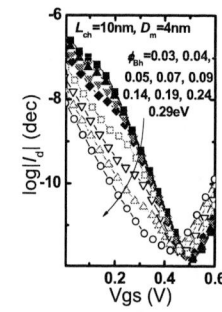

Fig.3 Principles and formulas of SB-NWTs. Tunneling current (J_{tunnel}) is focused in our work since in ON-state total current $J_{\text{total}}\approx J_{\text{tunnel}}$.

Fig.4 Convert region that shifts from OFF- to ON-state under different φ_{Bh}.

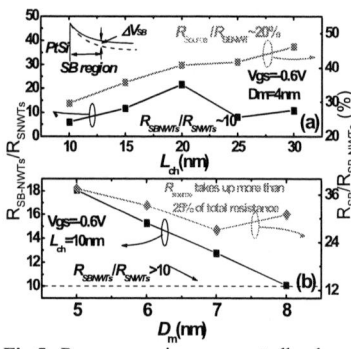

Fig.5 $R_{\text{lin-SB-NWTs}}$ is unexpectedly larger than $R_{\text{lin-SNWTs}}$ under different L_{ch} (a)/D_m (b) and R_{source} as source R_{SB} is a significant component. R_{SB} is extracted by the method in the inset of (a) as $R_{\text{SB}}=\Delta V_{\text{SB}}/\Delta I_d$.

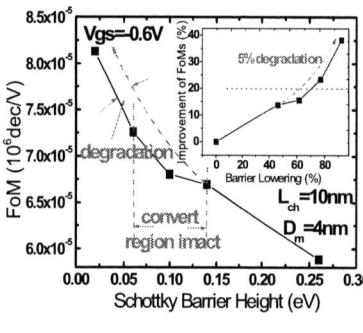

Fig.6 FoM under BL is limited by the convert region, which induces 5% degradation.

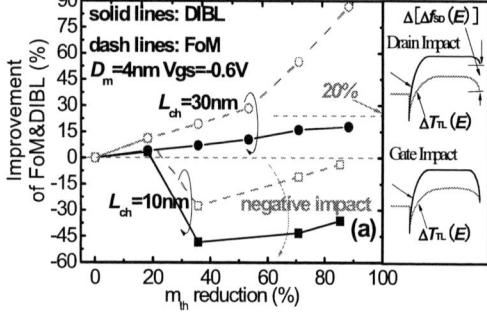

Fig.7 DIBL and FoM under MR at L_{ch}=10, 30nm. Under shorter L_{ch} severe degradation of DIBL is induced due to the enlarged V_d impact which causes unpromising FoM properties. The comparison between the drain&gate impact on J_{tunnel} is shown on the right.

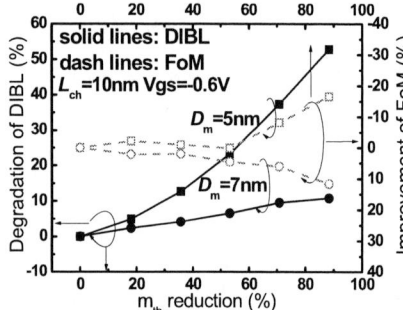

Fig.8 DIBL and FoM under MR at D_m=5, 7nm. Under shorter D_m degradation of DIBL and FoM is induced due to the shift in the extracting point of V_{th}, which degrades device scalability.

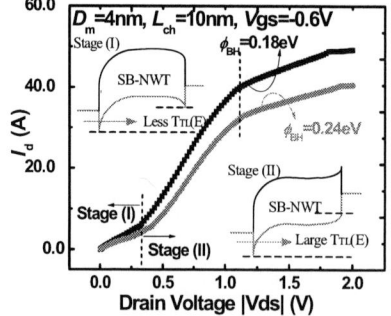

Fig.9 output curve of SB-NWTs undergoes two stage (I, II): in (I) $T_{\text{TL}}(E)$ is limited due to drain SB backscattering, while in (II) no scattering barrier exists at drain junction.

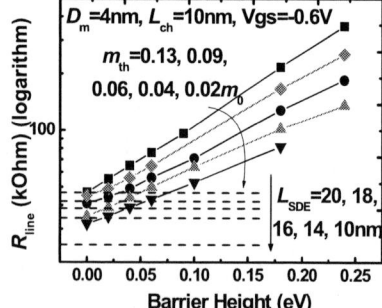

Fig.10 R_{lin} under different φ_{Bh} and m_{th}. BL and MR both bring better R_{lin} but $R_{\text{lin-SB-NWTs}}>R_{\text{lin-SNWTs}}$ is inextricable once $L_{\text{SDE}}<10\text{nm}$.

978-1-4244-2784-0/09 $25.00 © 2009 IEEE

HIGH MOBILITY SIGE SHELL-SI CORE OMEGA GATE pFETs

HEMANT ADHIKARI[1], HARLAN R. HARRIS[2], CASEY E. SMITH , JI-WOON YANG, BRIAN COSS, SRIVATSAN PARTHASARATHY ,
BICH-YEN NGUYEN[3], PAUL PATRUNO[4], TEJAS KRISHNAMOHAN[5], IAN CAYREFOURCQ [3], PRASHANT MAJHI[6], AND RAJ JAMMY

SEMATECH, [1] AMD assignee, [2]Texas A&M, [3] SOITEC USA, [4] Formerly SOITEC USA, [5]Intel, SantaClara, Stanford [6]Intel assignee
2706 Montopolis Drive, Austin, Texas 78741, U.S.A.
Tel: 1-512-356-7012 Fax: 1-512-356-7640; e-mail: hemant.adhikari@sematech.org

ABSTRACT

Omega gate type pFETs with SiGe shell-Si core are demonstrated that show 30% mobility enhancement for (110) oriented fins and 46% mobility enhancement for (100) oriented fins compared to Si omega gate devices. Performance improvement is demonstrated because of higher mobility and inherent epitaxial strain, while the external resistance in the two SiGe and Si omega FETs is comparable. Performance can further be improved by uniaxial compressive stress.

INTRODUCTION

Planar Si channel devices may not be able to provide for the high performance needed in future MOSFETs beyond 22nm node. [1] Two categories of advanced performance enhancements currently dominate the literature: alternate channel materials [2] and Multi-Gate FETs (MuGFETS) [3]. The simplest of these to introduce from a materials perspective is MuGFETS, while both methods have integration challenges. For these reasons, it may be speculated that MuGFETS will be introduced first, possibly as early as the 22nm node. This is because gate length scaling has severely saturated, as the composite data in Fig. 1 indicate. While the advent of High-k/metal gates for gate stack scaling [4] may increase the rate of scaling, future advances in MuGFET performance may require that addition of alternate channel materials. Ge based channels have been reported to demonstrate performance improvement over Si [2]. This work is the first demonstration of a SiGe shell-Si core omega gate type of pFET.

DEVICE FABRICATION

The SiGe/Si transistors were formed by standard gate-first CMOS flow. Thin (~15nm) epitaxial $Si_{0.7}Ge_{0.3}$ is selectively grown by CVD on Si fins etched on SOI substrates. As Fig 2-4 show, uniform epitaxy on (110) and (100) oriented fins is observed. $HfSiO_x$ dielectric and TiN metal gate was deposited using atomic layer deposition (ALD). Both (110) and (100) oriented fins show >90% gate coverage.

RESULTS AND DISCUSSION

Figure 5 shows the Id-Vg of SiGe shell-Si core (110) omega gate devices compared to Si omega gate devices for saturation and linear drain voltages. The higher drain current is expected because of higher mobility of SiGe channel and the biaxial stress in the fins along the channel direction. Figure 6 shows that a threshold shift of ~200mV is observed because of the SiGe channel. A 300mV Vt shift between SiGe and Si

devices is expected for similar Ge % [5], which suggests that the SiGe in these Omega FETs is partially relaxed. Figure 7 and 8 show 46% mobility enhancement for (100) oriented fins and 30% mobility enhancement for (110) oriented fins in SiGe based omega FETs compared to Si based FETs. These experimentally observed mobility trend matches that expected by simulations for relaxed Si and biaxialy strained 30% SiGe FinFET, as shown in Fig 9 and 10. The difference in experimental and simulation observations for SiGe can be attributed to several reasons like partial relaxation of the SiGe layer or scattering in SiGe fins because of either alloy scattering or surface roughness scattering or the presence charge traps. The temperature dependence of peak mobility between SiGe and Si fins, Fig 11, shows similar slopes are observed, suggesting that alloy scattering is not a significant contributor to mobility degradation.

Fig 12 shows the Ion-Ioff plot for (110) oriented SiGe and Si channels. The Id-Vg plot for 140nm gate length (as drawn) device shows higher on current for the SiGe based devices compared to Si fin devices. Fig 13 shows that one of the contributing factors for the higher off-state current for SiGe/Si devices is the gate leakage which is approximately two orders of magnitude larger in SiGe based devices compared to Si based devices. Fig. 14 and 15 show the drain current variation with drain bias for various gate overdrives and the Capacitance voltage curves for (110) oriented fins. A performance improvement of ~19% at Vd=0.9V and (Vg-Vt)=0.6V. Figure 16 shows that a comparable external resistance of 357 ohm-um and 253 ohm-um is extracted for SiGe and Si fins respectively.

A flexure based 4-point bending setup [6] was used for applying uniaxial compressive stress (up to 300 MPa) on p-finFETs of varying channel lengths to understand the performance additivity of external stress. The linear drive-current enhancement plotted as a function of stress is shown in Fig 17 and 18. As shown in Figure 17, SiGe/Si devices have the same enhancement as observed in Si FinFETs. Hence the enhancement shown by SiGe based devices because of enhanced mobility and stress in channel is additive to the external stress applied.

CONCLUSION

SiGe shell-Si core pFETS with an omega gate are demonstrated. A mobility enhancement of up to 46% and current improvements of ~19 % are observed compared to Si devices. We have presented a first study and demonstration of advanced devices beyond the introduction of FinFETs.

References: [1] R. Chau, Proc. *IEEE. Nanotechnology,* 2004; [2] T. Yamamoto, IEDM 2007; [3] J. Kavalerios, VLSI 2006; [4] K. Mistry IEDM 2007; [5] Harris, VLSI 2007 ; [6] S. Suthram, IEDM 2007;

978-1-4244-2784-0/09 $25.00 © 2009 IEEE

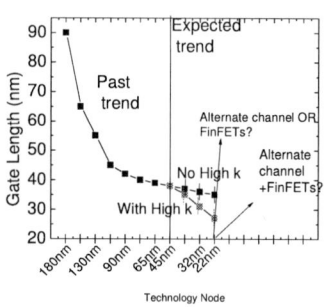

Figure 1. Past gate length scaling and possible future technology solutions for various nodes.

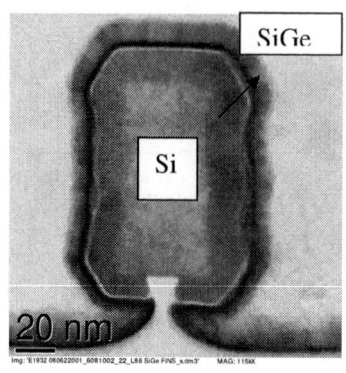

Figure 2. (110) oriented fins showing good epi. Angled Epi on fin sidewalls.

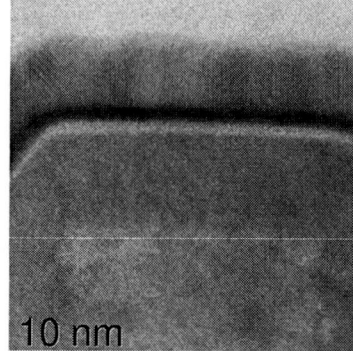

Figure 2. (100) oriented fins showing continuation of lattice between Si and SiGe. ~3.6nm of dielectric layer.

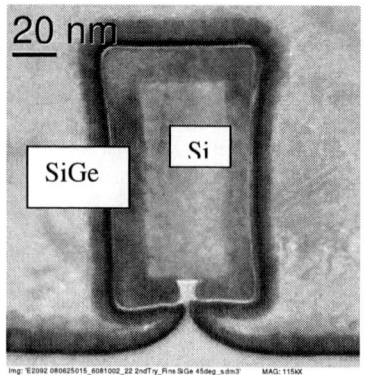

Figure 3. (100) oriented fins showing uniform growth of epitaxy in all directions

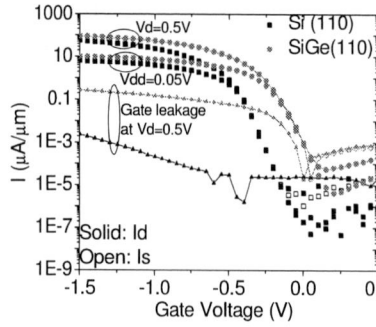

Figure 5. Id-Vg plot for SiGe and Si channel fins for long channel (2um) device.

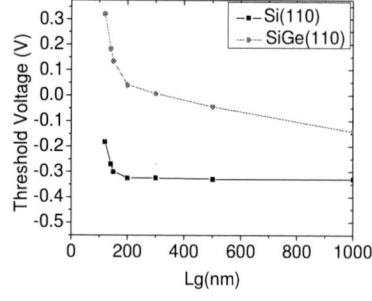

Figure 6. Threshold voltage shift because of SiGe

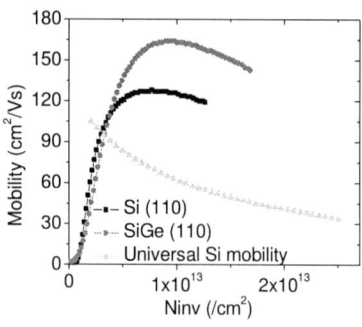

Figure 7. Mobility comparison between (110) oriented SiGe and Si fins. (100) universal Si mobility is shown for comparison.

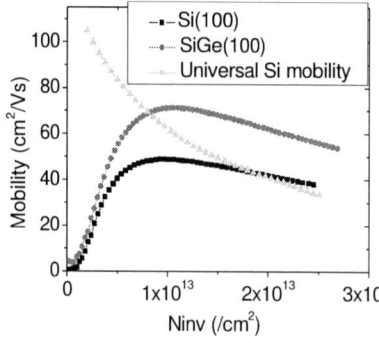

Figure 8. Mobility comparison between (100) oriented SiGe and Si fins. (100) universal Si mobility is shown for comparison.

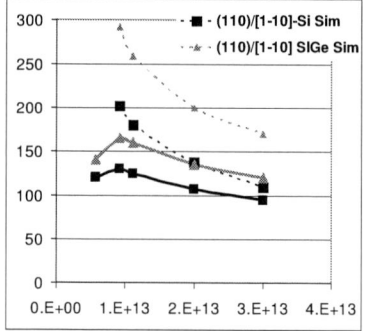

Figure 9. Experimental (solids lines) and simulation (dashed lines) mobility comparison for (110) oriented fins.

978-1-4244-2784-0/09 $25.00 © 2009 IEEE

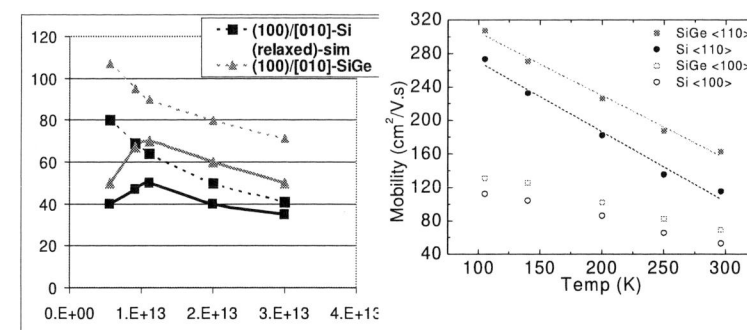

Figure 10. Experimental (solids lines) and simulation (dashed lines) mobility comparison for (100) oriented fins.

Figure 11. Temperature dependence of mobility curves show similar slopes for peak mobility between SiGe and Si.

Figure 12. Ion-Ioff comparison between SiGe and Si for (110) oriented fins.

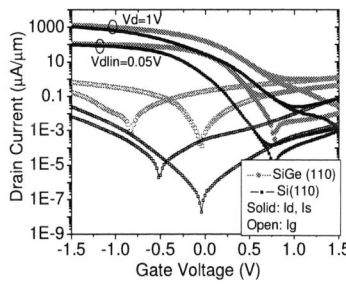

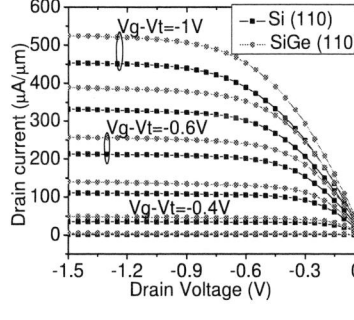

Figure 13. Id-Vg comparison between SiGe and Si based devices with Lg=140nm (as drawn).

Figure 14. Id-Vd comparison between SiGe and Si based devices with (110) orientation and Lg=250nm (as drawn) for Vg-Vt from 0 to -1V (in steps of -0.2V).

Figure 15. Capacitance Voltage curves for (110) oriented fins.

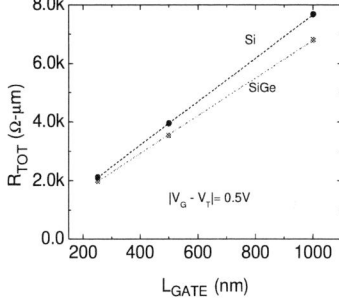

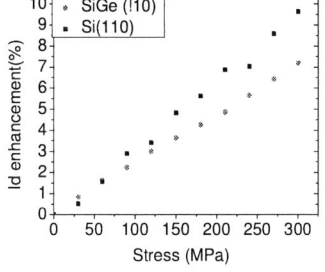

Figure 16. The external resistance comparison between SiGe and Si shell shows comparable external resistance.

Figure 17. SiGe shell-Si core omega gate FET and Si omega gate FET show similar performance enhancement with uniaxial bending stress.

Figure 18. Id enhancement for (110) oriented fins with applied uniaxial stress for various gate lengths.

978-1-4244-2784-0/09 $25.00 © 2009 IEEE

Metal-Oxide-Semiconductor devices with UHV-Ga$_2$O$_3$(Gd$_2$O$_3$) on Ge(100)

L. K. Chu[a], T. D. Lin[a], C. H. Lee[a], L. T. Tung[a], W. C. Lee[a], R. L. Chu[a], C. C. Chang[a], M. Hong[a*], and J. Kwo[b*]

Dept. [a]Materials Sci. and Eng., and [b]Physics, National Tsing Hua Univ, Hsinchu, Taiwan 30013

*Authors to whom the correspondence is addressed: mhong@mx.nthu.edu.tw (M. Hong) and Raynien@phys.nthu.edu.tw (J. Kwo)

ABSTRACT

Ultra-high vacuum (UHV)-deposited high Ga$_2$O$_3$(Gd$_2$O$_3$) was proved to passivate Ge effectively, as evidenced by comprehensive investigations including structural, chemical, and electrical analyses. The Ga$_2$O$_3$(Gd$_2$O$_3$)/Ge interface is revealed to be abrupt even being subjected to a 500°C anneal, a high κ value of 14.5, a low leakage current density of ~10^{-9}A/cm^2 with a Fowler-Nordheim tunneling behavior, and well-behaved C-V characteristics are achieved. Furthermore, Ge self-aligned pMOSFETs with Al$_2$O$_3$/ Ga$_2$O$_3$(Gd$_2$O$_3$) as the gate dielectrics have demonstrated a high drain current and a peak transconductance up to 252mA/mm and 143mS/mm, respectively, of 1μm-gate length.

INTRODUCTION

Beyond the 22-16 nm node technology, Ge is a promising channel candidate to replace Si since Ge offers higher electron and hole mobility comparing with Si. Moreover, the mobility may also be enhanced by strain. However, direct deposition of high κ dielectrics on Ge usually results in unwanted interfacial layer formation, leading to Fermi-level pinning and poor device performances. Feverish research efforts have concluded a necessity of utilizing an intentional buffer layer like GeO$_x$N$_y$ or Si to passivate Ge surface and prevent Ge in-diffusion into the gate dielectrics. Unfortunately, such a layer would decrease the permittivity of the overall gate stack and hampers further reduction of equivalent oxide thickness (EOT). UHV-deposited high κ dielectric Ga$_2$O$_3$(Gd$_2$O$_3$) (denoted as GGO) on Si and InGaAs are known to give atomically abrupt interfaces and excellent electrical (device) performances. [1-3] In this work, Ge MOSFETs were fabricated with GGO as a gate dielectric, along with studies of the properties of GGO/Ge gate stacks.

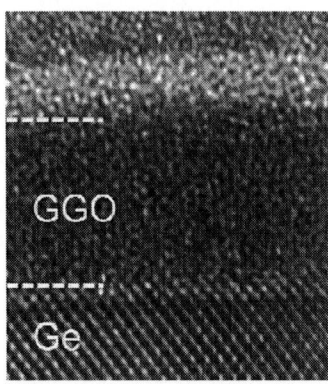

FIGURE 1. HR-TEM micrograph of the GGO/Ge interface after 450°C, 20min, N$_2$ anneal.

EXPERIMENT

Ge(100) wafers were dipped in diluted HF and rinsed in de-ionized water, followed by UHV annealing at ~450°C for removing native oxides. E-beam evaporated GGO film was then deposited on Ge and found to be amorphous as evidenced by RHEED patterns and high-resolution transmission electron microscopy (HR-TEM). Angle-resolved x-ray photoelectron spectroscopy (AR-XPS) and HR-TEM were also utilized to reveal the chemical and structural characteristics of the GGO/Ge interface. The wafers were also used to fabricate MOS capacitors and self-aligned MOSFETs with Ti/Au and TiN as the gate metal, respectively. The self-aligned process flow is listed in Table 1.

- Diluted HF dip + Deionized water rinse
- UHV annealing to 350°C~450°C
- UHV-Ga$_2$O$_3$(Gd$_2$O$_3$) deposition
- UHV-Al$_2$O$_3$ capping layer ~ 2nm
- TiN metal gate by sputtering
- Dry etching by ICP-RIE
- S/D implantation: B (25keV, 4E15cm^{-2})
- S/D activation: 450°C, 20min
- S/D region pattering
- Wet etching GGO
- Contact metal deposition

Table I. Process flow of self aligned pMOSFETs.

RESULTS AND DISCUSSION

The GGO/Ge hetero-structure is thermally robust, with its interface remaining intact after a 450°C-N$_2$-20min anneal, as shown in a cross-sectional HR-TEM picture (Fig. 1), in which an atomically abrupt interface between GGO and Ge is presented. In addition, AR-XPS data also indicate that there is no change in the spectra of Ge 3d, Ga 2p, and Gd 3d even withstanding an anneal up to 500°C for 5min, suggesting the absence of Ge oxide formation and no further Ge diffusion [4].

Furthermore, C-V curves without a notable shoulder feature are obtained for a GGO(10.8nm)/Ge capacitor (Fig. 2), which respond well to the applied frequencies from 1k to 1M Hz. A D$_{it}$ is calculated to be ~10^{12}cm^{-2}eV^{-1} near the mid-gap by Terman method. A very low gate leakage current density of 3×10^{-9}A/cm^2 at flat-band voltage +1V is obtained as shown in Fig. 3. The J-E curve also indicates Fowler-Nordheim tunneling through the oxide layer as well as the high-quality of GGO (inset of Fig. 3) [5]. A conduction band offset is then extracted to be ~2.27eV, suggesting a sufficient barrier to retard the leakage. Moreover, a valence band offset is also calculated to be 2.47eV as the band gap of GGO was known to be ~5.4eV, which is consistent with that obtained by XPS analyses.

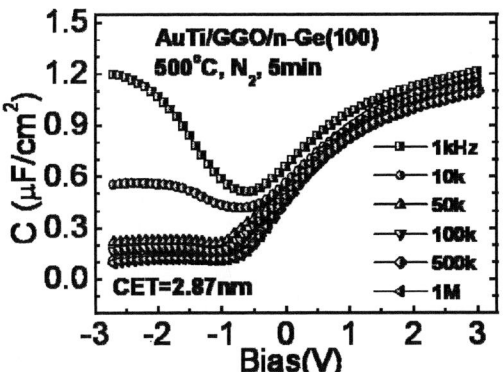

FIGURE 2. C-V characteristics of the AuTi/GGO(10.8nm) /Ge(100) capacitor.

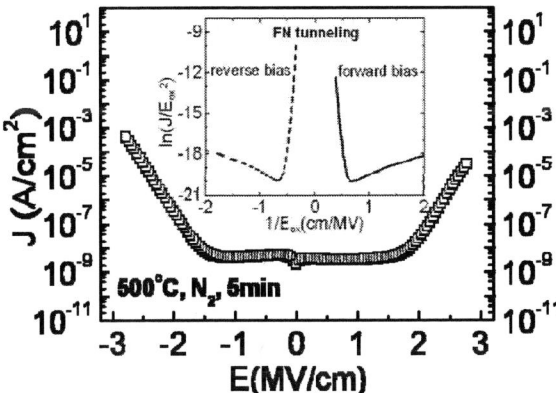

FIGURE 3. J-E characteristics of AuTi/GGO(10.8nm)/Ge(100) capacitor. The inset shows the F-N tunneling at high electrical field.

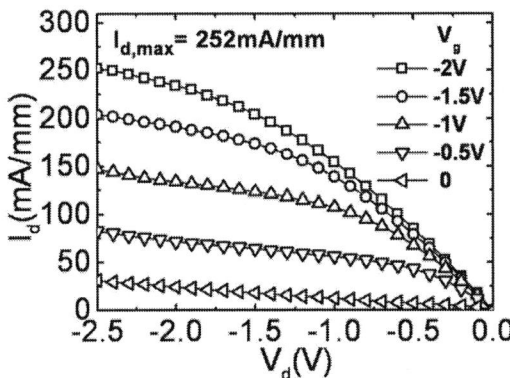

FIGURE 4. Drain current-voltage characteristics of the TiN/Al$_2$O$_3$/GGO/Ge pMOSFET.

Ge pMOSFETs of 1μm gate length (L$_g$) and 50μm gate width (W$_g$) exhibit excellent electrical performances with an EOT of 3.6nm (a 2nm-thick Al$_2$O$_3$ capping layer used to protect GGO increased the total EOT (6)). A maximum drain current density (I$_{d,sat}$) of 252mA/ mm at V$_g$=-2V (Fig. 4), and a peak transconductance (g$_{m,peak}$) of 143mS/mm (Fig. 5) are achieved, respectively. Systematic dependence of I$_{d,sat}$ and g$_{m,peak}$ on gate lengths varying from 20μm to

1μm is also observed, indicating much improved performance can be expected with further device down-scaling. The on-current can be further increased by improving the sheet resistance and contact resistivity, which is ~10^{-4} Ω-cm^2, extracted by TLM analysis. In addition, a peak hole mobility of ~260cm^2/V-s, which is about ~1.5× of the universal hole mobility for Si, is also obtained. Besides, Ge nMOSFETs using Al$_2$O$_3$/GGO as the gate oxide also displayed good I$_d$-V$_d$ characteristics (not shown). However, the performances are not as good as that of our pMOSFET counterparts presented in this work. The reason for the disparity was given previously [7].

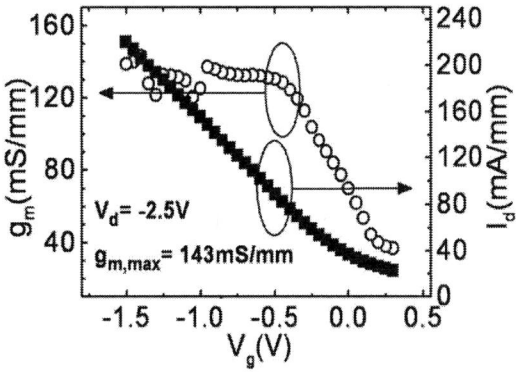

FIGURE 5. Corresponding I$_d$ vs V$_g$ and g$_m$ vs V$_g$ characteristics at saturation region.

CONCLUSION

Without the need for an interfacial layer, a high-quality high κ dielectric Ga$_2$O$_3$(Gd$_2$O$_3$) directly deposited on Ge substrate, has demonstrated excellent thermodynamic stability and electrical performances, such as a high κ value and extremely low leakage current. Moreover, although the processing steps are yet to be fully optimized, excellent output performances of the pMOSFETs such as high drain current density and transconductance were demonstrated with the strain-free Ge substrate, suggesting significant headroom for further improvement in the future.

ACKNOWLEDGEMENTS

We would like to acknowledge the support from National Science Council, Taiwan.

REFERENCES

[1] W. C. Lee *et al*, J. Cryst. Growth, **278**, 619 (2005).

[2] J. Kwo, *et al*, Appl. Phys. Lett., **75**, 1116, (1999).

[3] T. D. Lin, *et al*, Appl. Phys. Lett. **93**, 033516 (2008).

[4] C. H. Lee, *et al*., J. Vac. Sci. Technol. B **26**, 1128 (2008).

[5] T. S. Lay, *et al*, Solid-state electronics **45**, 1679 (2001).

[6] K. H. Shiu, *et al*, Appl. Phys. Lett. **92**, 172904 (2008).

[7] A. Dimoulas, *et al*, Appl. Phys. Lett., **89**, 252110 (2006).

Self-aligned Inversion Channel In$_{0.53}$Ga$_{0.47}$As N-MOSFETs with ALD-Al$_2$O$_3$ and MBE-Al$_2$O$_3$/Ga$_2$O$_3$(Gd$_2$O$_3$) as Gate Dielectrics

H. C. Chiu[a], T. D. Lin[a], P. Chang[a], W. C. Lee[a], C. H. Chiang[a], J. Kwo[b]*, Y. S. Lin[c], Shawn S. H. Hsu[c], W. Tsai[d], and M. Hong[a]*

[a] Dept of Materials Sci. and Eng., [b] Dept of Physics, and [c] Electrical Eng., Inst of Eletronics Eng., National Tsing Hua Univ., HsinChu 30012, Taiwan, R.O.C. and [d] Intel Corporation, SC1-05, 2200 Mission College Blvd, Santa Clara, CA 95052
*Authors to whom the correspondence is addressed: mhong@mx.nthu.edu.tw (M. Hong) and Raynien@phys.nthu.edu.tw (J. Kwo)

ABSTRACT

Self-aligned inversion-channel In$_{0.53}$Ga$_{0.47}$As n-MOSFETs with *ex-situ* atomic-layer-deposited Al$_2$O$_3$ and *in-situ* ultra-high-vacuum deposited Al$_2$O$_3$/Ga$_2$O$_3$(Gd$_2$O$_3$) as gate dielectrics have been demonstrated. Both devices exhibit excellent DC characteristics, including high drain currents and transconductances. In addition, RF characteristics of both devices were analyzed; without using any isolation, non de-embedded current gain cutoff frequency (f_T) and maximum oscillation frequency (f_{max}) of ~ 3.1 and 1.1 GHz (ALD-Al$_2$O$_3$) and of ~ 17.9 and 11.2 GHz (MBE-Al$_2$O$_3$/Ga$_2$O$_3$(Gd$_2$O$_3$)), respectively, have been obtained.

INTRODUCTION

The quest for technologies beyond the 22-16 nm node CMOS devices has now called for research on high κ gate dielectrics on channel materials with high carrier mobility. In$_{0.53}$Ga$_{0.47}$As has long been used as a backbone for nearly all high-speed electronic devices. The first inversion-channel (non-self-aligned) In$_{0.53}$Ga$_{0.47}$As MOSFET was demonstrated a decade ago by employing molecular beam epitaxy (MBE) grown Ga$_2$O$_3$(Gd$_2$O$_3$) [GGO] 40 nm thick as a gate dielectric, giving a maximum drain current (I$_{ds}$) of 375 mA/mm (L_g = 1 μm) and a transconductance (G_m) of 190 mS/mm (L_g= 0.75 μm) [1]. Recently, non-self-aligned In$_{0.53}$Ga$_{0.47}$As MOSFETs (L_g= 0.5 μm) with atomic-layer-deposited (ALD) grown Al$_2$O$_3$ dielectrics were fabricated, showing a maximum I$_{ds}$ of 430 mA/mm, and a G_m of 160 mS/mm [2]. Nonetheless, the non-self-aligned process is impractical for device integration. By utilizing an improved Al$_2$O$_3$/GGO bi-layer gate dielectric [3], we recently have achieved a self-aligned inversion-channel In$_{0.53}$Ga$_{0.47}$As MOSFET (L_g= 1 μm), reaching a record high I$_{ds}$ of 1.05A/mm and a G_m of 714 mS/mm [4]. In this work, the self-aligned process was also applied for fabricating In$_{0.53}$Ga$_{0.47}$As MOSFETs with ALD-Al$_2$O$_3$. In$_{0.53}$Ga$_{0.47}$As MOSFETs with ALD-Al$_2$O$_3$ and MBE-Al$_2$O$_3$/GGO both demonstrated excellent DC and RF characteristics.

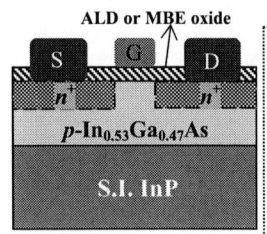

ALD or MBE oxide

Fig. 1 Schematic view of self-aligned inversion-channel In$_{0.53}$Ga$_{0.47}$As n-MOSFET with ALD-Al$_2$O$_3$, or MBE-Al$_2$O$_3$/GGO dielectrics and a TiN metal gate.

1. TiN gate formation by lift-off or dry-etching
2. S/D region patterning, Si implantation, and activation
3. S/D contact patterning, wet-etching of oxide, contact metal deposition
4. ohmic contact alloying, and pad metal deposition

Table 1. Process flow of self-aligned inversion-channel In$_{0.53}$Ga$_{0.47}$As MOSFETs.

EXPERIMENT

A schematic cross-section of the self-aligned inversion-channel In$_{0.53}$Ga$_{0.47}$As MOSFET is shown in Figure 1. The p-type In$_{0.53}$Ga$_{0.47}$As channel layer 3000Å-thick with a Be doping of 5×10^{16} cm^{-3}, and unintentionally doped In$_{0.53}$Ga$_{0.47}$As buffer layer 5000Å thick were grown on semi-insulating InP substrates by molecular beam epitaxy. The wafers were subsequently *ex-situ* loaded into an ALD reactor with a transfer time of less than 10 minutes. An ALD-Al$_2$O$_3$ film with a thickness of ~10 nm was then deposited on top of the In$_{0.53}$Ga$_{0.47}$As channel layer. Also, an Al$_2$O$_3$/GGO dual-layer gate dielectric was *in-situ* e-beam evaporated onto In$_{0.53}$Ga$_{0.47}$As surface. The detailed process flow of device fabrication was shown in Table 1.

RESULTS AND DISCUSSION

Firstly, the self-aligned ALD-Al$_2$O$_3$/In$_{0.53}$Ga$_{0.47}$As MOSFET of 1μm×100μm (W×L) showed well behaved I-V and G_m characteristics with a maximum I$_{ds}$ of 288 mA/mm (Fig. 2a), and a G_m of 93 mS/mm (Fig. 2b), comparable to, and better than those obtained with a non-self-aligned ALD-Al$_2$O$_3$ device with a 0.5 μm gate length [2]. An extremely low leakage current (<10^{-12} A at V$_{gs}$ = 4 V) was measured between gate and source (not shown here), demonstrating the good thermal stability of the hetero-structure withstanding RTA at 750°C for 5 s in N$_2$ ambient and excellent insulating property of ALD-Al$_2$O$_3$. Furthermore, the mid-gap interfacial density of states (D_{it}) ~2.5×10^{11} cm^{-2}eV^{-1} was determined by the charge pumping method on another similar device structure after a dopant activation at RTA 650°C in N$_2$ [5].

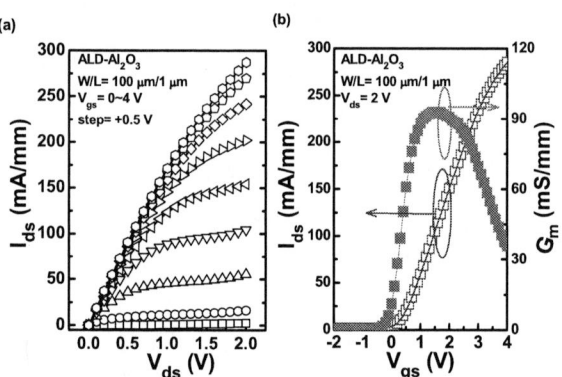

Fig. 2 DC (a)I-V characteristics and (b)drain current density and transconductance versus gate bias of a 1μm×100μm ALD-Al$_2$O$_3$/In$_{0.53}$Ga$_{0.47}$As n-MOSFET.

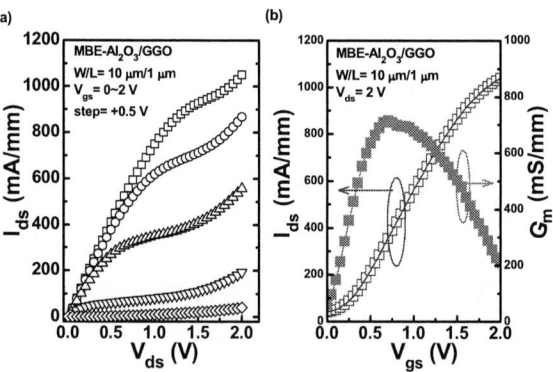

Fig. 3 DC (a)I-V characteristics and (b)drain current density and transconductance versus gate bias of a 1μm×10μm MBE-Al$_2$O$_3$/GGO/In$_{0.53}$Ga$_{0.47}$As n-MOSFET.

978-1-4244-2784-0/09 $25.00 © 2009 IEEE

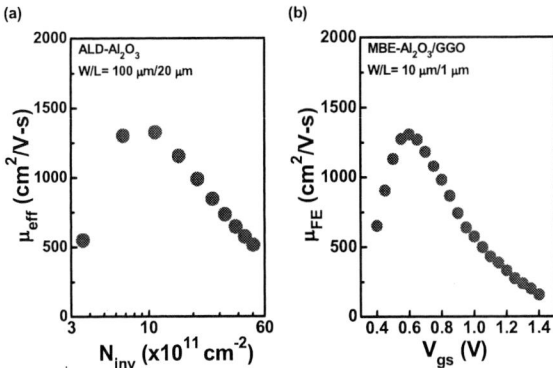

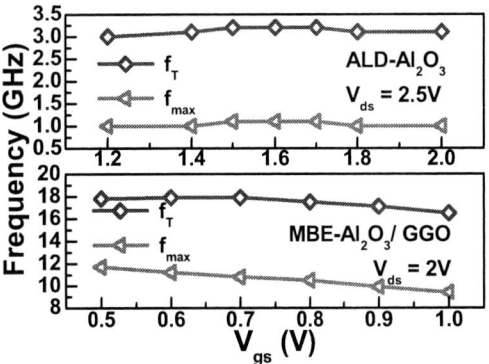

Fig. 4 Plot of (a)effective electron mobility (μ_{eff}) as a function of inversion charge density (N_{inv}) for a ALD-Al_2O_3/$In_{0.53}Ga_{0.47}As$ n-MOSFET extracted by split-CV and (b) field effective electron mobility (μ_{FE}) as a function of gate bias for a MBE-Al_2O_3/GGO/$In_{0.53}Ga_{0.47}As$ n-MOSFET.

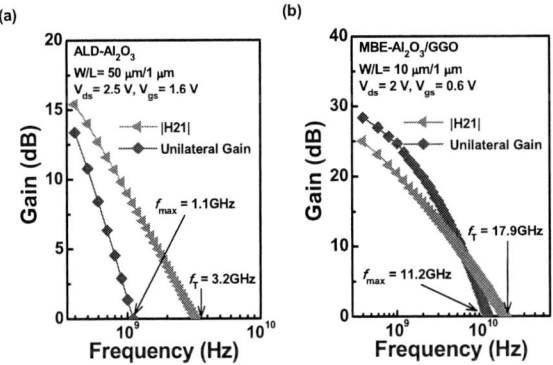

Fig. 5 RF performance of (a) a $1\mu m \times 50\mu m$ ALD-Al_2O_3/$In_{0.53}Ga_{0.47}As$ n-MOSFET and (b) a $1\mu m \times 10\mu m$ MBE-Al_2O_3/GGO/$In_{0.53}Ga_{0.47}As$ n-MOSFET, respectively.

Secondly, the self-aligned MBE-Al_2O_3/GGO/$In_{0.53}Ga_{0.47}As$ MOSFET of $1\mu m \times 10\mu m$ showed a maximum I_{ds} of 1.05 A/mm at V_{gs} = 2V and V_{ds} = 2V (Fig. 3a), and a transconductance of 714 mS/mm at V_{ds} = 2V and V_{gs} = 1V (Fig. 3b). The peak transconductance is the highest ever reported among all types of E-mode III-V MOSFETs, fabricated by self-aligned or non-self-aligned processes.

For ALD-Al_2O_3/$In_{0.53}Ga_{0.47}As$ n-MOSFETs, Fig. 4(a) shows the effective mobility (μ_{eff}) versus inversion charge density (N_{inv}) extracted by the split-CV method; a peak value of 1330 cm^2/V-s is derived from the integral of the gate-to-source capacitance measured at 100 kHz. Moreover, for MBE-Al_2O_3/GGO/$In_{0.53}Ga_{0.47}As$ MOSFETs, the field-effect electron mobility reaches its peak value of 1300 cm^2/V-s, obtained by I_{ds}-V_{gs} method, at a V_{gs} = 0.6V (Fig.4b). It should be noted that the mobilities mentioned here are extrinsic values without any correction.

The RF performances of these $In_{0.53}Ga_{0.47}As$ MOSFETs were characterized. The results are shown in Figure 5(a) for a $1\mu m \times 50\mu m$ ALD-Al_2O_3/$In_{0.53}Ga_{0.47}As$ device biased at V_{ds} = 2.5V and V_{gs} = 1.6V, giving a f_T = 3.2 GHz and a f_{max} = 1.1 GHz. Fig. 5(b) shows the RF performance of a $1\mu m \times 10\mu m$ MBE-Al_2O_3/$Ga_2O_3(Gd_2O_3)$/$In_{0.53}Ga_{0.47}As$ device, for which the f_T and f_{max} reach 17.9 GHz and 11.2 GHz, respectively, when biased at V_{ds} = 2V and V_{gs} = 0.6V. Fig. 6 shows the frequency response of each device as a function of the gate voltage. Both the f_T and f_{max} maintain good performance over a wide bias range, excellent characteristics for circuit applications. Note that the RF performance of the devices is

Fig. 6 Frequency response of inversion-channel $In_{0.53}Ga_{0.47}As$ n-MOSFET with ALD- and MBE-gate dielectrics as a function of gate voltage.

underestimated since no RF de-embedding procedure was applied. With the simple structure employed here, there was no device isolation and thus the de-embedding standard is difficult to define. As a consequence, an additional conduction path surely exists in the substrate leading to undesired parasitic components and seriously degraded f_T and f_{max}. The transistor isolation in our next devices is essential for improving the RF characteristics, following the Si CMOS technology in which the shallow-trench-isolation (STI) is used.

CONCLUSION

In summary, self-aligned inversion-channel $In_{0.53}Ga_{0.47}As$ MOSFETs based on both *ex-situ* ALD-Al_2O_3 and *in-situ* MBE-Al_2O_3/GGO as gate dielectrics, have shown excellent DC and RF device characteristics, particularly, the high drain current, transconductance, and current gain cutoff frequency. The advances of the InGaAs MOSFETs demonstrated in this work help enable future CMOS technology.

ACKNOWLEDGEMENTS

The authors wish to thank National Science Council under grants of NSC-97-2120-M-007-008 and NSC-96-2628-M-007-003-MY3, Taiwan, the support from the Asian Office of Aerospace Research and Development of the U.S. Air Force, and Intel Corporation.

REFERENCES

[1] F. Ren, et al, IEEE Electron Device Lett. **19**, 309 (1998).

[2] Y. Xuan, et al, IEDM 2007, p637.

[3] K. H. Shiu, et al, Appl. Phys. Lett. **92**, 172904 (2008).

[4] T. D. Lin, et al, Appl. Phys. Lett. **93**, 033516 (2008).

[5] H.C. Chiu, et al, Appl. Phys. Lett., *in press*.

Inversion-Type Surface Channel In$_{0.53}$Ga$_{0.47}$As Metal-Oxide-Semiconductor Field-Effect Transistors with Metal-Gate/High-k Dielectric Stack and CMOS-Compatible PdGe Contacts

Hock-Chun Chin, Xinke Liu, Leng-Seow Tan, and Yee-Chia Yeo.

Silicon Nano Device Lab., Dept. of Electrical and Computer Engineering, National University of Singapore (NUS), 117576 Singapore.

Phone: +65 6516-2298, Fax: +65 6779-1103, E-mail: yeo@ieee.org

ABSTRACT

We report the first demonstration of a surface channel inversion-type In$_{0.53}$Ga$_{0.47}$As n-MOSFET featuring gold-free palladium-germanium (PdGe) ohmic contacts and self-aligned S/D formed by silicon and phosphorus co-implantation. A gate stack comprising TaN/HfAlO/In$_{0.53}$Ga$_{0.47}$As is also featured. Excellent transistor output characteristics with high drain current on/off ratio of 10^4, high peak electron mobility of 1420 cm^2/Vs and peak transconductance of 142 mS/mm at gate length of 2 μm were demonstrated. In addition, the integration of low resistance PdGe ohmic contacts on In$_{0.53}$Ga$_{0.47}$As alleviates contamination concerns associated with the common use of gold-based contacts on In$_{0.53}$Ga$_{0.47}$As.

INTRODUCTION

High mobility III-V compound semiconductors, particularly InGaAs, have received renewed interest as potential channel materials to replace conventional Si or strained Si channels [1]-[18]. However, the lack of high quality and thermodynamically stable gate dielectrics on III-V semiconductors hinders the development of well-tempered III-V MOSFET. In addition, gold-based contact technologies that are commonly used in III-V devices cannot be employed for CMOS device integration. The formation of low resistance CMOS-compatible ohmic contacts and the achievement of high dopant activation [19]-[20] remain as important integration challenges. For the In$_{0.53}$Ga$_{0.47}$As n-MOSFET, there has not been any demonstration of CMOS-compatible contact technology.

In this work, we report the integration of multiple advanced process modules to form a novel self-aligned surface channel In$_{0.53}$Ga$_{0.47}$As n-MOSFET, featuring 1) a TaN/HfAlO/InGaAs gate stack; 2) CMOS-compatible palladium-germanium (PdGe) S/D ohmic contacts; and 3) silicon and phosphorus co-implanted source and drain (S/D) regions for high dopant activation.

EXPERIMENTAL WORK

P-type Zn-doped ($N_A = \sim 1 \times 10^{18}$ cm^{-3}) InP wafers were used as the starting substrates. The fabrication process flow is depicted in Fig. 1. Be-doped ($N_A = \sim 1 \times 10^{15}$ cm^{-3}) In$_{0.53}$Ga$_{0.47}$As layer (1 μm) was first grown by molecular beam epitaxy. The wafers then underwent a three-step pre-gate cleaning process comprising native oxide and excess elemental arsenic removal using hydrochloric acid (HCl) and ammonium hydroxide (NH$_4$OH) respectively, and surface passivation using ammonium sulfide (NH$_4$)$_2$S solution. After wet cleaning, the samples were quickly loaded into a MOCVD reactor for HfAlO high-k gate dielectric deposition. Post-deposition anneal (PDA) at 500˚C for 60 s was carried out to improve the quality of the as-deposited HfAlO film before reactive sputter deposition of the TaN metal gate. After gate patterning, self-aligned n$^+$ S/D regions of the transistors were formed by Si$^+$ and P$^+$ co-implantation and dopant activation at 600˚C to 800˚C for 10s to 60s. Forming gas anneal at 300˚C to 400˚C was also performed. Low resistance CMOS-compatible PdGe S/D ohmic contacts were integrated to complete the fabrication.

RESULTS AND DISCUSSION

Fig. 2 shows a schematic of the cross-section of the In$_{0.53}$Ga$_{0.47}$As n-MOSFET fabricated in this work. Key features are highlighted. The sheet resistance of the n$^+$ regions formed by Si$^+$

and P$^+$ co-implantation and under different dopant activation conditions was extracted by using TLM test structures and summarized in Fig. 3. Fig. 4 reveals the excellent ohmic behavior of the PdGe contacts on n$^+$ In$_{0.53}$Ga$_{0.47}$As regions. The contact resistance R_C and specific contact resistivity ρ_C of the PdGe contact are 0.63 Ω.mm and 1.2×10^{-4} Ω.cm^{-2}, respectively. Fig. 5 plots the I_D-V_G transfer characteristics of an In$_{0.53}$Ga$_{0.47}$As n-MOSFET. This surface channel device with a gate length of 4 μm demonstrates good transistor performance with promising drive current and high I_{on}/I_{off} ratio of 10^4. Subthreshold swing SS and threshold voltage V_t of the transistor are 274 mV/decade and 8.9 mV, respectively. Transconductance G_m characteristics of the same device at V_D of 0.1 and 1.2 V are shown in Fig. 6. Fig. 7 shows the I_D-V_D characteristics of the In$_{0.53}$Ga$_{0.47}$As n-MOSFET at various gate overdrives. This gold-free In$_{0.53}$Ga$_{0.47}$As MOSFET demonstrates excellent saturation and pinch-off characteristics with reasonably low series resistance.

The C-V characteristic in Fig. 8 indicates the inversion mode operation of the surface channel In$_{0.53}$Ga$_{0.47}$As n-MOSFET. The equivalent-oxide-thickness (EOT) of the transistor is ~ 4.7 nm. The effective carrier mobility is estimated by using linear I_D-V_G characteristic at low V_{DS} and the total inversion charge density, which is obtained by integrating the measured C-V curve. In$_{0.53}$Ga$_{0.47}$As n-MOSFET demonstrates much higher electron mobility than GaAs MOSFET with similar gate stack at high E_{eff} [Fig. 9]. Peak electron mobility exceeding 1420 cm^2/Vs can be realized in the In$_{0.53}$Ga$_{0.47}$As MOSFET in this work. The effect of FGA on subthreshold swing SS was also investigated, as illustrated in Fig. 10. FGA is useful to improve the device performance by passivating the interfacial dangling bonds. Drain currents at V_D of 1.2 V and gate-overdrives V_G-V_t of 1.0 V, 1.5 V and 2.0 V were plotted as a function of gate length L_G in Fig. 11. Fig. 12 compares the peak transconductance G_m achieved in this work with published data from 1965. Maximum G_m as high as 142 mS/mm at V_D of 2 V was realized here. Much higher drive current and transconductance are expected by further L_G and EOT scaling.

CONCLUSION

Surface channel In$_{0.53}$Ga$_{0.47}$As n-MOSFETs with TaN/HfAlO/In$_{0.53}$Ga$_{0.47}$As gate stack, gold-free PdGe S/D ohmic contacts and Si$^+$ plus P$^+$ co-implanted S/D regions were demonstrated. The gate-first inversion-type MOSFETs exhibit good transistor output characteristics with attractive drive current and transconductance, high I_{on}/I_{off} ratio, and maximum carrier mobility of 1420 cm^2/Vs. The successful integration of PdGe contacts alleviates the contamination concerns arising from the use of gold-based contacts in III-V MOSFETs.

REFERENCES

[1] H. Becke et al., SSE 8, pp. 813, 1965. [2] T. Mimura et al., IEEE TED 25, pp. 573, 1978. [3] G. G. Fountain et al., IEDM Tech. Dig., pp. 887, 1989. [4] M. Passlack et al., IEDM Tech. Dig., pp.383, 1995. [5] F. Ren et al., SSE 41, pp. 1751, 1997. [6] F. Ren et al., IEEE EDL 19, pp. 309, 1998. [7] M. Hong et al., Science 283, pp.1897, 1999. [8] Y. Sun et al., IEEE EDL 28, pp. 473, 2007. [9] I. Ok et al., IEDM Tech. Dig., pp.829, 2006. [10] K. Rajagopalan et al., IEEE EDL 27, pp. 959, 2006. [11] K. Rajagopalan et al., IEEE EDL 28, pp. 100, 2007. [12] R. J. W. Hill et al., IEEE EDL 28, pp. 1080, 2007. [13] H. C. Lin et al., Appl. Phys. Lett. 91, pp. 212101, 2007. [14] Y. Xuan et al., IEEE EDL 28, pp. 935, 2007. [15] Y. Xuan et al., Appl. Phys. Lett. 91, pp. 232107, 2007. [16] Y. Xuan et al., IEDM Tech. Dig., pp. 637, 2007. [17] Y. Xuan et al., IEEE EDL 29, pp. 294, 2008. [18] H.-C. Chin et al., IEEE EDL 29, pp. 553, 2008. [19] J. P. de Souza et. al., IEEE Trans. Elec. Dev. 39, pp.166, 1992. [20] F. Hyuga et. al., Appl. Phy. Lett. 50, pp.1592, 1987.

○ In$_{0.53}$Ga$_{0.47}$As Epitaxy Process
○ Pre-gate Clean
○ High-*k* Dielectric Deposition and PDA
○ TaN Metal Deposition and Gate Patterning
○ Si$^+$ + P$^+$ S/D Implantation and Dopant Activation
○ Forming Gas Anneal
○ Contact Patterning and Pd/Ge Deposition
○ Lift-off and Contact Formation

Fig. 1. Process sequence employed in the fabrication of inversion-type surface channel In$_{0.53}$Ga$_{0.47}$As MOSFETs.

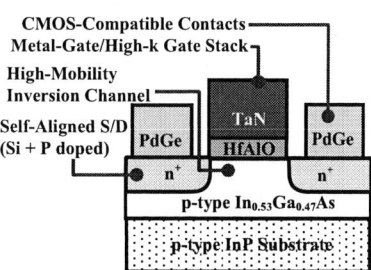

Fig. 2. Schematic of a InGaAs MOSFET showing key features. Gold-free ohmic contacts was integrated in a In$_{0.53}$Ga$_{0.47}$As-channel MOSFET for the first time.

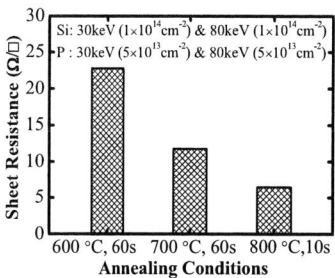

Fig. 3. Sheet resistance as a function of various dopant activation conditions. The sheet resistance was evaluated by using TLM test structures.

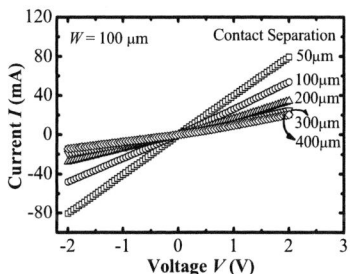

Fig. 4. The PdGe contacts demonstrate excellent ohmic behaviors on In$_{0.53}$Ga$_{0.47}$As at different contact separations.

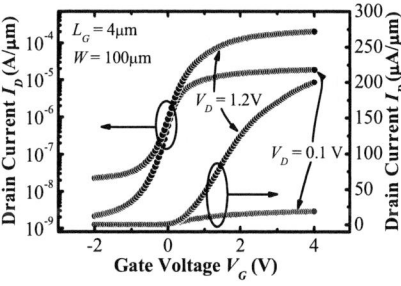

Fig. 5. Transfer characteristics of an In$_{0.53}$Ga$_{0.47}$As MOSFET with L_G of 4 μm. The threshold voltage and subthreshold swing of the device are 8.9 mV and 274 mV/decade, respectively.

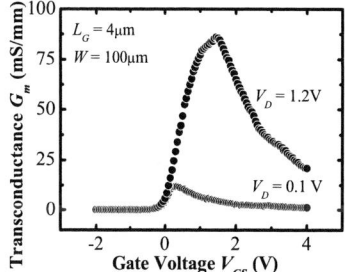

Fig. 6. Transconductance G_m versus gate voltage V_G of an In$_{0.53}$Ga$_{0.47}$As MOSFET at drain bias of 0.1 V and 1.

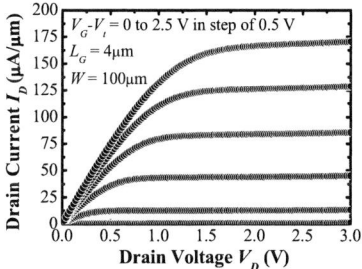

Fig. 7. I_D-V_{DS} output characteristics of an In$_{0.53}$Ga$_{0.47}$As transistor, demonstrating good saturation and pinch-off behaviors.

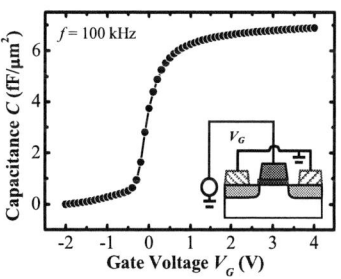

Fig. 8. Inversion C-V characteristic of the In$_{0.53}$Ga$_{0.47}$As n-MOSFET. Equivalent-oxide-thickness (EOT) of the transistor is ~ 4.7 nm.

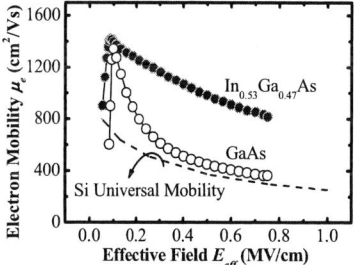

Fig. 9. Effective carrier mobility of In$_{0.53}$Ga$_{0.47}$As and GaAs surface channel transistor as a function of effective electric field E_{eff}. The In$_{0.53}$Ga$_{0.47}$As n-MOSFET demonstrates much higher electron mobility than GaAs MOSFET at high E_{eff}.

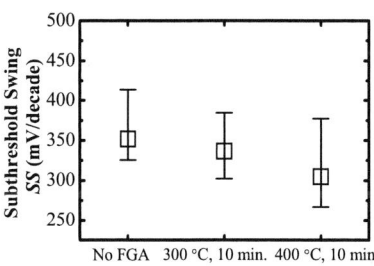

Fig. 10. Subthreshold swing SS of the In$_{0.53}$Ga$_{0.47}$As MOSFETs at different FGA conditions. FGA is helpful to improve SS by passivating interfacial dangling bonds.

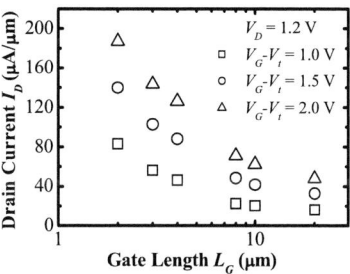

Fig. 11. Drain current I_D as a function of gate length L_G. at drain voltage of 1.2 V and gate overdrives of 1.0, 1.5 and 2.0 V.

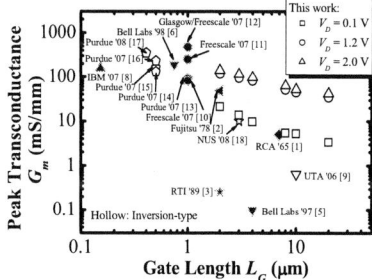

Fig. 12. Comparison of this work with published peak transconductance G_m of n-channel III-V MOSFETs from 1965. Peak transconductance exceeding 142 mS/mm can be realized in the In$_{0.53}$Ga$_{0.47}$As MOSFET with L_G of 2 μm in this work.

978-1-4244-2784-0/09 $25.00 © 2009 IEEE   133

Sub-100nm High-K Metal Gate GeOI pMOSFETs performance:
Impact of the Ge Channel Orientation and of the Source Injection Velocity

C. Le Royer, A. Pouydebasque, K. Romanjek, V. Barral, M. Vinet, J.-M. Hartmann,
E. Augendre, H. Grampeix, L. Lachal, C. Tabone, B. Previtali, R. Truche, F. Allain

"CEA, LETI, MINATEC", F38054 Grenoble, FRANCE
e-mail : cyrille.leroyer@cea.fr

ABSTRACT

We report here experimental investigations on GeOI pMOSFET: Besides the +65% mobility enhancement in narrow channel GeOI pMOSFETs as compared to wide channels, attributed to improved sidewall transport properties, <100> channel orientation transport is investigated for the first time in Ge (001): unlike Si, no current gain is observed compared to <110> channel orientation. Finally, ballisticity rates (BR) and source injection velocities (v_{inj}) were extracted, demonstrating 22% higher v_{inj} in Ge than in Si.

INTRODUCTION

Germanium or Germanium-On-Insulator (GeOI) pMOSFET is considered as a serious contender to Silicon due to its higher hole mobility [1-4]. However, large pMOS performance improvements have been demonstrated in Si using process induced strain (SiGe S/D, CESL, etc.), (110) surface operation as in hybrid orientation substrates or FinFET devices, or <100> channel orientation [5-7]. In addition, when reaching the end of the roadmap, quasi-ballistic transport is expected to be predominant and intrinsic material properties, such as the source injection velocity, will be the drivers of the performance. In this paper, we report on the better sidewall conduction properties of GeOI pMOSFETs with mesa isolation. Moreover, Ge <100> and <110> channel orientations are compared for the first time, without significant current gain for <100> oriented devices. Finally, ballisticity rates BR and injection velocities v_{inj} extracted in GeOI devices are discussed and compared to SOI devices data.

DEVICE FABRICATION

The device fabrication process is summarized in Fig. 1. Starting substrates are 200mm (001) oriented GeOI wafers with a 60-80nm thick Ge active layer obtained using the Smart Cut™ technology [1].

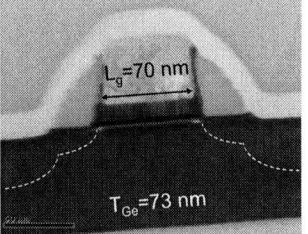

- GeOI Smart Cut™ (T_{Ge}~60-80nm)
- Channel Implant
- Mesa Isolation
- Si capping epi (1nm) + wet oxidation
  ~1nm SiO_2 on ~0.5nm Si
- HfO_2 ALCVD 3.5nm + TiN PVD 10nm
- Poly 50nm + SiO_2 Hard mask
- Gate photo-lithography + etch
- SDE + Halo Implant
- HfO_2+SiN Spacer formation
- S/D Implant + Activation anneal
- Contact formation + Metallization

Figure 1. Summary of the GeOI pMOSFET process (left). TEM image of GeOI pMOSFET with 70nm physical gate length (right).

The Ge layer was counter-doped with As implantation. Ge mesa structures were then patterned, followed by Si passivation and oxidation. 3.5nm HfO_2 ALCVD, 10nm TiN PVD, Poly-Si were deposited. Conventional DUV photo-lithography and etch led to devices with gate length down to 70nm as shown in Fig. 1. To our knowledge, this is among the shortest L_G reported for GeOI MOSFETs [1,8]. The measured EOT is 1.8nm. After BF_2 Source/Drain Extension formation, n-pockets were implanted with P and low temperature spacers were deposited and etched. BF_2 S/D were then implanted and annealed (without germanidation) followed by a standard Ti/TiN/W contact formation and Al metallization..

DEVICE PERFORMANCE

With the As counterdoping of the Ge film combined with the Pockets implants, we have demonstrated functional GeOI pMOSFETs with improved Short Channel Effects [1]: the threshold voltage becomes negative and constant (-0.1V) with respect to gate length even at L_{eff} = 90nm. The parameters (μ_0, V_{th}, L_{eff}...) were extracted using the Y-function.

For narrow width devices, we have observed larger I_{LIN} current density values (@ V_G-V_{th}=-0.8V): +40% I_{LIN} enhancement is observed at W_{eff}=0.29µm compared to 10µm [9]. The I_D gain when decreasing W_{eff} is similar for all L_{eff} investigated (L_{eff} 2.4µm-90nm).

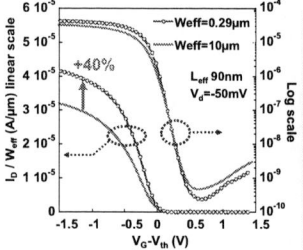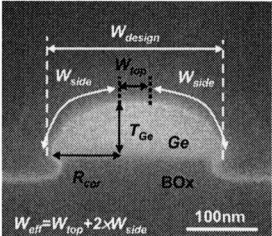

Figure 2. $I_D(V_G)$ characteristics (V_D=-50mV) of L_{eff}=90nm GeOI pMOS with a large and narrow W_{eff} (left). SEM cross-section along the gate of the edge of a narrow width GeOI mesa. Due to surface migration, the edge of the mesa is rounded, with a width W_{side} given by the quarter of perimeter of an ellipsoid. The total effective width is given by W_{eff}=W_{top}+2xW_{side} (right).

Even for the smallest L_G, there is no degradation of the sub-threshold current (Fig. 2-a). As discussed in [9] we attribute the positive impact of W_{eff} reduction on the mobility to larger mobility at the sidewalls [2,12,13] rather than mesa edge effects induced by the strained metal gate [14,15] or strain relaxation in narrow width devices [16]. Remarkably, the extracted μ_{top} and μ_{side} values exhibit no dependence on L_{eff} (μ_{top}=125cm²/V.s, μ_{side}=240cm²/V.s) with +90% mobility improvement at the sides.

The strong I_{LIN} & I_{ON} enhancements are obtained without compromising on the sub-threshold characteristics. Excellent on-state performances are demonstrated at L_{eff}=90nm (W_{eff}=0.29µm): I_{ON}=385µA/µm at V_g-V_{th}=-0.8V, V_D=-1.2V (estimated I_{ON}=485µA/µm using germanidation [1]).

INFLUENCE OF <100> CHANNEL ORIENTATION

For the first time <100> channel orientation transport is investigated in Ge devices. Fig. 3 shows the distribution of the linear

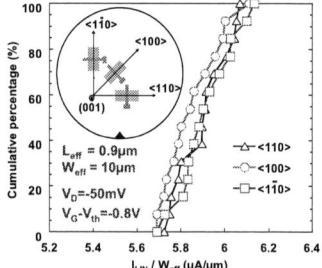

Figure 3. I_{LIN} distribution (L_{eff}/W_{eff}=0.9/10µm) for the three channel orientations indicated in the inset

978-1-4244-2784-0/09 $25.00 © 2009 IEEE

current I_{LIN} measured in long ($L_{eff} = 0.9\mu m$) and large ($W_{eff} = 10\mu m$) devices for three channel orientations (<100> + equivalent <110> & <1$\bar{1}$0>, inset Fig. 3). Contrary to what has been observed in Si [7], <100> channel orientation do not bring any current improvement in Ge, as predicted in [17]. The same behavior is observed whatever L_{eff} is (Fig. 4-a). When decreasing W_{eff}, a slight I_{lin} increase is observed in <100> channel devices, that is much smaller than for the equivalent <110> and <1$\bar{1}$0> oriented devices as shown in Fig. 4-b.

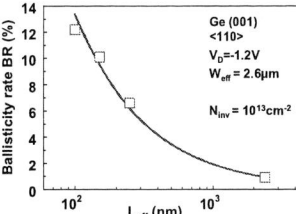

Figure 6. Extracted ballisticity rate (BR) vs. L_{eff} showing a continuous improvement. BR can be enhanced by a further reduction of L_{eff}.

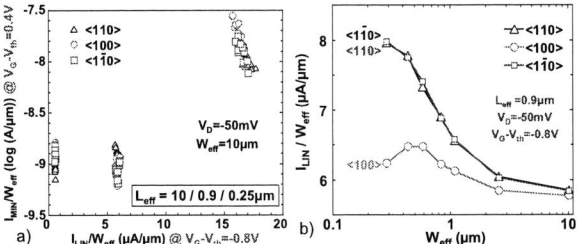

Figure 4. I_{LIN} vs. I_{MIN} trade-off for L_{eff} ranging from 10μm to 0.25μm (left). Evolution of I_{LIN} as a function of W_{eff} for the three channel orientations. A much smaller gain is observed for <100> oriented devices (right).

Fig. 5-a represents μ_0 as a function of W_{eff} for different L_{eff}. No significant impact of L_{eff} is observed and μ_0 at the smallest W_{eff} is 35% higher with <110> channel orientation compared to <100>. We assume this is due to less favorable surface orientations at the edge of the mesa for <100> channel devices. Top and side mobilities were extracted, showing improved μ_{side} vs. μ_{top} for <100> oriented devices, that is however 40% lower than μ_{side} extracted for <110> oriented devices (Fig. 5-b). Comparable top mobility values are extracted for both device orientations.

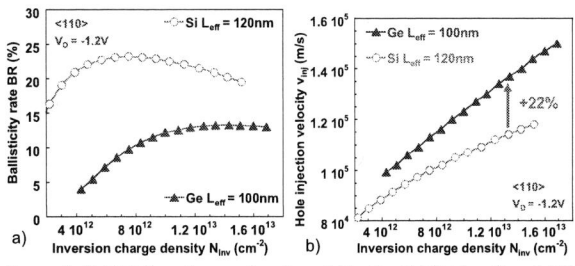

Figure 7. Ballisticity rate as a function of N_{inv} for GeOI (doped) and SOI (undoped) devices with similar L_{eff} and gate stack (left). Hole injection velocity vs. Inversion charge density (right).

CONCLUSION

We have demonstrated improved transport properties at the sides of mesa-isolated (001) GeOI pMOSFETs: +40% I_{LIN} and +65% μ_0 are reported for $W_{eff} = 0.29\mu m$ compared to 10μm. The gain is found to be independent of L_{eff}, and is attributed to surface orientations with higher hole mobility at the sides of the mesa. It is thus expected that GeOI-based "Ω-FET" devices can benefit from improved transport properties. Besides, <100> channel oriented GeOI devices were investigated for the first time: unlike Si, no current gain is observed in large width Ge devices. In narrow width devices, a smaller mobility improvement was extracted in <100> vs. <110> oriented devices, that was attributed to less favorable surface orientations at the edge of the mesa. Finally, quasi-ballistic transport was investigated in short channel GeOI devices. If the ballisticity rate was found to be affected by the non-optimized process conditions, +22% v_{inj} enhancement was reported for the first time in GeOI vs. SOI pMOSFETs, that underlines the potential of GeOI devices for the future CMOS technologies.

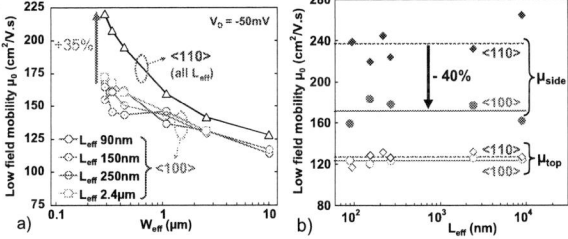

Figure 5. Low field mobility vs. W_{eff} for <100> and <110> channel devices. For the shortest W_{eff}, μ_0 is 35% higher in <110> oriented devices (left). Extracted μ_{top} and μ_{side} in <100> or <110> channel devices. A similar μ_{top} is found, but μ_{side} is 40% lower in <100> channel devices (right).

BALLISTICITY RATES AND INJECTION VELOCITY

In the ultimate nodes, quasi-ballistic transport is expected to be predominant and the injection velocity will be one of the main drivers of the device performance. Due to its lower effective masses and larger relaxation time τ, Ge is believed to provide a higher v_{inj} compared to Si [17,18]. Using the methodology described in [19], the ballisticity rate BR and source injection velocity v_{inj} were extracted in large width <110> oriented GeOI pMOS devices. Fig. 6 represents BR as a function of L_{eff} showing a continuous improvement when decreasing the effective length.

Due to the non optimized process conditions (channel doping, large interface state densities, etc.), the short channel GeOI pMOSFET exhibits a smaller BR compared to a SOI reference device (undoped) with similar L_{eff} and gate stack (Fig. 7-a). Nevertheless, v_{inj} is found to be +22% higher in Ge than in Si at the same N_{inv} as shown in Fig. 7-b.

REFERENCES

[1] K. Romanjek et al., *ESSDERC*, pp. 75-78, 2008.
[2] D. Kuzum et al., *IEDM*, pp. 723-726, 2007.
[3] T. Yamamoto et al., *IEDM*, pp. 1041-1043, 2007.
[4] G. Nicholas et al., *IEEE TED*, vol. 54, pp. 2503-2511, 2007.
[5] S. E. Thomson et al., *IEEE TED*, vol. 53, pp. 1010-1020, 2006.
[6] N. Collaert et al., *VLSI Tech.*, pp. 108-109, 2005.
[7] H. Sayama et al., *IEDM*, pp. 657-660, 1999.
[8] H. Kuribayashi et al., *J. Vac. Sci. Technol. A*, vol. 21, p. 1279, 2003.
[9] S. Bedell et al., *IEEE EDL*, vol. 29, pp. 811-813, 2008.
[10] A. Poudebsaque et al., *IEEE EDL, submitted*.
[11] E. Dornel et al., *APL*, vol. 91, 233502, 2007.
[12] D. S. Yu et al., *IEEE EDL*, vol. 26, pp. 118-120, 2005.
[13] J. Feng et al., *IEEE EDL*, vol. 28, pp. 637-639, 2007.
[14] Z. Krivocapič et al., *IEDM*, pp. 445-448, 2003.
[15] F. Andrieu et al., *VLSI Tech.*, pp. 50-51, 2007.
[16] T. Irisawa et al., *IEDM*, pp. 727-730, 2005.
[17] M. Uchida et al., *SISPAD*, pp. 315-318, 2005.
[18] S. Takagi, *VLSI Tech.*, pp. 115-116, 2003.
[19] V. Barral et al., *VLSI Tech.*, pp. 128-129, 2007.

ACKNOWLEDGMENT

This work was supported by the French National Research Agency (ANR) through Carnot Institute funding.

978-1-4244-2784-0/09 $25.00 © 2009 IEEE

Tri-gated Poly-Si Nanowire SONOS Devices

Hsing-Hui Hsu[a], Ta-Wei Liu[a], Chuan-Ding Lin[b], Chiu Kuo-Jung[c], Tiao-Yuan Huang[a] and Horng-Chih Lin[a, b, *]

[a]Department of Electronics Engineering and Institute of Electronics, National Chiao Tung University,
1001 Ta-Hsueh Road, Hsinchu, Taiwan 300, ROC; *E-mail: hclin@faculty.nctu.edu.tw
[b]National Nano Device Laboratories; [c]MSSCORPS CO., LTD.

INTRODUCTION

Si nanowire (NW) SONOS devices have recently been demonstrated as a good candidate for high-density non-volatile memory application [1][2]. Owing to the high surface-to-volume ratio of the NW channel, the programming and erasing (P/E) operation of the device could be performed at a lower voltage and much faster speed over the planar counterpart [2]. However, the fabrication of NW devices typically requires advanced lithographic tools and/or complicated process flow. These are not compatible with the manufacturing of flat-panel products where the device feature size is generally several microns or larger. In this work, we propose a simple and cost-effective approach to integrate planar poly-Si thin-film transistors (TFTs) and tri-gated poly-Si NW SONOS devices without resorting to advanced lithographic tools. Greatly enhanced P/E speed with the use of NW structure is clearly demonstrated.

DEVICE FABRICATION

Figure 1 shows the process sequence to fabricate poly-Si TFTs together with the tri-gated SONOS devices. In addition to the cross-sectional view of the completed device structure at each step, the top view of device layout at some key steps are also shown to help understand the process features. The fabrication began on Si wafers capped with thermal oxide used to simulate the glass substrate. First a 50nm-thick TEOS oxide and a 50nm-thick SiN were deposited on the wafer surface sequentially by LPCVD (Fig.1(a)). Next, an anisotropic etch step was performed to etch the SiN and TEOS oxide and form a dummy structure, as shown in Fig.1(b). Then an HF-based wet etching step was performed to selectively remove the TEOS layer and form cavities underneath the nitride at the sidewalls of the dummy structure (Fig.1(c)), followed by the deposition of an amorphous silicon (a-Si) layer by LPCVD (Fig.1(d)). In this step the cavities were fully filled with the a-Si layer which was subsequently transformed into polycrystalline phase with solid-phase crystallization treatment (at 600 $^{\circ}$C in N_2 for 12 hours). Then a lithographic step was used to define the active region including the source/drain and channel for both planar and NW devices, followed by an anisotropic etch step to define the active region (Fig.1(e)). Afterwards, the Si layers remaining in the cavities formed snugly the NW channels once the top hardmask and side TEOS oxide were selectively removed (Fig.1(f)). Then, an ONO (4nm/7nm/13nm) stacked layer and an n^+ poly-Si layer were sequentially deposited. Gate formation was subsequently performed. Self-aligned P^+ ion implant (30 keV, $10^{15}cm^{-2}$) was executed to dope the S/D region (Fig.1(g)). A TEM image of a NW SONOS device is shown in Fig.2. Height and width of the NW are around 50nm and 25nm, respectively.

RESULTS AND DISCUSSION

The use of ONO as the gate dielectric for the planar TFTs has been described previously [3]. Well-behaved device characteristics are confirmed in this work. Nevertheless, the NW devices show steeper subthreshold swing and better on/off current ratio than the planar counterparts, thanks to the better gate controllability of the NW channels. In addition, much better device fluctuation is demonstrated with the NW structure.

For P/E operations, a high voltage is applied to the gate while the source and drain are both grounded. Figure 3 shows the V_{TH} shift versus programming time of NW SONOS with gate bias of 11V, 12V, 13V, and 14V, respectively. For a fixed period, the V_{TH} shift increases with increasing gate bias. It can be seen that memory window of 3V can be achieved within 1 ms with applied voltage of 14V (Fig.4). We apply this operation condition to program fresh devices for investigation of erasing characteristics. Figure 5 is the result of V_{TH} shift as a function of erasing time with gate bias of -9V, -10V, -11V, respectively. The V_{TH} shift increases with increasing gate voltage and/or operation time. It can be seen that the rate of V_{TH} shift slows down when erasing time is longer than 100ms with gate bias of -11V.

Figure 6 compares the programming characteristics of planar and NW SONOS devices. To compare the programming capability, we apply the same gate bias of 14V to both devices. For NWs, V_{TH} shift is over 1V when the programming time is merely 1μs. On the contrary, V_{TH} shift of planar devices is less than 0.1V even when the programming time increases to 10ms. In order to obtain the same V_{TH} shift as the NW device at 10ms, a gate bias of 20V is needed for the planar device. Similar situations also occur in erasing characteristics, as shown in Fig.7. To attain comparable erasing speed to that of the NW device biased with gate voltage of -11V, the gate bias of planar devices must be raised to -20V.

Data retention characteristics of fresh devices measured at various temperatures are shown in Fig.8. If we program the devices by gate bias of 13V for 1ms, and -11V with 50ms for erasing, the memory window can be larger than 0.6V after 10 years at room temperature. The data lost paths of the fabricated SONOS devices are identified to be mainly through Frenkel-Poole emission and tunneling effect. Hence the temperature dependence is not evident. The endurance characteristics are shown in Fig.9. It is seen that the stressing cycles can be more than 10^4 with acceptable memory window. Nevertheless, window opening is observed in our measurements. As stressing cycles increase, the windows extend from 1.39 and 2.29V to 1.98 and 2.96V, respectively. To inspect the origins, we measured and inspected the evolution of subthreshold characteristics during the P/E operations. Figure 10 is the I_D-V_G characteristics of the device coinciding to the measured points shown in Fig.9. In the figure we conclude that the increase in subthreshold swing (SS) due to the generation of interface states is responsible for the increase in V_{TH} of the programmed state. In Fig.10, we also observe a negative shift in V_{TH} of the erased state, implying the occurrence of hole-trapping in the gate dielectric. To make it clear, we also characterized the retention characteristics of the device after 10^4 P/E cycles. The results are shown in Fig.11. It can be seen that, in the erased state, most of the negative V_{TH} shift recovers within the first 1000 sec. This indicates that most of the trapped holes are located close to the oxide/channel interface.

In summary, the method developed in the work for integration of planar TFTs and NW SONOS devices can be easily implemented in modern flat-panel manufacturing without resorting to costly lithography. Based on the results obtained in this work, the proposed method appears to be very promising for the realization of system-on-panel (SOP).

Acknowledgment −This work was supported by the National Science Council under contract No. NSC 95-2120-E-009-003.

REFERENCES

[1] S. D. Suk et al., Symp. VLSI Technol., p.142, 2007.
[2] J. Fu et al., IEEE EDL, vol.29, p.518, 2008.
[3] Szu-I Hsieh et al.,JJAP, vol.45, p.3154, 2006.

978-1-4244-2784-0/09 $25.00 © 2009 IEEE

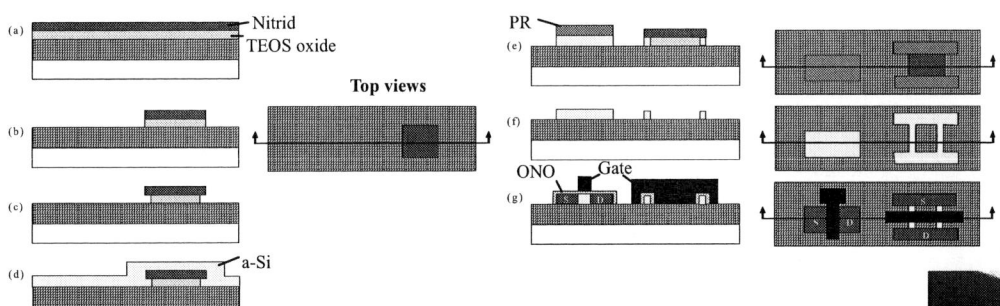

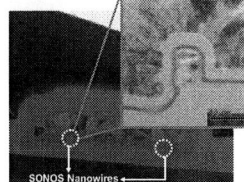

Fig. 1 Fabrication flow of the planar and NW devices. Top views at some key steps are shown in the right side. The straight lines denote the direction corresponding to the cross-sectional views shown in the left.

Fig. 2 Cross-sectional TEM image of a NW SONOS device.

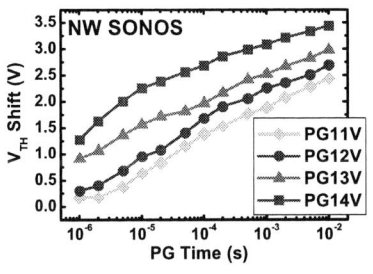

Fig. 3 V_{TH} shift as a function of programming (PG) time for NW SONOS devices stressed under different voltage. (L=0.7μm)

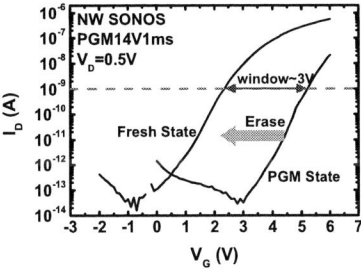

Fig. 4 Fresh and programmed subthreshold characteristics of a NW SONOS device. (L=0.7μm)

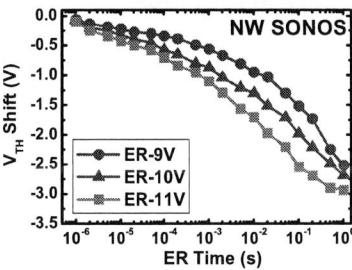

Fig.5 V_{TH} shift as a function of erasing (ER) time for NW SONOS devices stressed under different voltage. (L=0.7μm)

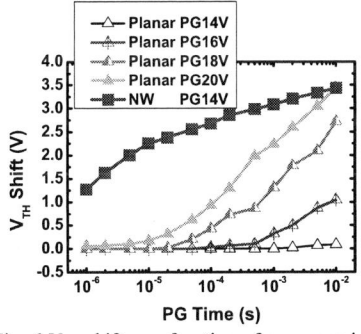

Fig. 6 V_{TH} shift as a function of programming time for NW (L=0.7μm) and planar (L/W=0.7/5μm) SONOS devices stressed under different voltage.

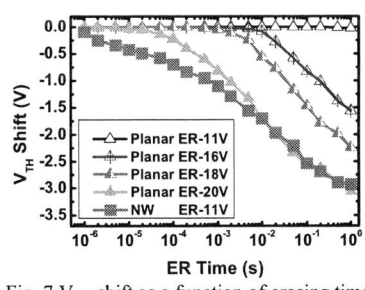

Fig. 7 V_{TH} shift as a function of erasing time for NW (L=0.7μm) and planar SONOS (L/W=0.7/5μm) devices stressed under different voltage.

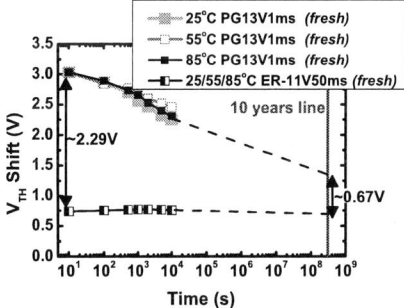

Fig.8 Retention characteristics of NW SONOS devices characterized at different temperature. (L=0.7μm)

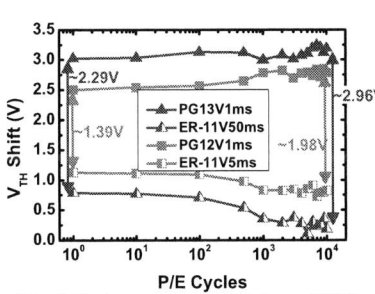

Fig. 9 Endurance characteristics of NW SONOS devices under two P/E conditions. (L=0.7μm)

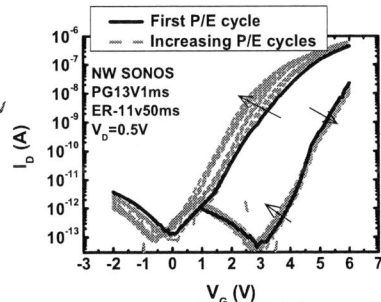

Fig. 10 Evolution of subthreshold characteristics of NW SONOS devices with increasing P/E cycles. The arrows indicate the evolution direction. (L=0.7μm)

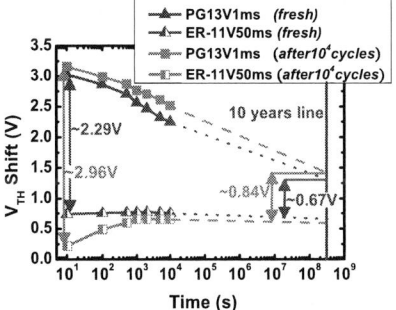

Fig.11 Retention characteristics of NW SONOS devices after 10^4 P/E cycles. Results for fresh device are also included for comparison. (L=0.7μm)

978-1-4244-2784-0/09 $25.00 © 2009 IEEE

Overall Operation Considerations for a SONOS-based Memory

C. H. Lee[1,2], W. H. Tu[2], L. H. Chong[1], S. H. Gu[1], K.F. Chen[1], Y. J. Chen[1], J. Y. Hsieh[1], I. J. Huang[1], N. K. Zous[1], T. T. Han[1], M. S. Chen[1], W. P. Lu[1], K. C. Chen[1], Tahui Wang[1,2], and C. Y. Lu[1]

[1]Macronix International Co., Ltd, No. 16, Li-Hsin Road, Science-Based Industrial Park, Hsin-Chu, Taiwan
[2]Dept. of Electronics Engineering, National Chiao-Tung University, Hsin-Chu, Taiwan
Phone: +886-3-5786688-78088 E-mail: nkzou@mxic.com.tw, shku@mxic.com.tw, twang@cc.nctu.edu.tw

ABSTRACT

Erase characteristics of a SONOS-based structure are emulated not only for n^+-poly and p^+-poly gates but also for TaN-gate+Al_2O_3 combination. By incorporating our previous studies, performances including program, erase, and read disturb can be reviewed for both SONOS and TANOS devices. Unsurprisingly, it is hard to satisfy all requirements by using a SONOS device. In a TANOS device, an optimal bottom oxide thickness can be specified with the consideration of the three factors simultaneously. Moreover, it is found that conventional extrapolation methodology is inadequate to predict the lifetime of a TANOS device and tends to under-estimate the tolerable read bias.

INTRODUCTION

Unlike a floating gate device, the structures of a SONOS-based device are still controversial. Various materials are tried to replace the poly-gate, the top oxide, and the nitride layer to improve the performances including program, erase, and read disturb [1-5]. Even for a TANOS device, which is the most mature structure among them, its optimal thickness of the three stacks is not concluded yet [1-2]. An accurate model is thus essential to predict operation performances of a SONOS based structure. In our previous studies, a transient model is developed and both program and read disturb behaviors have been well emulated [8-9]. In this paper, an erase model with the cogitation on electrons back tunneling (EBT) is developed. This unwanted gate injection process restricts the erase capability in a SONOS cell [1], [5]. Although much works have been demonstrated to improve erase performance by replacing n^+-poly with p^+-poly as gate [5-7] or using a TANOS structure [1-2], a consistent erase model for them had not been accomplished. In this paper, with the help of developed model, process margins for either a SONOS cell or a TANOS cell can be examined according to its program speed, read disturb, and erase performance inclusively.

RESULTS AND DISCUSSIONS

According to the schematic energy band diagram illustrates in Fig.1, a competition between gate electrons and substrate holes can be identified as a key factor to an erase characteristic. Other possible factors such as capture/emission currents contributed from nitride traps and currents tunneled out from a nitride layer are also indicated. However, in Fig.2, it is found that temperature dependence of an erase transient is irrelevant. Possible explanations are as follows; first, the hole barrier between nitride and SiO_2 [10] is large enough to retard out-going holes (J_{h-out}) from a nitride layer to the gate. Second, during erase operation, electric field across the bottom oxide decrease and out-going electrons (J_{e-out}) from nitride thus become less important. So, we rule out these components during calculation. Next, to check if electron (J_{EBT}) and hole (J_h) tunneling components can be described by simplified tunneling equations [10-11], erase characteristics of a SONOS cell with an n^+-poly gate and a 1.5nm bottom oxide are then emulated in Fig.3 and acceptable results are obtained.

Using a p^+-poly gate to reduce a gate injection current is a popular way [5-7]. Comparing with an n^+-poly gate, no inverted electrons can be supplied but only valence electrons, which are 1eV lower than that of an inverted one. Unfortunately, early erase saturation is still observed in literatures [5-7] and in our experiments

(Fig. 4(a)). According to the literature [12], the band-to-band tunneling process, which is shown in Fig. 4(b), could be a possible source for EBT. By taking this factor into account, in Fig. 4(a), a good agreement is found between simulation and measurement of a p^+-poly gate SONOS. With the proposed model, in Fig.5, V_T shifts before erase saturation occurs are compared for different ϕ_B under three applied gate biases with a same ONO stack (8nm/6nm/3nm). It is clear that the effect of using a p^+-poly gate does not increase the tunneling barrier of back tunneling electrons by 1eV but only about 0.4eV. That is why the reduction of erasure saturation is always limited by using a p^+-poly gate.

By incorporating our previous studies [8-9], performances including program, erase, and read disturb can now be reviewed for a SONOS based structure. In Fig. 6, the T_{ox} thickness, which meets both program speed and read disturb criteria, is ranged from 4.7nm to 5.1nm. However, it is shown in Fig.7 that erasing a cell to achieve ΔV_T=4V within 2ms, T_{ox} of 2.3nm or thinner is a must even with a p^+-poly gate. The other way is to increase the tunneling barrier of back tunneling electrons. In Fig. 8, erase capability versus the band offset ϕ_B is simulated under T_{ox}=4.8nm. To fulfill the erase requirement, ϕ_B should be as high as 4.25eV. To our best knowledge, suitable metal material is not found in semiconductor related publications.

Using a high κ material as a top-blocking layer can help to relieve the demand for a high work-function metal gate. A TANOS (TaN+Al_2O_3) cell is the most successfully case. In Fig.9, the data of reference [2] is quoted and our simulations fit its program and erase behaviors well. In simulation, a smaller electron tunneling probability at the top layer, which is due to a high κ dielectric always suffer a lower electric field, is the reason that the erase ability is enhanced. Moreover, in Fig. 10(a), the read retention time is also emulated and compared with reference [2]. Discrepancy is observed when the read bias is lower than 5V. In Fig. 10(b), the simulated time evolution at V_{read}=5V is plotted. This ΔV_T transient shows two distinct slopes, which means the tunneling mechanism of read disturb varies from direct tunneling (a trapezoid oxide barrier only) to modified FN tunneling (a trapezoid oxide barrier with a triangular nitride barrier) with time. As shown in Fig. 10(b), the existence of such phenomenon may lead to a large error when an extrapolated method is used. Cares should be taken during extrapolation. Fig. 11 shows the bottom oxide thickness effect on program, erase, and read disturb for a TANOS cell under EOT= 11.7nm. At this case, 5.75/6.0/5.6nm is suggested for an $Al_2O_3/Si_3N_4/SiO_2$ stack. Unfortunately, an Al_2O_3 film of 5.75nm may raise another issue such as anomalous leakage [12].

CONCLUSION

An erase model is developed and verified. For a p^+-poly gate, the gate injection procedure includes band-to-band tunneling in a poly Si followed by electrons tunneling through a top-blocking layer and the effective barrier increment is merely 0.4eV. Hence, it is hard to find an optimal thickness combinations of ONO stack to satisfy program, read disturb, and erase performances at a same time. Using a TANOS cell can help to relieve the demand for a high work-function metal. A 5.75/6.0/5.6nm is suggested for an $Al_2O_3/Si_3N_4/SiO_2$ stack under EOT=11.7nm. However, a very high quality Al_2O_3 is required at this case. By the way, cares should be taken when a read disturb is extrapolated at a TANOS cell.

978-1-4244-2784-0/09 $25.00 © 2009 IEEE

References

[1] C. H. Lee et al., *IEDM*, pp.613-616, 2003.
[2] C. H. Lee et al., *VLSI*, pp.21-22, 2006.
[3] Y. Q. Wang et al., *TED*, pp.2699-2705, 2007.
[4] Y. Q. Wang et al., *IEDM*, pp.162-165, 2005.
[5] H. Reisinger et al., *VLSI*, pp.113-114, 1997.
[6] C. W. Kuo et al., *MTDT*, pp.77-79, 2006.
[7] C. Friederich et al., *IEDM*, pp.963-9663, 2006.
[8] C. H. Lee et al., *NVSMW*, pp.109-110, 2008.
[9] C. H. Lee et al., *SSDM*, pp.826-827, 2008.
[10] H. Bachhofer et al., *JAP*, pp.2791-2800.
[11] S. M. Sze ,In *Physics of Semiconductor Devices* ,John Wiley, New York, 1981, p.520
[12] C. H. Lin et al., *IEEE TED*, pp.2125-2130, 2001.
[13] A. Furnemont et al., *Sol. State Elec.*, pp.577-583, 2008

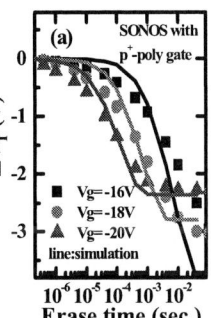

Fig.1 Schematic energy band diagram and current flows during erase operation for a SONOS cell.

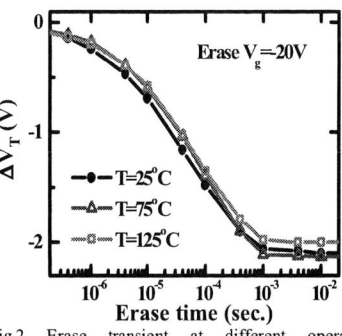

Fig.2 Erase transient at different operation temperatures (T=25°C, 75°C, and 125°C, respectively) for a SONOS cell with ONO stack is 8nm/6nm/3nm.

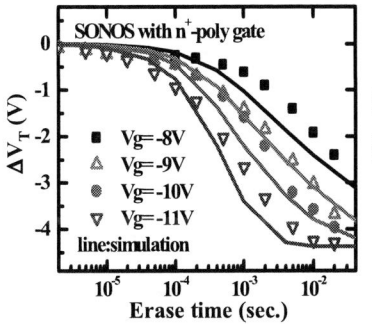

Fig.3 Measured (symbol) and simulated (line) erase transient at various V_g in a SONOS cell with n+ poly gate. ONO stack is 6nm/6nm/1.5nm.

Fig.4 (a) Measured (symbol) and simulated (line) erase transient at various V_g in a SONOS cell with p+ poly gate. ONO stack is 8nm/6nm/3nm; (b) Schematic energy band diagram of band-to-band tunneling mechanism.

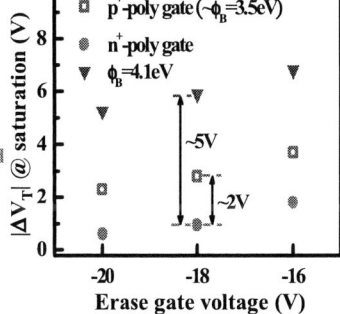

Fig.5 Compare the erase capability under various gate biases and ϕ_B. ϕ_B is the band offset between gate material and blocking oxide.

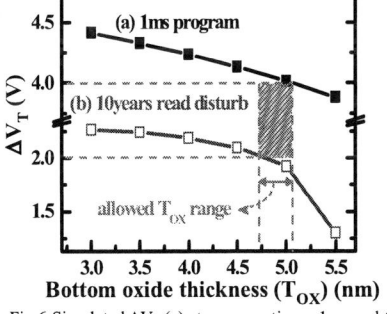

Fig.6 Simulated ΔV_T (a) at program time =1ms and (b) at disturb time=10years for different T_{ox} but under fixed EOT. Program voltage=17V and disturb voltage=5.7V.

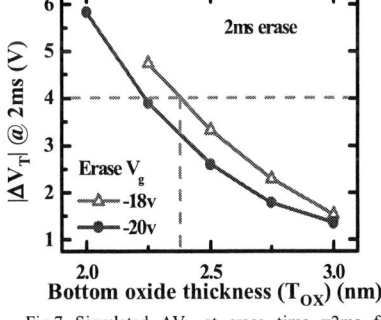

Fig.7 Simulated ΔV_T at erase time =2ms for different T_{ox} but under fixed EOT. Erase voltage is -18V and -20V.

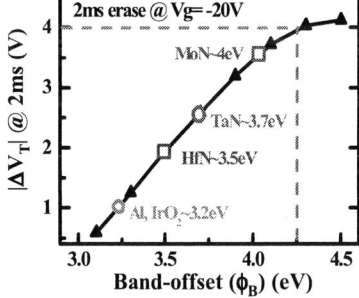

Fig.8 Simulated ΔV_T at erase time =2ms under erase voltage=-20V for various ϕ_B. ONO stack now is 6.2nm/6nm/4.8nm to satisfy the program and disturb criteria. Several metal materials are also labeled.

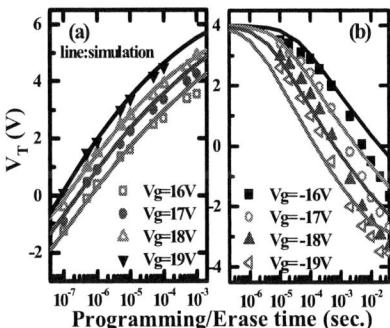

Fig.9 (a) Program and (b) erase characteristics for a TANOS cell. Symbols are reported by [2] and lines are our simulation results.

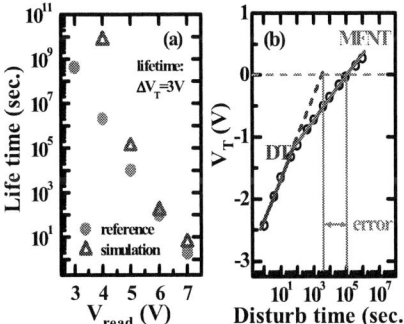

Fig.10 (a) Read retention of a TANOS cell. Solid symbols are quoted from [2] and open symbols are simulated results based on our previous work[9].

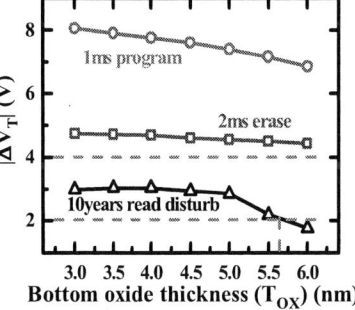

Fig.11 Optimal bottom oxide thickness range extraction for a TANOS cell with EOT=11.7nm. Program V_g= 17V. Erase V_g= -18V. Read V_g=4.25V.

978-1-4244-2784-0/09 $25.00 © 2009 IEEE

Reliability Study of MANOS with and without a SiO$_2$ Buffer Layer and BE-MANOS Charge-Trapping NAND Flash Devices

Chien-Wei Liao, Sheng-Chih Lai, Hang-Ting Lue, Ming-Jui Yang [a)], Chin-Yen Shen [a)], Yi-Hsien Lue [b)], Yu-Fong Huang, Jung-Yu Hsieh, Szu-Yu Wang, Guang-Li Luo [a)], Chao-Hsin Chien [a)], Kuang-Yeu Hsieh, Rich Liu and Chih-Yuan Lu

Macronix International Co., Ltd., Hsinchu, Taiwan, R.O.C., (E-mail: cwliao@mxic.com.tw)

[a)] National Nano Device Laboratories, [b)] Department of Electro-physics, National Chiao-Tung University, Hsinchu, Taiwan, R.O.C.

ABSTRACT

The reliability of MANOS devices with an oxide buffer layer (MAONOS) in between SiN trapping layer and high-K Al$_2$O$_3$ top dielectric is extensively studied. We conclude that the primary function of high-K Al$_2$O$_3$ is to suppress the gate electron injection during erase instead of increasing the P/E speed. As a result, inserting a buffer oxide only changes EOT but does not change the P/E mechanisms. On the other hand, the buffer oxide can greatly improve data retention by suppressing leakage through Al$_2$O$_3$. However, owing to the slow erase performances with a thick bottom oxide, both MANOS and MAONOS erase slowly and very high erase voltages must be used. Also, both MANOS and MAONOS devices show very fast endurance degradation below P/E<10, which is inherent due to electron de-trapping mechanism. Moreover, the large erase voltage also causes severe degradation of tunnel oxide after many P/E cycling. To get both speed and reliability performances, it is necessary to introduce bandgap engineered tunneling barrier (BE-MANOS) to solve the fundamental problems of MANOS.

I. INTRODUCTION

The difficulties in scaling floating gate NAND Flash will eventually require using the charge-trapping devices [1-2]. One of the most popular charge-trapping devices is the MANOS (or TANOS) device [3-7], where a high-K Al$_2$O$_3$ material is used to replace the top oxide of MONOS. However, there is often a misunderstanding [4] that the high-K top dielectric will increase the P/E efficiency of charge-trapping devices, an analogy drawn from the floating gate memory with a high-K IPD. In fact, for a planar charge-trapping device, the bottom oxide electric field is simply determined by $E_{O1}=(V_G-V_{FB})/EOT$. Therefore, if the EOT of MANOS is the same as that for the MONOS, the bottom tunnel oxide will have the same electric field, hence the P/E speed should be identical. In principle, Al$_2$O$_3$, with a higher K should give a lower EOT when it is used to replace the SiO$_2$ in MONOS. However, because of its higher leakage almost twice the thickness is used and the EOT stays the same.

We have previously found that the actual role of high-K top dielectric is to suppress the gate electron injection during –FN erase [8-9]. Due to the higher dielectric constant, the electric field in Al$_2$O$_3$ is reduced, thus the gate injection can be suppressed. Since FN tunneling barrier at high electric field is a narrow triangular barrier, it depends only on the interfacial region of top dielectric and metal gate. Therefore, it doesn't require very thick Al$_2$O$_3$, and by inserting a buffer oxide in between Al$_2$O$_3$ and SiN similarly low gate injection can also be achieved.

In this work, the impact of the buffer oxide for MANOS device is studied extensively and the performance and reliability are compared with BE-MANOS [8-9] using the bandgap engineered tunneling barrier.

II. SAMPLE PREPARATION

The processing for MANOS, MAONOS and BE-MAONOS samples was previously reported [8-9]. Before Al₂O₃ deposition, various oxide buffer layers were formed by wet oxidation of nitride. A typical TEM micrograph of MAONOS is shown in Fig. 1.

III. BASIC DEVICE CHARACTERISTICS

Figure 2 shows the retention of MAONOS with various buffer oxide thickness. For the same Al$_2$O$_3$ top dielectric (60 Å), thicker buffer oxide shows apparently improved data retention. This suggests that Al$_2$O$_3$ is leaky, and buffer oxide is efficient in blocking the leakage. On the other hand, Fig. 3 shows that a thicker buffer oxide decreases the erase speed at the same voltage. This is simply due to the larger EOT. We can use the transient analysis method [10] to extract the erase current density (J) versus the bottom oxide field (E), as shown in Fig. 4. It clearly shows that these samples have very similar J-E curves. This suggests that the buffer oxide thickness does not change the erase mechanism.

Figure 5 shows that the two MAONOS device with similar EOT (~14 nm) have identical erase speed at –20 V. This further proves that the

erase speed is independent of Al$_2$O$_3$ and buffer oxide thickness.

Figure 6 compares the breakdown characteristics. For a fair comparison of different EOT, the x-axis is plotted as electric field. It shows that both MANOS and MAONOS have breakdown field at ~20MV/cm (in O1), independent of the buffer oxide thickness.

In Fig. 3, for a sufficiently fast erase speed (J>10^{-3} A/cm^2), the required erase field should be greater than 18 MV/cm, which is quite close to the breakdown field. Thus MANOS inherently has serious reliability issues by operating near breakdown region.

Figure 7 compares the erase characteristics of various devices. It clearly shows that with the help of bandgap engineered tunneling barrier, BE-MAONOS shows a much faster erase speed at lower voltages, which enable the practical applications in Flash memory devices.

We further compare the programming characteristics using the Incremental Step-Pulse Programming (ISPP) method [11]. We find that the ISPP of MAONOS device is quite linear, and the slope is close to 0.85. This suggests that MAONOS have a good electron capturing efficiency. On the other hand, when the buffer oxide is thin (20Å), the programming has apparent degradation at higher V_{FB}. This suggests that a thin buffer oxide cannot stop the out tunneling of electron toward the gate, leading to degraded capture efficiency. Therefore, increasing the buffer oxide thickness or Al$_2$O$_3$ thickness is necessary to enhance the capture efficiency at higher injection level.

IV. Cycling Endurance and Retention Comparison

Figure 9 compares the cycling endurance of MAONOS, MANOS and BE-MAONOS. BE-MAONOS shows the best endurance. On the other hand, MAONOS has fast endurance degradation even below P/E<10, while MANOS has significant V_{FB} roll-off after 1K cycling.

Figure 10 further studies the endurance degradation of MAONOS. The results show that the endurance degradation is independent of the gate material or erased state. Interestingly, the endurance degradation is almost linear, quite different from Flash devices for which degradation rolls in after high numbers of cycling.

The reason for such fast endurance degradation of MAONOS is explained by Fig. 11, where the "refill" characteristics [12-13] are examined. The MAONOS device has gradually decreased erase speed after the refill operation. The refill characteristics are typically associated with the electron de-trapping process – the electron trapping energy spectrum shifts higher (blue shift) [12-13] and this in turn reduces the erase speed. On the other hand, BE-MAONOS does not show refill characteristics, because it is erased by hole tunneling through the bottom.

Figure 12 shows the CV curves of erased states after cycling. For MAONOS and MANOS devices, it shows significant distortion, indicating that the tunnel oxide is seriously damaged with D$_{it}$ generation. On the other hand, BE-MAONOS shows a much smaller distortion, owing to the much smaller voltages needed for P/E cycling.

The retention after various P/E cycling for MAONOS and BE-MAONOS are compared in Fig. 13. BE-MAONOS clearly shows better post-cycling retention because of much less device damage.

Finally, the schematic band diagrams to explain the erase mechanism of various devices are shown in Fig. 14. MONOS device has a large gate electron injection, leading to high erase saturation. Using Al$_2$O$_3$ top dielectric makes the gate injection barrier wider, such that erase saturation is lower. Inserting a buffer oxide does not change the gate injection at the same field. Both MANOS and MAONOS are erased by electron de-trapping out of nitride, which is very slow [14]. On the other hand, BE-MAONOS allows hole-tunneling injection, thus a faster erase speed is obtained.

V. SUMMARY

The reliability of MANOS devices with buffer oxide is studied extensively. Our results suggest that using BE-MAONOS is necessary to solve the fundamental issues of MANOS.

978-1-4244-2784-0/09 $25.00 © 2009 IEEE

References:

[1] K. Kim, *IEDM,* pp. 323-326, 2005.
[2] H.T. Lue *et al, IEDM,* pp. 547-550, 2005.
[3] Y. Shin *et al, IEDM,* pp. 327-330, 2005.
[4] C. H. Lee, *et al, IEDM,* pp. 613-616, 2003.
[5] E.S. Choi *et al, NVSMW,* pp. 83-84, 2007.
[6] T. Melde *et al, NVSMW,* pp. 130-132, 2008.
[7] L.Breuil *et al, NVSMW,* pp. 126-127, 2008.
[8] S.C. Lai *et al, NVSMW,* pp. 88-89, 2007.
[9] S.C. Lai *et al, NVSMW,* pp. 101-102, 2008.
[10] H.T. Lue *et al, EDL,* pp.816-818, 2004.
[11] H. T. Lue, *el al, IRPS,* pp.693-694, 2008.
[12] H.T. Lue *et al, IRPS,* pp.168-174, 2005.
[13] S.C. Lai *et al, EDL,* pp. 643-645, 2007.
[14] S.C. Lai *et al, VLSI-TSA,* pp. 14-15, 2007.

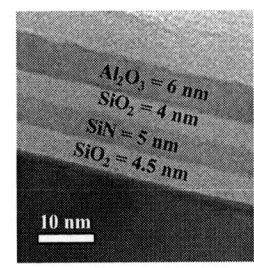

Fig. 1 TEM micrograph of MAONOS (MANOS with a SiO_2 buffer layer between SiN and Al_2O_3).This TEM micrograph was taken before metal gate deposition.

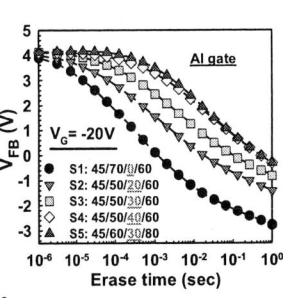

Fig. 2 Fresh-state (P/E=1) retention of MANOS with various thicknesses of SiO_2 buffer layer. Thicker buffer oxide can significantly improve data retention.

Fig. 3 Erase characteristics of MAONOS. Thicker buffer oxide causes "apparently" slower erase speed because of increased EOT.

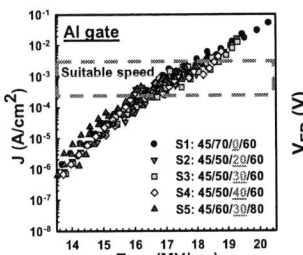

Fig. 4 Erase current density (J) versus the bottom tunnel oxide electric field (E), calculated from Fig. 3 using the transient analysis method [11]. All *J-E* curves fall onto a single consistent curve for MANOS and MAONOS devices.

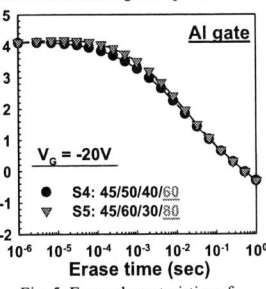

Fig. 5 Erase characteristics of MAONOS with different $O2/Al_2O_3$ thickness. EOT of these two devices are almost identical (~14nm).

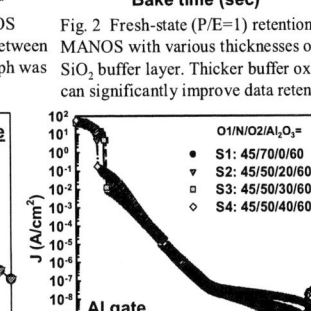

Fig. 6 The –Vg breakdown of MANOS and MAONOS. The x-axis is normalized to electric field for fair comparison of different EOT.

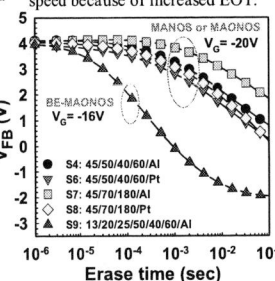

Fig. 7 Erase characteristics of MANOS, MAONOS and BE-MAONOS with different gate materials (Al or Pt).

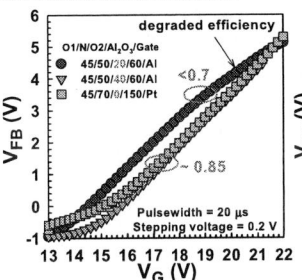

Fig. 8 Incremental stepping pulse programming (ISPP) of MANOS and MAONOS. ISPP is to constantly increase the programming bias after each step.

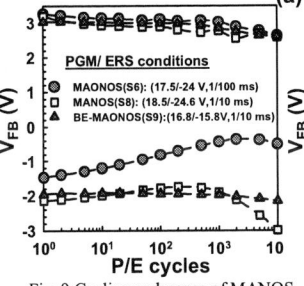

Fig. 9 Cycling endurance of MANOS, MAONOS, and BE-MAONOS. BE-MAONOS shows the best endurance. For the erased state, MAONOS has increased V_{FB}, while MANOS has decreased V_{FB} after 1K cycling.

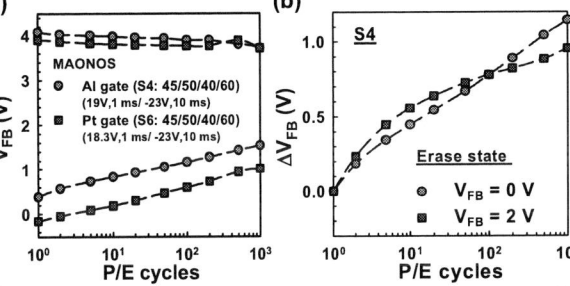

Fig. 10 (a) Endurance degradation of MAONOS with different gate material (Al and Pt). Both devices shows fast endurance degradation below P/E<10. (b) Endurance degradation of MAONOS with different erased state (V_{FB}=0 or –2 V) of Al-gate MAONOS (S4). Only erased state is shown. The programmed state is V_{FB}=4V.

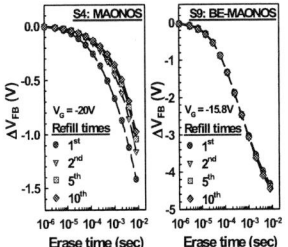

Fig. 11 Refill characteristics of MAONOS and BE-MAONOS. MAONOS shows decreased erase speed after repeated refill sequence, while BE-MAONOS shows no such refill effect. Also note the much lower erase voltage for BE-MAONOS.

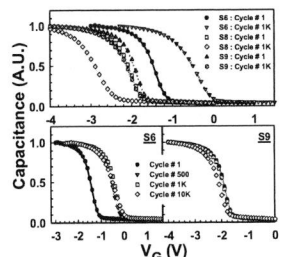

Fig. 12 C-V characteristics of MAONOS(S6), MANOS (S8) and BE-MAONOS (S9) during P/E cycling. MAONOS and MANOS shows large CV distortion after P/E cycling, owing to more severe stressing because of the need for higher voltage.

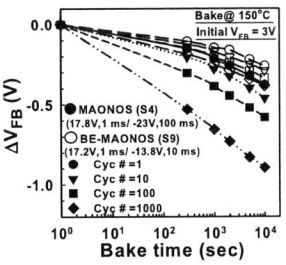

Fig. 13 Retention of MAONOS and BE-MAONOS after different P/E cycling. Cycling conditions for MAONOS and BE-MAONOS are adjusted to keep the same memory window (0 to 3V). Before 150 °C baking, all the devices are programmed to V_{FB}=3V.

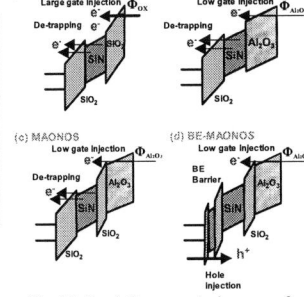

Fig. 14 Band diagrams during erase for (a) MONOS, (b) MANOS, (c) MAONOS and (d) BE-MAONOS. Due to the higher dielectric constant of Al_2O_3, the gate injection barrier is wider, thus gate injection is suppressed. Using a buffer oxide does not change the gate injection.

978-1-4244-2784-0/09 $25.00 © 2009 IEEE

RELIABILITY OF PLANAR AND FINFET SONOS DEVICES FOR NAND FLASH APPLICATIONS – FIELD ENHANCEMENT VS. BARRIER ENGINEERING

Tzu-Hsuan Hsu[*], Hang-Ting Lue, Sheng-Chih Lai, Ya-Chin King[*], Kuang-Yeu Hsieh, Rich Liu, and Chih-Yuan Lu

Emerging Central Lab., Macronix International Co., Ltd., 16 Li-Hsin Road, Hsinchu Science Park, Hsinchu, Taiwan.
[*]Also with Institute of Electronics Engineering, National Tsing-Hua University, Hsinchu, Taiwan
E-mail: Brucehsu@mxic.com.tw

ABSTRACT

The reliability of sub-40nm SONOS NAND devices with various tunnel oxide thickness and FinFET structures are studied for future NAND Flash application. SONOS intrinsically has slow erase speed and high erase saturation for tunnel oxide ranging from 25 to 45 Å. Furthermore, the endurance degradation occurs very early at low P/E<10, owing to the nature of electron de-trapping mechanism at tunnel oxide > 20A. Thus planar SONOS is not suitable for NAND Flash applications. On the other hand, when SONOS is applied to FinFET structure, significantly faster erase speed is obtained, owing to the field enhancement effect. However, it is still hard to erase below the initial Vt. We conclude that barrier engineering, such as BE-SONOS is more efficient in providing faster erase speed at lower erase voltages without endurance degradation. We also estimated the large density (4Mb) array distribution of sub-40 nm SONOS and BE-SONOS devices, and found that the distribution width is quite insensitive to the tunnel oxide thickness. This suggests that for future scaled NAND devices the edge effect is more important in determining the P/E distribution than the tunnel oxide thickness variation.

I. Introduction

SONOS-type devices are forecasted to be the promising solution of NAND Flash beyond 40 nm. However, SONOS devices easily suffer edge effect [1-3] which greatly changes the P/E performance and reliability properties. This effect was rarely considered in many previous literatures, leading to scattered results and equivocal conclusions.

In this work, we first characterize the intrinsic SONOS capacitors, since it is completely immune to any edge effect. It should be mentioned that N$^+$ source/drain are still necessary for the capacitors to avoid any possible substrate avalanche injection during programming [4].

Next, bulk FinFET structures are compared. It was found that FinFET structures naturally introduce a field enhancement effect [1,2,3,6,7] around the fin tip due to the curvature. Thus the enhanced erase performance creates a larger memory window for MLC applications. Since this corner edge effect causes faster speed, it also introduces more variations in P/E distribution. A 4Mb (one sector) distribution is estimated to understand the issues of FinFET SONOS devices. Finally, BE-SONOS devices are briefly compared with SONOS.

II. Intrinsic Characteristics of SONOS capacitors

We have fabricated the sub-40nm near-planar and FinFET SONOS NAND devices using the hard-mask trimming technique [2], as shown in Fig. 1. P$^+$-poly gate is used to reduce the gate injection during –FN erase. The total EOT of gate stacks for various samples are around 13 to 17 nm to meet the criterion of sub-40 nm NAND Flash operations.

We first investigate the capacitors to study the intrinsic characteristics. In Fig. 2(a), SONOS capacitor has a high erase saturation level at $V_{FB} \sim 3V$ even for a 25 Å O1. This is because when O1>25 Å hole injection is negligible and electron de-trapping is the major erase mechanism [8]. Electron de-trapping is slow and gate electron injection is even larger, leading to the high erase saturation [5]. On the other hand, Fig 2(b) shows that BE-SONOS has a much faster erase speed at lower erase voltages. This indicates that hole injection through the bandgap engineered O1/N1/O2 barrier is much more efficient for the erase [8].

Next, incremental-step-pulse programming (ISPP) [7] is studied. ISPP is to constantly increase the programming voltages after each step. ISPP is often plotted as V_T versus V_{PGM}, which is often a linear curve with a constant slope. Figures 3 shows that the ISPP slope of SONOS is constant (~0.8) and independent of the O1 thickness. In Fig. 4, BE-

SONOS also shows a nearly ideal ISPP slope (~0.9) for various O1 or EOT. The constant ISPP slope is very important in controlling a tight Vt distribution, since it provides a constant Vt shift during ISPP steps, irrespective of thickness variations.

Figure 5 compares the P/E endurance of SONOS and BE-SONOS capacitors. We find that SONOS has a fast endurance degradation even at low P/E cycling (<10). This fast endurance degradation happens for all tunnel oxide thickness (not shown here), and it is not related to the tunnel oxide damages. On the other hand, BE-SONOS does not show such fast endurance degradation.

The reason of the fast endurance degradation of SONOS is due to the nature of electron de-trapping mechanism. A simple experiment is shown in Fig. 6, where "refill" method [5,8] is tested. Because shallowly trapped electrons are reduced owing to the spectrum blue shift [5] SONOS becomes hard-to-erase after only a few cycles. On the other hand, hole injection does not shift the trapped electron energy thus there is no endurance degradation for BE-SONOS.

III. The sub-40 nm SONOS NAND Devices

Figure 7 shows that the ISPP slope is generally degraded (<0.6) in a NAND devices. Previous analysis indicates that the ISPP slope degradation originates from the edge effect that changes the local electric field and causes non-uniform injection in NAND devices [1-3].

Figure 8 shows that the erase speed can be enhanced when more STI recess is introduced (bulk FinFET structure). This suggests that with the help of field enhancement, SONOS can be erased to near the initial Vt, as shown in Fig. 9(a). On the other hand, BE-SONOS shows a much faster erase speed (Fig. 9(b)), thus offering much more memory window (peak-to-peak window ~8V) than SONOS.

Read disturb is compared in Fig. 10. It shows that thicker O1 has a larger read disturb immunity. BE-SONOS read disturb is in between SONOS with O1=35 and 45 Å.

IV. P/E distribution Analysis

Several hundreds of cells are collected to estimate the Gaussian distribution in a 4Mb array (one sector). Figure 11 shows that the dumb-mode P/E distribution is generally wide, and some tail distribution happens for SONOS. Although FinFET SONOS has a much faster erase speed than planar SONOS, the erase distribution is very wide and higher than the initial Vt distribution such that overall window is small.

O1=25 Å shows very poor data retention and can not be considered in Flash applications, while O1=45 Å shows a reasonable retention for Flash applications.

It is interesting that the P/E distribution width is relatively insensitive to the O1 thickness, as shown in Fig, 11(a)(b). In order to clarify it, we compare the dumb-mode programmed state distribution for FinFET SONOS and BE-SONOS with various tunnel barrier, as shown in Fig. 12. Interestingly, the programmed distribution is almost identical (~2.5V) for various tunnel barrier. This suggests that for very scaled devices the edge effect and geometry variation plays more significant role in determining the total P/E distribution than the ONO thickness variation.

V. Summary

SONOS with various tunnel oxide thickness and FinFET structures are studied extensively in this work. BE-SONOS shows superior erase and endurance performance, and is much more feasible in NAND Flash applications. Furthermore, the edge effect variation will become the important concern in the future scaled NAND Flash devices.

References:

[1] H.T. Lue et al, IEDM, pp. 161-164, 2007.[2] T.H. Hsu et al, *IEDM*, pp. 913-916, 2007. [3] H. T. Lue et al, VLSI, pp. 116-117, 2008. [4] T. Ishida et al, IRPS, pp. 516-522, 2006. [5] H.T. Lue et al, IRPS, pp. 168-174, 2005. [6] T.H. Hsu et al, *NVSMW*, pp. 115-116, 2008. [7] H. T. Lue et al, IRPS, pp. 693-694, 2008. [8] P. Y. Du, et al, IRPS, pp. 399-405, 2008.

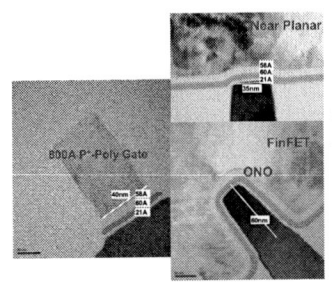

Fig. 1 TEM pictures of the sub-40nm SONOS NAND devices with near-planar or FinFET (with STI recess~600Å) structure.

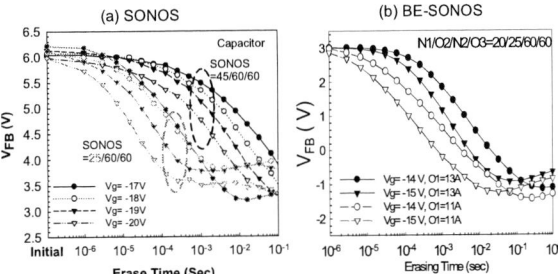

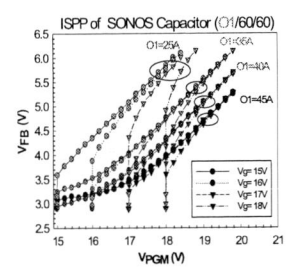

Fig. 3 ISPP characteristics of SONOS capacitor with various O1 thickness. All the devices are programmed from erased sate (V_{FB}~3V). Thinner O1 requires lower programming voltages, but the ISPP slope are all quite similar (~0.8).

Fig. 2 (a) Erase characteristics of SONOS capacitors with O1=45 and 25Å. Although thinner O1 provides better erase speed, SONOS has a very high erase saturation (V_{FB}~3V) even with a 25Å O1. Therefore, SONOS must be programmed to higher V_{FB}~6V for the erase. (b) Erase characteristics of BE-SONOS. Compared with SONOS, BE-SONOS shows a much faster erase speed and lower erase saturation, and can be operated at much lower erase voltages.

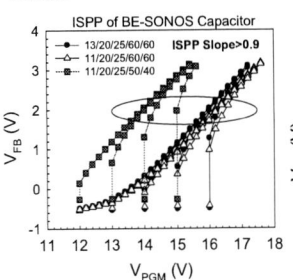

Fig. 4 ISPP characteristics of BE-SONOS capacitors with various thickness. All the devices are programmed from erased state. All the devices have similar ISPP slope ~0.9, independent of O1 or total thickness. Thinner EOT requires lower voltages.

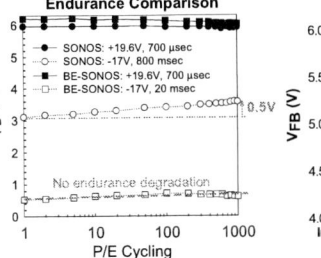

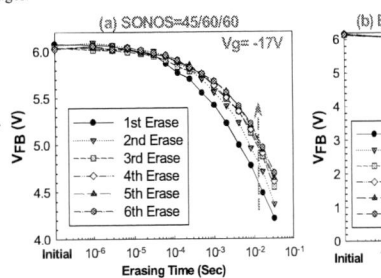

Fig. 5 Endurance comparison of (a) SONOS (45/60/60) and (b) BE-SONOS (13/20/25/60/60) capacitors. SONOS shows fast endurance degradation even at low P/E cycling (<10), while BE-SONOS does not show any endurance degradation.

Fig. 6 Refill characteristic comparison of (a) SONOS (45/60/60) and (b) BE-SONOS (13/20/25/60/60). Refill method is to repeatedly erase and then program back to the same PV level [5]. SONOS shows gradually decreased erase speed after refill method, because the erase mechanism is the electron de-trapping and shallowly trapped electrons are reduced after repeated erase [5]. On the other hand, BE-SONOS shows repeated erase performance, because the erase mechanism is hole injection [8].

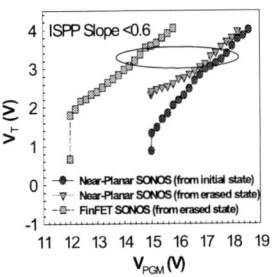

Fig. 7 The ISPP characteristics of 3X-nm SONOS NAND devices. In general, the ISPP slope (<0.6) has significant degradation than the intrinsic capacitors (~0.8), owing to the STI edge effect [5]. FinFET SONOS shows lower programming voltages due to the field enhancement effect [1,2,4].

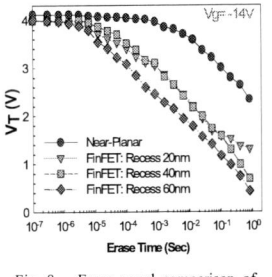

Fig. 8 Erase speed comparison of SONOS (25/60/60) with various STI recess. The near-planar structure shows the slowest erase speed, while FinFET with more STI recess shows faster erase speed. The effect of STI recess saturates at larger recess depth.

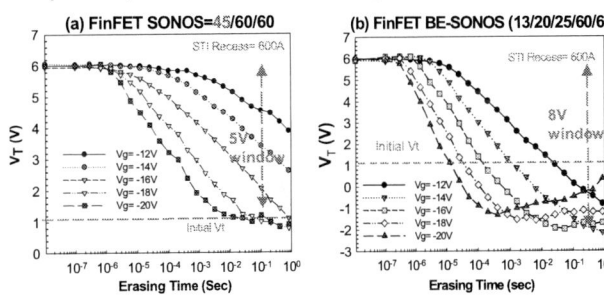

Fig. 9 Erase comparison of (a) FinFET SONOS and (b) FinFET BE-SONOS. All the devices have the same 600Å STI recess. With the help of field enhancement, SONOS with a thick O1 can be erased to nearly initial Vt. On the other hand, BE-SONOS has a much faster erase speed and can be erased far below initial Vt.

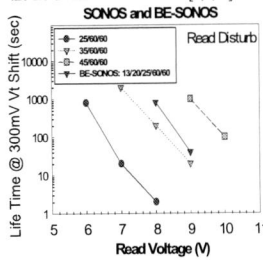

Fig. 10 Read disturb comparison of SONOS and BE-SONOS. For SONOS, thinner O1 shows larger read disturb. The read disturb of BE-SONOS is in between O1=35 and 45Å SONOS.

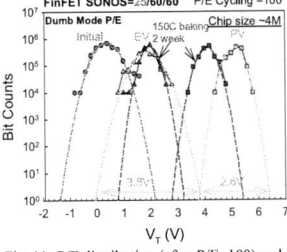

Fig. 11 P/E distribution (after P/E=100) and retention comparison of FinFET SONOS with (a) O1 = 25Å and (b) O1=45Å. A few hundred cells are collected to simulate a 4Mb sector Gaussian distribution. Both programming and erasing are carrier out by dumb-mode operation without any algorithm and verify. For all the devices, programmed state is slightly narrower than the erased state. The erased state is hard-to-erase lower than the initial Vt even with the field enhancement effect in FinFET structure. O1=25Å show very poor data retention. Interestingly, the P/E distribution width is quite similar for different O1 thickness.

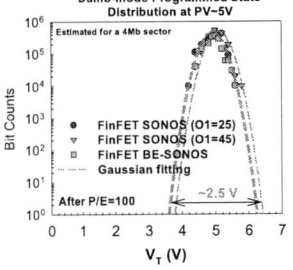

Fig. 12 Dumb-mode programmed state distribution at PV~5V for FinFET SONOS and FinFET BE-SONOS (13/20/25/80/60). All the devices are similar sub-40nm FinFET structures (but not optimized). Interestingly, the distribution width (~2.5V) is insensitive to the tunnel barrier.

978-1-4244-2784-0/09 $25.00 © 2009 IEEE   143

Band Engineered Tunnel Oxides for Improved TANOS-type Flash Program/Erase with Good Retention and 100K Cycle Endurance

David C. Gilmer, [a]Niti Goel, [b]Sarves Verma, Hokyung Park, Chanro Park, Gennadi Bersuker, Paul D. Kirsch, [b]Krishna C. Saraswat and Raj Jammy

SEMATECH 2706 Montopolis Drive, Austin, TX 78741, U.S.A., [a]Intel, [b]Stanford University, Email: david.gilmer@sematech.org

ABSTRACT

We demonstrate for the first time improved program, erase, and endurance for charge trap flash TaN-Al_2O_3-Si_3N_4-"Tunnel-oxide (TO)"-Si MOSFETs through band engineered tunnel oxides (BE-TO). Several high-K dielectrics (HfO_2, HfSiO, Al_2O_3, Si_3N_4) and tunnel stack sequences (SiO$_2$-high-k, SiO$_2$-high-k-SiO$_2$) are compared. New results are as follows: SiO_2/Al_2O_3 (OA) BE-TO and $SiO_2/Si_3N_4/SiO_2$ (ONO) BE-TO ΔVth windows improve >300% vs. standard SiO_2-TO. Both OA and ONO stacks endure P/E cycles to at least 100K cycles maintaining a window >4V. Results are consistent with a model based on high-k conduction/valence band offsets. Increased erase efficiency for BE-TO enables improved endurance without sacrificing P/E window due to lower P/E voltage stressing. These large, enduring windows are favorable for multi-level cell application and may extend TANOS flash beyond the 20nm node.

INTRODUCTION

TANOS flash devices[1,2] may replace floating gate structures for sub 30nm node technology[3]. Typical TANOS SiO_2-TO must be sufficiently thin for high electric fields to induce electron tunneling into the silicon nitride (SiN) charge storage layer, but thick enough to prevent "trapped" electrons from escaping when the bias is removed. TO thickness controls the trade-off between charge retention and P/E efficiency. With a thin TO, P/E speed is high, while charge retention is low. With thicker TO, P/E speed is slow, but charge retention is improved. This TO trade-off can force higher P/E bias stress, which then compromises retention and/or endurance. To address this trade-off, BE-TO have been proposed [4,5] and demonstrated in SONOS[6]. However, a systematic comparison of HfO_2, Al_2O_3, Si_3N_4, HfSiO in the BE-TO has not been done. In this work, we experimentally demonstrate the BE-TO in TANOS MOSFETs and compare the previously mentioned high-k materials, demonstrating P/E window and endurance benefits.

EXPERIMENT

TANOS flash MOSFETs (NMOS) were formed using 40A SiO_2-TO as an experimental control. Various BE-TO structures of symmetric OXO (SiO$_2$/high-K/SiO$_2$) or asymmetric OX (SiO$_2$/high-K) layers were also fabricated into a TANOS type device (Fig. 1) using Hf(Si)O$_2$, Al_2O_3, or Si_3N_4 as the high-K and labeled as OHO, OAO, ONO or OA respectively. All devices used the same fabrication steps for the 6nm SiN charge trap, 11nm Al_2O_3 blocking layer, and TaN electrode. Dopant activation was done with a 1020°C-spike anneal. All P/E measurements were done using FN methods.

DISCUSSION

Given the band offsets vs. Si for HfO_2, Si_3N_4, and Al_2O_3,[7] the programming efficiency for these high-K dielectrics as "X" in OXO should follow HfO_2>Si_3N_4>Al_2O_3. Ease of hole injection (erase) should follow Si_3N_4>HfO_2>Al_2O_3. Figure 1a,b shows OXO BE-TO TANOS stacks. MOSCAP results in Fig. 2 show P/E efficiency improves as expected based on the OXO band offset arguments made above. The larger P/E window comes at the cost of lower retention for ONO, and significant retention loss for OHO (Fig. 3). The larger retention loss in OHO may be due to shallow traps in

Hf(Si)O$_2$ dielectrics[8]. The layer thicknesses and band offsets are important levers to modulate P/E.

As shown in Fig. 4, OAO-TO and OA-TO program more efficiently than SiO_2-TO consistent with band offset arguments. The OA device programs >4V with only a 17V 1μs pulse. Program Vth shift as a function of effective field shows similar behavior for OAO and OA, but lower efficiency for SiO_2-TO. This similarity suggests lower SiO_2 plays a dominant program role. The band offsets also suggest OAO and OA should erase less efficiently than SiO_2-TO. However, Fig. 5 shows OA erases faster than SiO_2-TO. The speed of OA erase vs. OAO also suggests trapped electrons are primarily in the SiN storage layer, not the Al_2O_3. Figure 6 shows 100K cycle endurance for OA where a P/E window >4V is maintained while the window and endurance for SiO_2-TO is much poorer failing near 2K cycles. When larger bias/time was used for SiO_2-TO to achieve a similar P/E window as the AO device, it survived less than 100 cycles.

The ease of erase suggested by band offsets for ONO structures, and shown in Fig. 2, are also shown in the MOSFET data for Fig. 5. After programming the ONO device to ΔVth=6V, the ONO can be 100% erased in just 1μs and -19V. This erase characteristic results in superior ONO endurance vs. SiO_2-TO. 100K cycles and >4V P/E window is achieved in Fig. 7. The ONO program speed does not improve vs. SiO_2-TO, but the speedy erase compensates resulting in larger effective P/E window. Figure 4, 5 compare the P/E speeds for ONO, OHO, OAO, OA , and SiO_2-TO. The greater erase efficiency for ONO and OA BE-TO leads to 300% and 400% larger effective P/E window with minimal retention penalty vs. SiO_2-TO as seen in Fig. 8. For the +/-17V P/E, the SiO_2-TO 2V window with 11% charge loss would reduce to 1.8V. However, even with higher charge loss of 14% and 27% for ONO and OA respectively, they still have much larger remaining windows after charge loss at 5.2V or 5.8V respectively.

CONCLUSION

We demonstrate for the first time improved program, erase, and endurance for CTF MOSFET devices through BE-TO engineering. The use of various dielectrics (Al_2O_3, HfO_2, Si_3N_4) verify the band engineering model. A 400% P/E Vth window improvement was realized using OA BE-TO, and 300% improvement for ONO BE-TO. Both OA and ONO stacks endure P/E cycling past 100K cycles maintaining a window >4V with good retention. Increased erase efficiency for BE-TO is responsible for the increased P/E window and endurance cycles. This concurrent improvement in erase and endurance with OA and ONO may address the key issues limiting charge trap flash (erase and endurance) and enable multi-level cell application and voltage scaling beyond the 20nm node.

REFERENCES

[1] Y-C. Shin, et al., *IEDM*, p. 327, 2005.
[2] C.H. Lee, et al., IEDM, p.26.5.1, 2003.
[3] K. Prall IEEE NVSMW, p. 5, 2007.
[4] K. K. Likharev, Appl. Phys Lett., P.2137, 1998.
[5] B. Govoreanu, et al., IEEE EDL, Vol. 24, No. 2, p.99, 2003.
[6] H. T. Lue, et al., IEDM, 2005.
[7] J. Robertson, JVST-B 18 p.1785, 2000.
[8] D. Hei, et al. IEEE TED p. 1338, 2007.

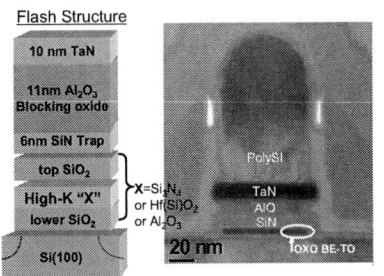

Figure 1(a) BE-TO TANOS schematic, and 1(b) TEM for a fabricated OHO BE-TO TANOS MOSFET.

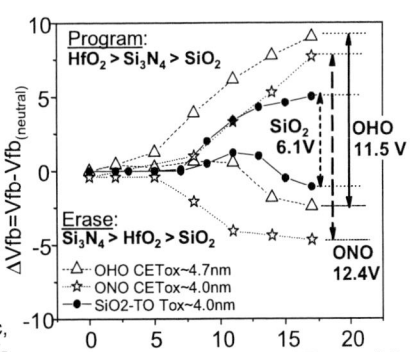

Figure 2. P/E window comparing various OXO BE-TO with X=Si₃N₄ or HfO₂ with SiO₂-TO in TANOS MOSCAP devices.

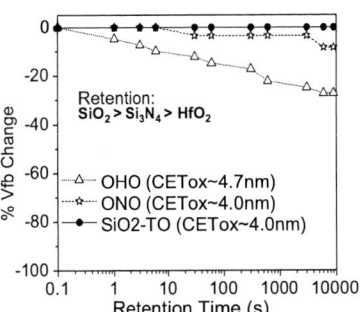

Figure 3. Retention measurements comparing various OXO BE-TO with X= Si₃N₄ or HfO₂ with SiO₂-TO in TANOS MOSCAP devices.

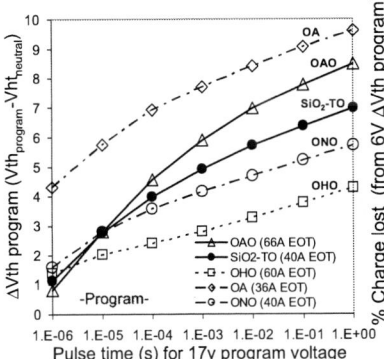

Figure 4. Program speed for 17V pulses comparing OA, OAO, ONO, OHO BE-TO with SIO2-TO in TANOS type Flash MOSFETs.

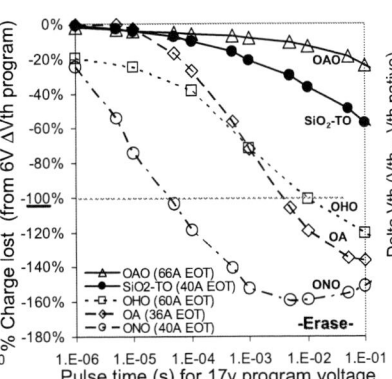

Figure 5. Erase speed for 17V pulses comparing OA, OAO, ONO, OHO BE-TO with SiO2-TO in TANOS type Flash MOSFETs. Only OHO, ONO, and OA BE-TO can 100% erase in 10ms or less.

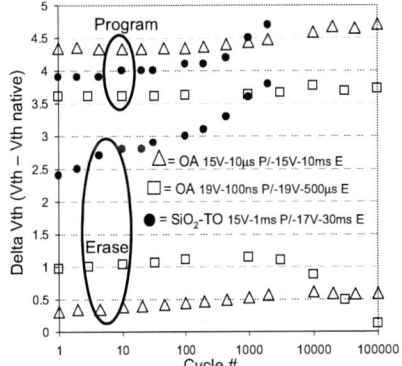

Figure 6. Endurance comparing OA BE-TO and SiO2-TO MOSFET TANOS showing much longer cycling and larger window for OA BE-TO. The high P/E efficiency for OA allows even nano-s/micro-s P/E speeds and also reduces bias stress leading to >100K endurance with large remaining window.

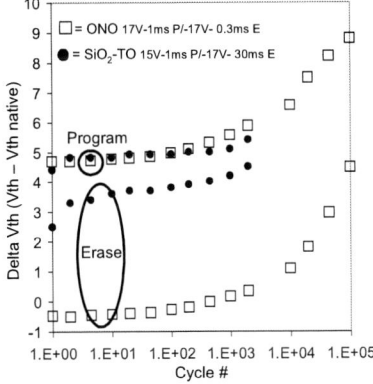

Figure 7. Endurance comparing ONO BE-TO and SiO2-TO MOSFET TANOS showing much longer cycling and larger window for OA BE-TO which is a direct benefit of improved erase efficiency.

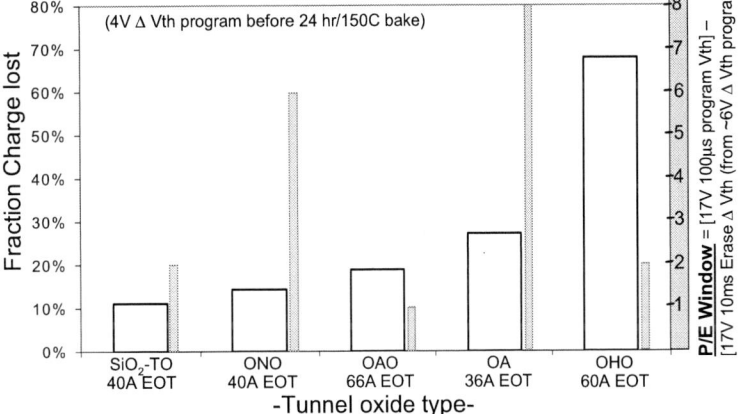

Figure 8. Retention as fractional charge lost from a 4V ΔVth program after 24 hours at 150C (open bars) and Effective +/-17V P/E window in volts (grey bars) comparing OA, OAO, ONO, OHO BE-TO with SIO2-TO (with approximate tunnel oxide EOT listed). Over 400% P/E window improvement realized using OA BE-TO, and >300% improvement for ONO BE-TO vs. SiO2-TO.

978-1-4244-2784-0/09 $25.00 © 2009 IEEE

AUTHOR INDEX

Absil, P. 43, 63
Adelmann, C. 63
Adhikari, Hemant 125
Afzalian, Aryan 115
Akasaka, Yasushi 19
Allain, F. 134
Aoki, H. 9, 11
Aoulaiche, M. 43, 63
Arikado, Tsunetoshi 19
Aritome, S. 31
Augendre, E. 134
Balan, V. 81
Bao, T. I. 45
Barnett, J. 53
Barral, V. 134
Baud, L. 81
Bécu, S. 102
Bersuker, G. 53, 55, 59, 144
Bidal, Grégory 100
Biesemans, S. 43, 63, 73
Boccardi, G. 51, 57
Boeuf, Frédéric 71
Bouvet, D. 109
Breitwisch, M. 27
Brut, Hugues 100
Bryant, Andreas 117
Bulle-Lieuwma, C. W. T. 51
Burenkov, A. 15
Cai, Jin 77
Carron, V. 81
Cayrefourcq, Ian 125
Chan, Kevin 117
Chang, C. C. 128
Chang, Josephine 117
Chang, P. 130
Chang, S. M. 88

Chang, S. Z. 63
Chang, W. H. 121
Chang, Y. C. 121
Chang, Y. H. 121
Chao, Der-Sheng 35
Cheek, R. 27
Chen, C. Y. 7, 31
Chen, Chih-Wei 35
Chen, F. 29, 33, 35
Chen, H. C. 45
Chen, H. H. 7
Chen, J. 83, 88
Chen, K. C. 138
Chen, K. F. 138
Chen, M. S. 138
Chen, Pang-Shiu 33
Chen, Wei-Chen 111
Chen, Wei-Su 35
Chen, William P. N. 75
Chen, X. 83
Chen, Y. J. 138
Chen, Y. S. 29, 33
Chiang, C. H. 130
Chiang, P. J. 31
Chiang, W. T. 79
Chiarella, T. 73
Chien, Chao-Hsin 140
Chin, Hock-Chun 132
Chiu, H. C. 121, 130
Cho, Jin 117
Cho, M. J. 43
Choi, Donghun 113
Chong, L. H. 138
Chor, Eng Fong 67
Chowdhury, M. 83
Chu, L. K. 128

AUTHOR INDEX

Chu, R. L. 128

Chuang, Ching-Te 69, 106

Chudzik, M. 83

Colinge, Jean-Pierre 115

Collaert, N. 73

Collonge, M. 102

Coss, Brian 125

Cousin, Bastien 98

Cros, Antoine 100

Cueto, O. 81

De Keersgieter, A. 73

De Michielis, L. 109

De Vries, René Penning 3

Dehdashti, Nima 115

Deleonibus, S. 81

Dennard, Robert. H. 77

Descombes, S. 81

Devi, Sivasubramaniam Nandini 67

Dip, Anthony 19

Divakaruni, R. 83

Dobrosz, P. 109

Dokumaci, Omer 117

Eller, M. 83

Ernst, T. 102

Escanes, P. 51

Fan, J. J. 7

Fan, Ming-Long 106

Fang, T. Y. 88

Faynot, O. 81

Feng, Philip 42

Ferain, I. 73

Fleury, Dominique 100

Fühner, T. 15

Fujita, H. 7

Fukunaga, Tetuya 13

Gehring, Andreas 117

Ghibaudo, G. 100, 102

Gilmer, David C. 144

Goel, Niti 144

Goethals, Mieke 90

Grampeix, H. 134

Grogg, Daniel 119

Gu, S. H. 138

Guan, Ximeng 113

Guillorn, Michael 117

Haensch, Wilfried 77, 117

Haetty, Jens 83

Han, J. P. 83

Han, S. 83

Han, T. T. 138

Hara, M. 9, 11

Harada, Y. 53

Harris, Harlan R. 125

Harris, James S 113

Hartmann, J.-M. 134

Hashimoto, K. 23

Hatzistergos, M. 83

Hendrickx, Eric 90

Heng, C. H. 104

Hoffman, T. 73

Hoffmann, T. 43, 63

Holt, Judson 117

Hong, J. M. 121, 128, 130

Hsiao, Yi-Hsuan 94

Hsieh, J. Y. 138, 140

Hsieh, Kuang-Yeu 94, 140, 142

Hsu, C. C. 7

Hsu, Hsiao-Hsuan 61

Hsu, Hsing-Hui 136

Hsu, Shawn S. H. 130

Hsu, Tzu-Hsuan 142

Hsu, Yen-Ya 35

AUTHOR INDEX

Hu, Chenming 96
Hu, Jenny 113
Hu, Vita Pi-Ho 106
Huang, C. H. 7
Huang, I. J. 138
Huang, J. 53, 59
Huang, R. M. 21, 79
Huang, Ru 123
Huang, Tiao-Yuan 111, 136
Huang, Yu-Fong 140
Huguenin, J. L. 51
Hung, C. H. 31
Hussain, M. M. 53
Hwang, H. P. 31
Hwang, Wei 69
Ikeda, K. 23
Ionescu, A. M 109, 119
Iyer, Subramanian 108
Jaeger, D. 83
Jammy, Raj 19, 53, 59, 125, 144
Jang, Simon 25
Jaud, Marie-Anne 98
Jeng, P. R. 25
Jiang, J. D. 7
Joe, Raymond 19
Jomaah, Jalal 98
Jonckheere, Rik 90
Josse, Emmanuel 100
Jurczak, M. 73
Kaczer, B. 63
Kaitsuka, Takanobu 19
Kampen, C. 15
Kanarsky, T. 83, 117
Kang, C. Y. 53, 59
Kang, Zhaoyi 123
Kao, M. J. 29

Kao, Ming-Jer 35
Kase, M. 23
Kauerauf, T. 43
Kelkar, P. 43
Kerner, C. 73
Khanal, P. 59
Khare, M. 83
Khater, Marwan 117
Kim, K. 83
Kim, N. 83
Kim, S. 83
Kimura, C. 9, 11
King, John 117
King, Ya-Chin 142
Kirsch, P. 53, 59, 144
Knoch, Joachim 40
Kobayashi, H. 7
Koester, Steven 77
Krecinic, Faruk 88
Krishnamohan, Tejas 125
Kubicek, S. 43
Kumar, Arvind 77, 117
Kuo, Jack J. Y. 75
Kuo, T. F. 21
Kuo-Jung, Chiu 136
Kurata, H. 23
Kwo, J. 121, 128, 130
Lachal, L. 134
Lai, C. S. 29
Lai, Sheng-Chih 140, 142
Lam, C. 27
Lattanzio, L. 109
Lauwers, A. 43
Le Royer, C. 134
Lee, B. H. 53
Lee, C. H. 121, 128, 138

AUTHOR INDEX

Lee, C. J. 45

Lee, Chain-Ming 35

Lee, Chi-Woo..115

Lee, Heng-Yuan 33

Lee, M-H. 27

Lee, Rinus T. P. 67

Lee, T. L. 25

Lee, W. C. 128, 130

Lee, Y. 83

Li, C. I. 21

Li, W. 83

Li, Xi 117

Li, Z. 57

Liao, Chien-Wei 140

Liao, M. H. 25

Lin, Burn 86, 88

Lin, C. H. 29

Lin, C. T. 57

Lin, Cha-Hsin 33

Lin, Chuan-Ding 111, 136

Lin, Chung-Hsun 117

Lin, Horng-Chih....................................136

Lin, S. J. 88

Lin, T. D. 128, 130

Lin, Y. M..31

Lin, Y. S. 130

Lina, Horng-Chih..................................111

Lindsay, R..83

Liu, C. H. 31

Liu, C. S. 51, 57

Liu, E. C 79

Liu, P. W. 21, 79

Liu, Rich 140, 94, 140, 142

Liu, Ta-Wei 136

Liu, Wen-Hsing 35

Liu, Xinke 132

Lo, Guo-Qiang 67

Loesing, R. 83

Lorenz, J. 15

Lu, Chih-Yuan 94, 138, 140, 142, 140

Lu, H. H. 45

Lu, J. C. 25

Lu, W. P. 138

Lucas, Kevin 87

Lue, Hang-Ting 94, 140, 142

Lue, Yi-Hsien 140

Lung, H-L. 27

Luo, Guang-Li 140

Lysaght, P. S. 55

Ma, G. H. 21, 79

Maekawa, H. 23

Maikap, S. 29

Majhi, Prashant 19, 125

Majumdar, Amlan.................................117

Masuzumi, T. 9, 11

Mayuzumi, S. 17

Melvin, C. 53

Mercha, A. 73

Mii, Y. J. 63

Moon, Yongsik 117

Morand, Y..81

Moselund, K. E. 109

Moumen, N. 83

Müller, M. 51, 57

Naeem, Munir 117

Nagashima, N. 17

Najmzadeh, M. 109

Nara, Yasuo 65

Narayanan, V. 83

Nemouchi, F. 81

Nguyen, Bich-Yen 125

Nishikawa, M. 23

AUTHOR INDEX

O'Neill, A. 109

Oh, Jungwoo 19

Ohno, T. 17

Okabe, K. 23

Olsen, S. 109

Osada, Kosei 13

Ostermayr, M. 83

Ouyang, Christine 117

Ouyang, Qiqing 77

Pacha, C. 83

Parat, Krishna 92

Park, C. 53, 59, 144

Park, D. G. 83, 117

Park, Hokyung 144

Park, Jemin 96

Parthasarathy, Srivatsan 125

Parvais, B. 73

Patruno, Paul 125

Pedini, J.-M. 102

Petry, J. 51, 57

Pinto, Mark 1

Pittikoun, S. 31

Poiroux, T. 81

Pouydebasque, A. 134

Previtali, B. 81, 102, 134

Price, J. 55

Rabaey, Jan 37

Ragnarsson, L-Å 43

Rahaman, S. Z. 29

Rajendran, B. 27

Ren, Zhibin 117

Reznicek, Alexander 77

Ribeiro, M. 102

Röhr, E. 43

Romanjek, K. 134

Ronse, Kurt 90

Rooyackers, R. 73

Rozeau, Olivier 98

Saenger, Katherine L. 77

Sakamoto, Y. 31

Samavedam, S. 83

Samudra, G. S. 104

Saraswat, Krishna 113, 144

Schepis, D. 83

Schram, T. 43

Schruefer, K. 83

Sebaai, F. 43

Selmi, L. 109

Shen, C. 104, 140

Shen, Tzermin 79

Sherony, M. 83

Shi, Leathen 77

Shibahara, Kentaro 13

Shih, Y-H 27

Shue, S. L. 45

Singanamalla, R. 51

Sinha, Mantavya 67

Sivasubramani, P. 53

Skotnicki, Thomas 71

Smith, Casey E. 125

Solomon, Paul. M. 77

Steegen, A. 83

Stein, K. 83

Su, Pin 75, 106

Subramanian, V. 73

Sugawara, Takuya 19

Sugino, T. 9, 11

Tabone, C. 134

Tamura, N. 23

Tan, Leng-Seow 132

Tang, T. 83

Tateiwa, K. 53

AUTHOR INDEX

Tateshita, Y. 17

Theon, V. Y. 83

Toh, E. H. 104

Tomoyasu, Masayuki 19

Tosti, L. 81

Truche, R. 134

Tsai, C. C. 121

Tsai, C. H. 21, 79

Tsai, C. T. 21, 79

Tsai, Jonas 79

Tsai, M. J. 29, 33, 35

Tsai, S. H. 79

Tsai, W. 130

Tseng, Fang-Churng 2

Tseng, H. H. 53, 55, 59

Tseng, J. 51

Tsui, Bing-Yue 61

Tsukamoto, M. 17

Tu, W. H. 138

Tung, L. T. 121, 128

Tzeng, P. J. 29, 33

Uchida, Ken 5

Umeyama, M. 23

Van Nimwegen, L. 51

Vandenberghe, Geert 90

Verma, Sarves 144

Vinet, M. 81, 102, 134

Von Arnim, K. 83

Voogt, F. 51

Wada, M. 43

Wakabayashi, H. 17

Wang, Cindy 117

Wang, H. Y. 57

Wang, L. T. 25

Wang, P. Y. 7

Wang, Runsheng 123

Wang, Tahui 138

Wang, Xinlin 117

Wang, Y. R. 21

Wann, C. 63

Watanabe, D. 9, 11

Wei, H. C. 31

Wei, Lan 71

Witters, L. 73

Wong, H. S. Philip 71, 113

Wong, W. Z. 7

Wu, J. 25

Wu, T. Y. 29, 33

Wu, Yu-Sheng 106

Xiong, K. 51

Xiong, Weize 115

Yamakawa, S. 17

Yan, Ran 115

Yan, W. 83

Yang, Hao-I 69

Yang, J. W. 53, 125

Yang, L. T. 104

Yang, Ming-Jui 140

Yang, Qingyun 117

Yang, R. J. 31

Yates, John 117

Yau, Jeng-Bang 77

Yeh, Lingyen 25

Yeo, Yee-Chia 38, 67, 104, 132

Yin, D. Y. 31

Yin, H. 83

Young, C. D. 59

Young, Rex 7

Yu, C. H. 45, 57

Yu, H. Y. 63

Yu, M. H. 25

Zhang, Liangliang 123

AUTHOR INDEX

Zhang, Ying 117

Zhao, C. 57

Zhuang, H. 83

Zous, N. K. 138

CURRAN ASSOCIATES INC.
proceedings
.com

9781424427840